# Construction Ecology

# Construction Ecology

## Nature as the basis for green buildings

Edited by
Charles J. Kibert, Jan Sendzimir,
and G. Bradley Guy

London and New York

First published 2002 by Spon Press
11 New Fetter Lane, London EC4P 4EE

Simultaneously published in the USA and Canada
by Spon Press
29 West 35th Street, New York, NY 10001

*Spon Press is an imprint of the Taylor & Francis Group*



Typeset in Garamond by
Prepress Projects Ltd, Perth, Scotland



British Library Cataloguing in Publication Data
A catalogue record for this book is available from the British Library

Library of Congress Cataloging in Publication Data

Kibert, Charles J.
Construction ecology: nature as the basis for green buildings / edited by Charles J. Kibert, Jan Sendzimir, and G. Bradley Guy.
p. cm.
ISBN 978-0-415-26092-3 (pbk. : alk. paper)
1. Building. 2. Building materials – Environmental aspects. 3. Green products. 4. Sustainable development. 5. Architecture and society. 6. Construction industry – appropriate technology. I. Sendzimir, Jan. II. Guy, G. Bradley. III. Title.

TH146.K53 2001
720′.47–dc21 2001034698

# Contents

# Figures

# Tables

# Boxes

# Contributors

**Timothy F.H. Allen** is Professor of Botany at the University of Wisconsin, Madison. He has been applying notions of complex systems and hierarchy theory to ecology for twenty-five years. His first book, *Hierarchy: Perspectives for Ecological Complexity* (Chicago: University of Chicago Press, 1982), established hierarchy theory and scaling in ecology. Four other books on hierarchy theory focus on ecosystem analysis or cover all types of ecology and beyond to the life and social sciences in general. He has published over fifty scholarly works in journals on community data analysis, agricultural systems, issues of scale, and sustainability. His latest work, *Supply Side Sustainability* (in press), written with Forest Service colleagues J. Tainter and T. Hoekstra, explores the emerging field of economic ecology and suggests that we must manage the whole ecosystem that makes resources renewable, not natural resources themselves. The historical analysis shows that failure to move beyond the information age to an age of quality global management invites a new dark age.

**Robert U. Ayres** is Sandoz Professor of Environment and Management, Professor of Economics and Director of the Centre for the Management of Environmental Resources at the European Business School INSEAD, in Fountainebleau, France. From 1979 to 1992 he was Professor of Engineering and Public Policy at Carnegie-Mellon University, Pittsburgh, Pennsylvania. He received a BA in mathematics from the University of Chicago and his PhD in mathematical physics from Kings College, University of London. From 1961 to 1966, he was a staff member of the Hudson Institute, where he worked on environmental problems. He worked at Resources for the Future, in Washington, DC. In 1968 he co-founded a research/consulting firm in Washington, DC, to perform studies for the US government and international agencies. He was also affiliated with the International Institute for Applied Systems Analysis (IIASA) in Austria in 1986–87 and 1989–90. He has published more than 200 journal articles and book chapters and has authored or co-authored twelve books on topics ranging from technological change, manufacturing, and productivity to environmental and resource economics. His most recent books are *Accounting for Resources 1: Economy-wide Applications of Mass-Balance Principles to Materials and Waste* (with L.W. Ayres, Cheltenham, UK: Edward Elgar, 1998), *Turning Point: the End of the Growth Paradigm* (London: Earthscan, 1998), *Information, Entropy, and Progress: A New Evolutionary Paradigm* (New York: Springer Verlag, 1994), and *Industrial Ecology: Towards closing the Materials Cycle* (Edward Elgar). He is also co-author of *Industrial Metabolism: Restructuring for Sustainable Development* (with U.E. Simonis, Tokyo: United Nations University Press, 1994) and editor-in-chief of a new volume, *Eco-Restructuring* (Tokyo: United Nations University Press, 1998).

**Fritz Balkau** is a chemist by training but has spent nearly all of his professional career working in the environmental field, including government and international organizations. He focused on land-use planning, pollution, and waste management, but has also worked in chemicals management, strategic policy development, and training. He is currently head of UNEP's Production and Consumption Unit, based in Paris, France.

**Jürgen Bisch** works as an architect for Seegy & Bisch Architects in Nuremberg, Germany. He has designed numerous administration, factory, and apartment buildings in Germany and has worked for many years on the revitalization and renovation of villages. His design principles are centered around the following concepts of reducing building components, avoiding the use of composite materials, reducing energy requirements by 50% compared with similar buildings, increasing productivity by promoting good indoor air quality, and reducing initial building costs by 15%. In addition to his architectural work, he taught industrial design as a professor at the Institute of Fine Art, University of Kassel, Germany.

**Stefan Bringezu** is head of industrial ecology research in the Department of Material Flows and Structural Change at the Wuppertal Institute, Germany. A biologist by training, his PhD was based on field research on regulating mechanisms within natural ecosystems. From 1987 to 1992 he was deputy leader of the Department of Biocidal Substances within the Chemicals Control Division of the German Federal Environment Agency. In 1992 he joined the Wuppertal Institute and from 1997 to 1998 was acting head of the Department for Supply Systems and Environmental Planning at Dortmund University. His main interests are the development of indicators for sustainability on different levels (local, regional, national), life cycle assessment methods, the analysis of industrial metabolism, integrated economic and environmental accounting, and sustainable resource management.

**G. Bradley Guy** is a research associate at the Center for Construction and Environment, M.E. Rinker, Sr. School of Building Construction, College of Architecture, University of Florida. His expertise is in the areas of sustainable architectural design and materials, urban planning, sustainability indicators, community development, and organizational activities. He is the principal designer and assistant project manager for the "Summer House" at Kanapaha Botanical Gardens, Alachua County, Florida, an environmental education facility and example of sustainable architecture. He is the project manager for a house deconstruction project that will provide materials to the "Summer House" as well as provide data for a cost–benefit analysis of this alternative approach to traditional demolition. Previous publications include contributions to the "Abacoa Sustainability Codes" for the 2,000-acre Traditional Neighborhood Development (TND) Abacoa development in South Florida and the "Greening Federal Facilities" for the Federal Energy Management Program (FEMP). He provided the technical assistance for the "Build Green and Profit" building contractors' CEU course manuals administered by the sixty-seven county offices of the Cooperative Extension Service of the State of Florida.

**James J. Kay** is an associate professor of environment and resource studies (with a cross-appointment in systems design engineering and the School of Planning) at the

University of Waterloo, Ontario, Canada. He studied physics at McGill University and systems design engineering at the University of Waterloo. His PhD thesis was entitled "Self-organization in living systems." His principal research focus is on the development of the ecosystem approach to ecological management and planning. His research activities span the full spectrum from the theoretical and epistemological basis for an ecosystem approach through the formulation of ecosystem-based environmental policy and the development of ecosystem monitoring programs to on-the-ground planning in the context of both urban and natural ecosystem and the greening of institutions. His theoretical focus is on the application of non-equilibrium thermodynamics, information theory, and systems theory to the problem of understanding the organization of ecosystems and to the development of the ecosystem approach to management. In particular, he has focused on the relationship between the second law of thermodynamics and ecosystem organization. This work is the subject of the article "Complexity and thermodynamics: towards a new ecology" in the journal *Futures* (August, 1994) and he is currently working on a book (*Order From Disorder: The Thermodynamics of Complexity and Ecology*) for the Columbia University Press series *Complexity in Ecological Systems.*

**Charles J. Kibert** is the Director of the Rinker School of Building Construction and former Director of the Center for Construction and Environment at the University of Florida. He is also a CSR/Rinker Professor in the M.E. Rinker School of Building Construction and a University of Florida Research Foundation Professor. He founded the Cross Creek Initiative, a non-profit organization for transferring sustainable building methods and technologies to construction industry. He organized and coordinates Task Group 39 of Conseil International du Batiment (CIB), an international research body that promotes building disassembly (deconstruction) and materials reuse as a high-priority strategy for sustainable construction. He is the editor of *Reshaping the Built Environment: Ecology, Ethics, and Economics* (Washington, DC: Island Press, 1999) and is the primary author of *Turning Brownfields into Vital Community Assets* (Washington, DC: Neighborhood Training Institute, 1999), *Greening Federal Facilities* (Washington, DC: Federal Energy Management Program, 1997), and over ninety papers on the sustainable built environment.

**Ernest Lowe** is Chief Scientist of RPP International's Sustainable Development Division and Director of this company's Indigo Development Center. He is one of the creators of the eco-industrial park concept and has worked with eco-park projects in the USA, South Africa, the Philippines, and Thailand. He is author of *Discovering Industrial Ecology* (Columbus, OH: Battelle Press, 1997) and the *Eco-Industrial Park Handbook* (prepared initially for US-EPA in 1995, Columbus, OH: Battelle Press). He has recently updated and revised the handbook for use in Asian developing countries under contract with the Asian Development Bank. In 1999, he presented a paper on sustainable new towns in the Rinker Eminent Lecture Series, which appears in *Reshaping the Built Environment: Ecology, Ethics, and Economics* (edited by Charles Kibert, Washington, DC: Island Press, 1999).

**Howard T. Odum** is Graduate Research Professor Emeritus in the Department of Environmental Engineering Sciences, University of Florida at Gainesville, founder of the University's Center for Wetlands and Center for Environmental Policy. While

Director of The University of Texas Marine Science Institute in 1957 with a Rockefeller Foundation project he developed the ecosystem metabolism concept of production–consumption–recycle comparing biomes of land and water including rainforests. The measured indices, thermodynamic concepts, and simulation models were extended to landscapes and human society. These helped start the fields of systems ecology, ecological economics, ecoenergetic energy analysis, and ecological engineering. His books include *Environment, Power and Society* (New York: Wiley-Interscience, 1970), *Ecological and General Systems (Systems Ecology)* (New York: John Wiley, 1983), *Ecological Microcosms* (with R.J. Beyers, New York: Springer Verlag, 1993), *Environmental Accounting, Emergy and Decision Making* (New York: John Wiley, 1996), and *Environment and Society* (with Elisabeth C. Odum and M.T. Brown, Boca Raton: Florida Lewis Publishers, 1997). He is a member of the Royal Swedish Academy of Science and co-recipient of its Crafoord Prize in 1987.

**Rob Peña** is an assistant professor of architecture at the University of Oregon, where he specializes in climate-responsive, energy-efficient design. Rob's efforts to integrate ecological design into architectural education include Outward Build, a summer program aimed at place-responsive, hands-on experiential learning. He was recently honored with the University of Oregon's Ersted Distinguished Teaching Award in recognition of these efforts. Rob also works with Van der Ryn Architects in Sausalito, California, where he is currently involved in ecological planning and housing projects in New Mexico, Arizona, and California. Rob holds a Bachelor of Science degree in architectural engineering from the University of Colorado and a Master of Architecture degree from the University of California, Berkeley.

**Garry Peterson** is a research associate at the Center for Limnology at the University of Wisonsin at Madison. He is currently working on the theory and practice of the ecological management of regional ecosystems. He was a post-doctoral fellow at the National Center for Ecological Analysis and Synthesis, at the University of California, Santa Barbara. He is currently working on a project entitled "Theories for Sustainable Futures: Understanding and Managing for Resilience in Human–Ecological Systems." In 1999, he received a PhD in zoology from the University of Florida.

**Jan Sendzimir** is a systems ecologist at the International Institute for Applied Systems Analysis in Laxenburg, Austria, and sits on the board of directors of the CEIBA Foundation and the Soils of Tomorrow Foundation. His research addresses environmental issues at global, regional, and ecosystem levels, particularly the resilience of human and natural communities in response to disturbances at different time and space scales. Of particular concern in addressing the interface between the human and natural worlds is the evolution of environmental management systems and policies which flexibly adapt to emerging conditions. He has taught ecology courses at Santa Fe Community College (1985), the Ecological Summer School of Central European University in Budapest (Soros Foundation) (1991), and the Academy of Mining and Metallurgy in Krakow (1998) and has published five papers. He holds a PhD and a MSc in systems and landscape ecology from the University of Florida, and an MA in teaching biology. He also was a member of the board of directors of the Kosciuszko Foundation from 1985 to 1996.

**Sim Van der Ryn** is President of the Ecological Design Institute and Van Der Ryn Architects in Sausalito, California, and is renowned as a leader in ecological design. For over thirty years, his design, planning, teaching, and public leadership has advanced the viability and acceptance of ecological principles and practice in architecture and planning. He is the author of over 100 articles, monographs, and book contributions, including *Ecological Design* (Washington, DC: Island Press, 1995), *The Toilet Papers: Recycling Waste and Conserving Water* (White River Junction, VT: Chelsea Green, 2000), *Sustainable Communities* (Sierra Club Books, 1986), and *The Integral Urban House* (San Francisco: Sierra Club Books, 1978).

**Malcolm Wells** is a pioneer of underground building and natural design. For 34 of his 45 years as an architect he has been an advocate of underground buildings as the answer to the paving over of America with asphalt and concrete. He is widely known for many books, including *How to Build an Underground House* (New York: Malcolm Wells, 1991) and *Gentle Architecture* (New York: McGraw Hill, 1981). His cartoons are featured in Rob Roy's *Mortgage Free! Radical Strategies for Home Ownership* (White River Junction, VT: Chelsea Green, 1998). He lives in the Underground Art Gallery in Brewster, Massachusetts.

**Iddo Wernick** is an associate research scientist at the Columbia Earth Institute and an adjunct professor in the Department of Earth and Environmental Engineering, Columbia University, New York. He received his PhD in applied physics in 1992 and has since devoted most time to the development of the emerging field of industrial ecology. He has written extensively in the areas of dematerialization, the metrics of national resource use, and the industrial ecology of specific materials including forest products and metals. He currently teaches a two-semester sequence in industrial ecology (industrial ecology of earth resources and industrial ecology of manufacturing) in the School of Engineering and Applied Science at Columbia University.

# Preface

Through the efforts of a wide array of public and private organizations around the globe, a strong and growing movement is beginning the process of transforming the systems that create the built environment from ones that pay no attention to resources and the environment to a new variety in which these considerations are the pre-eminent criteria for "good" construction. Although dating back only to the early 1990s, there is already ample evidence that the "green building" movement is affecting the design, construction, operation, and disposal of the built environment. It utilizes many of the same efforts used in the mid-1970s to reduce building energy consumption and promote a shift to renewable energy resources. Green buildings are designed and built with a sharp focus on the overall environmental and resource impacts of human habitation. In addition to significantly reducing energy use, the green building movement also attempts to reduce water consumption, minimize construction and operational waste, select materials that are recyclable and/or with recycled content and renewable resource content, optimize the siting of buildings, and insure healthy interior environments for the occupants.

In spite of the early success of this movement, much work needs to be done, particularly in deepening the understanding of the connections and interplay between the built environment and natural systems. At present, green building movements rely on a virtual smorgasbord of options that are no more than the best judgment of the designers and builders who are seeking to explore new practices. Building professionals are using a largely intuitive approach to creating a green built environment that, although fairly effective in decelerating the destruction of environmental and resource capacity, lacks an adequate understanding of the very systems it purports to protect. A strong philosophical and technical basis for the wide array of decisions that must be made in the selection of sites, materials, or energy systems does not currently exist. This is the purpose of a new discipline called construction ecology involving the development of an understanding of how the structures and behavior of natural systems, their organization, metamorphosis, resource utilization, pulsing, complexity, and other important aspects can be examined both as heuristic metaphors and for providing information for the life cycle of the built environment. This volume is the first attempt to develop an ecology of the built environment that explores how the built environment affects the human–natural system interface. This development must be an important part of the research agenda of the green building movement as it proceeds over the next few decades. The ecologists, industrial ecologists, and architects who wrote the chapters of this volume describe their thoughts on how nature can inform the built environment and lead to an era in which these significant human artifacts are integrated with, rather than made apart from, Nature.

Outside the building industry, similar struggles to link human and natural systems behavior have been taking place. One result has been the emergence of the field of industrial ecology, which, since 1989, has been investigating, with some degree of success, the application of the behavior of natural ecological systems to manufacturing and other industrial activities such as power generation and wastewater management. The initial efforts of industrial ecology have focused on closing materials loops by integrating industries in what has been called *industrial symbiosis*. Several eco-industrial parks have been designed and are achieving varying degrees of success around the world. The study of economics is also undergoing a transformation as a result of examining how the behavior of the economy has much in common with the ebb and flow of nature. Michael Rotschild, in his 1992 book *Bionomics*, described many of these similarities. The emergence of complexity theory has opened many new avenues for exploration of these connections and similarities. M. Mitchell Waldorp describes the application of complexity theory to economics in *Complexity: The Emerging Science at the Edge of Order and Chaos* (New York: Simon & Schuster, 1992). Michael Byrne, in his book *Complexity Theory and the Social Sciences: An Introduction* (New York: Simon & Schuster, 1998) explores how human systems behave in a chaotic/complex fashion. Meanwhile, ecologists are beginning to tease out the complex behavior of natural systems. The remarkable outcome of these disparate efforts is that Nature and the built environment seem to exhibit much the same types of behaviors. The growth, maturation, and decline of natural systems mirror the development, aging, and decay of cities. The resources and population of human settlements exhibit the same pulsing character as natural systems. The questions begging to be answered are: Does human habitat, its planning, content, and functioning behave in a complex/chaotic manner analogous to natural systems? What are the lessons that can be learned from the behavior and functioning of natural systems that the green building movement should adopt as its philosophical basis and incorporate into both its foundation philosophy and its design principles?

These questions stimulated the organization of a collaborative effort at the University of Florida in 1999 among architects, ecologists, industrial ecologists, and materials manufacturers. The Rinker Eminent Scholar Workshop on Construction Ecology and Metabolism brought together a group of the world's foremost authorities and practitioners from academia and the public and private sectors. This volume is a result of their post-workshop thoughts on, and responses to, these questions and others that were stimulated during the workshop. It is organized into three main parts. Part 1 contains the reflections of the ecologists who attended the workshop, and in each case they illuminate the current state of thinking about ecological systems as well as their notions about how the built environment may benefit from considering the behavior of natural systems. Part 2 provides the insights of leading thinkers in the field of industrial ecology, which for the past decade has been attempting to determine how industrial systems can adopt the elegant functioning of natural systems into their processes. Part 3 is a compilation of the responses of leading architects to the possibility of considering natural systems as both metaphor and model for the built environment. The final chapter, Conclusions, provides some initial thoughts on new directions for green building using construction ecology as its basis.

In organizing and carrying off the truly complex effort of bringing together a group of people representing the leading edge of thinking on the human–natural system interface, many acknowledgements are in order.

First, our gratitude must be expressed to the Marshall E. Rinker, Sr. Foundation and the Marshall and Vera Lea Rinker Foundation for their continuous support of the Rinker School of Building Construction. The endowments created in memory of "Doc" Rinker continue to provide opportunities to the faculty of the school to create a world-class construction program. The faculty of the Rinker School of Building Construction responded positively to a proposal for the workshop and provided the financial resources from the Rinker endowment necessary to support the expense of bringing this group to Gainesville and organizing their thoughts into a hopefully coherent volume. The then Dean of the College of Architecture, Wayne Drummond, fully supported this concept and provided much moral support in its execution. Thanks also to Jimmie Hinze, former Director of the Rinker School, for his support of this program and his efforts in bringing it to fruition.

I would also like to acknowledge the efforts of the Center for Construction and Environment in the Rinker School for their energy and enthusiasm in performing all the detailed tasks of making the workshop a reality. Gisela Bosch led the group which organized the workshop, and it is largely through her tireless work and dedication that it was a great success. G. Bradley Guy assisted in both the selection of the eminent attendees and the editing of this volume. Jan Sendzimir was responsible for selection of leading ecologists to attend the workshop and also assisted in editing the volume. Dottie Beaupied worked diligently to insure the details of travel and accommodation were thoroughly attended to and continued to assist the editors throughout the entire process of bringing this resulting book to press. Finally, we are grateful to Spon Press for deciding to publish this volume and for the efforts of Sarah Kramer in making it a reality.

Charles J. Kibert
Gainesville, Florida

# Foreword

My first direct experience of the need for adding significantly to the built environment was in the 1960s, when I interviewed farm workers in their shanty towns in California. Often, a dozen family members would be crowded into a two- or three-room tarpaper shack with corrugated iron roof and no running water. Then in 1997–98 my perception of this challenge became much more acute when I visited townships such as Alexandra and the Cape Flats in South Africa. For families there to be able to afford housing sufficient to their very deep needs, they also needed work places and training facilities – further additions to the built environment.

In the Philippines in 1999 I saw another version of this situation in the strip residential and commercial developments strung out between a freeway and factory walls. Ironically, people from these poverty-stricken neighborhoods would dress up in their finest to visit Glorietta Mall or MegaMall and enjoy the fantastic luxury of air conditioning! The multiplicity of such malls, filled with global identity shops such as Pierre Cardin and fast food chains such as Pizza Hut, are another indicator of a huge increase in construction activity, even in developing countries.

China plans twenty new towns a year in the west to accommodate the hundreds of thousands of farm families unable to compete with global prices when this country enters the World Trade Organization. The large cities in the east could not withstand even more migration into their overcrowded neighborhoods.

This escalating demand for built space makes dramatically clear that we cannot afford to continue designing and building our homes, commercial centers, and industrial and public facilities in the old unsustainable models. Chapter 1 in this volume gives detailed evidence of the enormous environmental costs of construction, operation, and deconstruction of our buildings. While green design has many success stories, such as the Herman Miller Phoenix Designs plant in Michigan, we must go beyond incremental and fragmented improvements in design and construction of the built environment. The disciplines of design and construction must go to a basic level of rethinking to achieve the level of change the challenges of sustainability demands. This publication is a powerful beginning of that process.

Construction ecology is a breakthrough in two fields of research and application – the design and construction of our built environment and the relatively new discipline of industrial ecology. The buildings in which we live and work and the infrastructure that supports our lives demand an amazing share of the natural resources we humans consume. As populations continue to grow and expectations rise, it is essential that we learn to design and build with a much higher level of efficiency and a much lower level of pollution and waste.

Industrial ecology has emerged in the last 15 years as a systems approach to design, development, and operation of human systems, both public and private. It aims to create the transition to a sustainable world in which our economic activities respect the limits of global and local carrying capacity. The construction ecology conference that was the source of this book was a very significant effort to use the concepts and methods of industrial ecology to begin charting the path to sustainable design and construction.

One of the most attractive concepts industrial ecologists have suggested is that we can learn how to design human systems from the dynamics of ecosystem behavior. While there is potentially great value in this approach, most of the efforts before this book have been naive and full of clichés. Relatively few industrial ecologists have written of "industrial ecosystems" with much insight into how ecosystems actually function. We hear repeatedly that the waste of one organism becomes the food for another and little more.

Construction ecology, and the conference that produced it, is the first large-scale attempt to learn about major human energy and materials flows by comparing them to the dynamics of natural systems. Ecologists, industrial ecologists, architects, construction researchers, and representatives of building products manufacturing have used the ecological metaphor in a profound way that other industry and academic clusters could learn from.

Ecosystems in Nature demonstrate many strategies beyond consumption of life's by-products (what we still call 'waste') that are relevant to design and construction. For instance:

- The sole source of power for ecosystems is solar energy.
- Concentrated toxic materials are generated and used locally.
- Efficiency and productivity are in dynamic balance with resiliency. Emphasis on the first two qualities over the third creates brittle systems, likely to crash.
- Ecosystems remain resilient in the face of change through high biodiversity of species, organized in complex webs of relationships. The many relationships are maintained through self-organizing processes, not top-down control.
- In an ecosystem, each individual in a species acts independently, yet its activity patterns cooperatively mesh with the patterns of other species. Cooperation and competition are interlinked and held in balance.

By exploring the implications of these and other ecological principles for design and construction the authors of the papers in *Construction Ecology* have created a context for the holistic rethinking required for a sustainable approach to the built environment.

As an industrial ecologist, I am pleased to see that we can apply the basic strategy of the ecological metaphor with real depth, not just as a cliché. As someone whose career focuses on supporting developing countries in their often daunting task of housing their poor and giving them employment to rise out of poverty, I feel that this volume may help them achieve these goals while preserving their environments.

Ernest Lowe

# Introduction

*Charles J. Kibert*

Construction ecology articulates the philosophical and technical foundations for the international movement most commonly referred to as *green building*. Linking the words "construction" and "ecology" into a single description may be a difficult concept for many people, and some may contend that it is an oxymoron. However, in attempting to respond to the need to connect human activities to Nature, many disciplines are beginning to explore their existing "ecology" and are trying to understand how it is related to the behavior of natural systems. Thus, one can now find references to urban ecology, social ecology, industrial ecology, and political ecology, to name a few.

The construction industry has much in common with the overall industrial subsystem of the economy. This industry is in fact tightly coupled to the industrial subsystem because modern buildings largely comprise products made in factories of materials extracted from the Earth's biological or mineral resources. It could be argued that the built environment is merely a portion of industrial output and that its so-called ecology could be studied wholly within the framework of industrial ecology, itself a relatively new discipline.

The built environment, however, is not merely an industrial product. Buildings are perhaps the most significant artifacts of human culture and can have historic meaning across millennia, something that can be said of few other human artifacts. The scale of the built environment is enormous, and it occupies a significant fraction of the Earth's surface. It replaces, at long time scales (typically 50–100 years), once productive natural systems with non-productive (in a natural ecological sense) structures. The alteration of existing natural systems is a primary effect of the built environment, and these systems are usually replaced in part by human-designed landscaping that may not bear any resemblance to the ecological systems that once occupied the site. In addition to the industrial products that constitute it, the built environment incorporates significant quantities of raw materials such as earth and gravel to create a structural boundary layer between buildings or infrastructure and the ground. During its operation, the built environment consumes energy, water, and materials and emits solid, liquid, and gas contaminants. At the end of their useful life, these structures contribute vast quantities of waste to the environment, on the order of 0.4–0.5 tons per capita per year in industrialized countries.

In spite of the differences between the built environment and other products of an industrial society, the construction industry can use the lessons learned from industrial ecology as a jumping off point to discover its existing metabolism of energy matter and its interactions with ecological systems. Examination of natural systems has revealed structures and behaviors, and their metaphors, that mutually reinforce the natural and built environments, and these lessons can benefit both industrial and construction ecology.

Although the international green building movement has enjoyed success during the 1990s, its initial decade, much work needs to be done, particularly in the area of understanding the connections between buildings and Nature and the effects of the built environment on natural systems. At present, green building movements rely on a wide array of options based on the intuition and judgment of professional planners, architects, engineers, and facility managers. Optimal approaches, techniques, and methods for producing green buildings do not currently exist because, although one of the stated purposes of green buildings is to protect the environment, the effects of specific design and operational decisions on the natural world are virtually unknown and clearly not able to be quantified in any meaningful way. Consequently, the professionals associated with the creation of the built environment are literally guessing at how best to reduce the effects on Nature of building construction, operation, and demolition. Although effective tools have been developed to examine financial trade-offs associated with shifting to low environmental impact options, methods to assess the direct effects of decisions on ecological systems are presently lacking. Understanding the effects of building decisions on the health of natural systems is absolutely essential to the long-term success of efforts to advance the state of ecological design and the construction of green buildings. Integrating natural systems function with the operation of buildings is another largely untapped source. For example, the possibility of buildings providing nutrients to natural systems while Nature does the work once performed by energy-intensive mechanical systems is a field of study ripe for both intellectual and commercial exploitation. A fundamental rethinking of green building strategies will be required for this to occur, and the integration of the study of both ecology and industrial ecology with architecture will be needed to begin this process of deep change.

## Current state of green building

A description of the current state of the movements to green the built environment would be useful in establishing a context for understanding the need to develop a sound basis for its future development. Many terms are used to describe these movements and, in addition to green building, terms such as sustainable construction, sustainable architecture, ecological architecture, ecologically sustainable design, and ecologically sustainable development have been used. The term "sustainable construction" seems to be the most comprehensive description of all the activities involved in trying to better integrate the built environment with its natural counterpart. Begun as an international movement in 1993, sustainable construction can be defined as "creating a healthy built environment based on ecologically sound principles." It looks at the entire life cycle of the built environment: planning, design, construction, operation, renovation and retrofit, and the end-of-life fate of its materials. It considers the resources of construction to be materials, land, energy, and water and has an established a set of principles to guide this new direction:

1. reduce resource consumption;
2. reuse resources to the maximum extent possible;
3. recycle built environment end-of-life resources and use recyclable resources;
4. protect natural systems and their function in all activities;
5. eliminate toxic materials and by-products in all phases of the built environment;
6. incorporate full-cost accounting in all economic decisions;
7. emphasize quality in all phases of the life cycle of the built environment.

Similar principles have been set out by many of the organizations involved in the greening of the built environment, all of them having much in common with the principles of sustainable construction. Progress in implementing these principles has been impressive. A comprehensive overview of programs in countries around the world would be lengthy and, for the sake of brevity, a review of progress in the USA will be used as indicative of how rapidly change is taking place worldwide. In the USA, there are several major entities driving the emergence of green buildings: the US Green Building Council, the National Association of Home Builders, and the federal and local governments.

### *The US Green Building Council*

The US Green Building Council (USGBC), established in 1993, represents a wide range of actors who concluded that construction industry must change course to be sustainable: architects, engineers, product manufacturers, academics, and public institutions. In the USA, the construction industry clearly has disproportionate impacts on the environment compared with other sectors of the economy. At present, although it represents just 8% of the country's gross domestic product, this industry is responsible for over 40% of total materials extraction to produce and alter buildings and infrastructure, and the operation of buildings consumes over 30% of the nation's primary energy. In a fashion similar to its counterparts in other major industrial countries, as its response to changing the playing field, the USGBC organized a system of rating buildings that would add new criteria for the siting, design, construction, and operation of new and renovated buildings in the USA. This rating system, known more commonly by its acronym of LEED (Leadership in Energy and Environmental Design), proposed to classify buildings into four categories depending on their level of performance with respect to energy and environmental issues: platinum (highest), gold, silver, and LEED-rated. In the short time since its proposal and subsequent piloting, the LEED standard must be declared to be a major success. Scores of buildings have been designed and built using its criteria, and many are more queuing up to employ it as perhaps the key focus for building design, ranking only behind the client's requirements for the building's function. The LEED standard is being expanded into other sectors of building construction, including residential housing. The beta testing of the standard started in 1998, and over thirty buildings received ratings based on version 1.0. In April 2000, the final standard, version 2.0, was issued and is now being used to rate commercial and institutional buildings.

### *The National Association of Home Builders*

The National Association of Home Builders (NAHB) is generally considered to be the most powerful construction industry organization in the USA, with over 200,000 members organized into 800 local chapters. According to the NAHB, in 1999 there were over 1.6 million single- and multi-family housing starts. In 1998, over $214 billion dollars of family housing was produced by the private sector, about one-third the value of total construction in the USA. Home ownership is a significant aspect of American culture. Home ownership is highly valued, and homes represent a significant portion of wealth. Home ownership accounted for approximately 44% of the nation's total net worth in 1993. The high level of home building also represents a significant proportion of the environmental impacts of construction, especially in terms of its land consumption. Fortunately, several homebuilder associations have actively engaged in determining how to build homes in an environmentally friendly manner. At least six of these associations, often in cooperation with local jurisdictions, have established a variety of green builder programs, and the NAHB now has an annual national conference devoted to green home building.

### *Federal and local government*

Of all the organizations involved in green building efforts in the USA, the federal government is both the largest customer and arguably its greatest proponent. A wide array of federal agencies have demanded better environmental and health performance for new buildings, among them the US Post Office, the National Park Service, and many of the military services. Many of the buildings that were rated by the first version of the LEED standard in the beta testing effort were federal buildings. The US Department of Energy has been a major supporter of the development and implementation of the LEED standard. Presidential executive orders have directed a variety of actions on the part of federal agencies that directly or indirectly support the construction of green buildings. Several highly visible federal building efforts, such as the "Greening to the White House" and the "Greening of the Pentagon," have been effective in publicizing green buildings in the USA.

Local government has also been a major force in the green building movement in the USA. The municipal government of Austin, Texas, initiated a green building program in the early 1990s. The Austin effort was initially directed at the procurement of city buildings and produced the first guidelines for municipal building, the *Sustainable Building Sourcebook*. The efforts of the city soon produced a parallel effort in the local homebuilding industry, and the Austin homebuilders association formed the first NAHB green residential construction program. The city of Seattle, Washington, now requires conformance to the LEED standard for all municipal buildings, and similar requirements for the use of the LEED standard are emerging from local government across the USA. The city of Boulder, Colorado, was the first municipality to require some level of green building measures for all housing constructed within city limits and enforces this requirement through the building permitting process.

## Organization

The purpose of this book, the investigation of how to base green buildings on the structure, behavior, and metaphor of natural systems, necessitated the collaboration of four groups of participants: ecologists, industrial ecologists, architects, and building product manufacturers. The workshop that was conducted to create the interaction of the selected individuals representing these fields occurred over a two-day period. Prior to the workshop the participants were provided with a notebook of information and papers to tutor them on the most recent developments in ecology and industrial ecology, as well as the state of the art in green building. The purpose was to cross-inform the participants about the others' disciplines. In the initial phase of the workshop all participants gave a presentation on their own work and provided some initial responses with respect to the goals of the workshop. The participants were then divided into two groups, with all disciplines represented in each group. The groups collaborated for several sessions to begin the process of applying ecology and industrial ecology to the built environment. The chapters in this book are the reflections of the participants on their work and its application to construction as affected by their interactions with colleagues in the other disciplines. In many cases, the authors provide background on their work in their chapters for two reasons. First, because this is a cross-disciplinary effort, there is the need to provide the reader with the vocabulary and state of the art in each discipline. Second, this background is needed to understand the conclusions reached by the authors with respect to the development of an ecology of construction. The chapters were written after the collaboration and interaction of the participants in the workshop and need to be understood in this context.

Chapter 1, an introduction to the concept of construction ecology and metabolism, was written by Charles Kibert in collaboration with Jan Sendzimir, an ecologist, and G. Bradley Guy, an architect. It presents some initial thoughts on the interplay of the built environment, ecology, and industrial ecology and is meant to lay the groundwork for the reader to understand the rationale for the development of construction ecology and metabolism. It discusses some of the specific difficulties facing the green building movement due to the lack of a coherent philosophy and design principles. Some additional thoughts on how buildings will have to change to more closely resemble nature are provided.

Part 1, The Ecologists (Chapters 2–5), contains the thoughts of four eminent ecologists: H.T. Odum, James Kay, Tim Allen, and Garry Peterson. It is introduced by Jan Sendzimir, who reviews the work of these contributors and provides the reader with insights into both current ecological theory and its application to the built environment.

In Chapter 2, H.T. Odum discusses the application of systems theory to the built environment. James Kay develops a powerful overview of how to consider the interface between human and natural systems in Chapter 3. The issues of surprise and emergence are discussed by Tim Allen in Chapter 4. In Chapter 5, Garry Peterson discusses ecological resilience, ecological change, and ecological scales and their relevance to the human built environment.

Developments in industrial ecology that may apply to the built environment are the subject of Part 2 (Chapters 6–9), which is introduced by Charles J. Kibert and contains the contributions of four eminent industrial ecologists: Robert Ayres, Iddo Wernick, Stefan Bringezu, and Fritz Balkau. In Chapter 6, Robert Ayres focuses the reader on the problem of creating a zero-emissions built environment from an industrial ecology point of view and suggests several materials and energy strategies to achieve this end. Iddo Wernick, in Chapter 7, suggests several strategies for the built environment that would reduce resource consumption and impacts on natural systems. In Chapter 8, Stefan Bringezu suggests that the key issue for the built environment, as for other industrial sectors, is maximizing resource efficiency. Fritz Balkau addresses how to manage the implementation of industrial ecology concepts in Chapter 9.

Part 3 (Chapters 10–12) is introduced by G. Bradley Guy and provides the thoughts of the architecture collaborators: Sim Van Der Ryn, Robert Peña, Jürgen Bisch, and Malcolm Wells. Sim Van Der Ryn and Robert Peña address the well-developed concepts of ecological design in Chapter 10. In Chapter 11, Jürgen Bisch describes a personal and historical journey toward architecture grounded in ecology and the evolution of a design process that is sensitive to nature and both respectful of and informative for his clients. His efforts to dematerialize buildings and couple them to the natural systems on the building site portray what may in fact be the cutting edge of applying ecology to construction. Malcolm Wells describes the total systems approach he has evolved over 40 years of architectural practice in Chapter 12. His conclusion is that the built environment should in fact impinge as little as possible on the planet's surface, and his fundamental strategy for achieving this is to place buildings underground, where ground contact and beneficial interaction with natural systems is maximized.

The final chapter, Conclusions, provides an overview of the outcomes of this initial collaboration of ecologists, Industrial Ecologists, and architects. This overview is organized into three major sections: (1) recommendations and agreements; (2) critical issues requiring further investigation; and (3) additional observations.

## Summary and conclusions

Developing an ecology of construction is an important step in the evolution of green building movements around the world. In almost every case, national and international movements indicate ecological design and the adoption of ecological principles as key elements of green building or sustainable construction. However, theory and practice developed in ecology have had little or no actual connection to green building and, as a consequence, the green building movement, although highly successful, stands a fair chance of ultimate failure because of the lack coherent underpinnings. This book suggests that for the green building movement to emerge successfully as standard practice, it must rely on ecology and industrial ecology to provide the coherent structure and science needed to serve as the basis for developing new theory, practices, and experiments; for decision making; and to serve as its general compass.

In addition to providing a badly needed basis for green building, construction ecology will require built environment professionals to dramatically increase their understanding of ecology. Because it could be said that the *raison d'être* for green building is the protection of natural systems, the re-education of planners, architects, builders, and policy-makers in ecological theory can only result in more sharply focused attention on the important issues of green building. The introduction of ecological education as a requirement for a professional design or construction license to impact natural systems in the dramatic fashion characteristic of the built environment would seem to be a highly beneficial and worthwhile outcome for society.

# 1 Defining an ecology of construction

*Charles J. Kibert, Jan Sendzimir, and G. Bradley Guy*

The construction and operation of the built environment has disproportionate impacts on the natural environment relative to its role in the economy. Although it represents about 8% of gross domestic product (GDP) in the USA, the construction sector consumes 40% of all extracted materials, produces one-third of the total landfill waste stream, and accounts for 30% of national energy consumption for its operation. The sustainability of this industrial sector is dependent on a fundamental shift in the way in which resources are used, from non-renewables to renewables, from high levels of waste to high levels of reuse and recycling, and from products based on lowest first cost to those based on life cycle costs and full cost accounting, especially as applied to waste and emissions from the industrial processes that support construction activity. Construction, like other industries, would benefit from observing the metabolic behavior of natural systems, in which sustainability is a property of a complex web of niche elements. The emerging field of industrial ecology, which is examining Nature for its lessons for industry, provides some insights into sustainability in the built environment or sustainable construction. This book proposes and outlines the concept of construction ecology, a view of construction industry based on natural ecology and industrial ecology for the purpose of shifting construction industry and the materials and manufacturing industries supporting it onto a path much closer to the ideals of sustainability. Additionally, construction ecology would embrace a wide range of symbiotic, synergistic, built environment–natural environment relationships to include large-scale, bioregional, "green infrastructure" in which natural systems provide energy and materials flows for cities and towns and the human occupants provide nutrients for the supporting ecological systems.

## Introduction

Ecosystems are the source of important lessons and models for transitioning human activities onto a sustainable path. Natural processes are predominantly cyclic rather than linear; operate off solar energy flux and organic storages; promote resilience within each range of scales by diversifying the execution of functions into arrays of narrow niches; maintain resilience across all scales by operating functions redundantly over different ranges of scale; promote efficient use of materials by developing cooperative webs of interactions between members of complex communities; and sustain sufficient diversity of information and function to adapt and evolve in response to changes in their external environment. A variety of approaches to considering the application of natural system design principles to the industrial subsystem of human activities is emerging to help redesign the conduct of a linear economy based largely on the consumption of non-renewable resources.

Industrial ecology is an emerging discipline that is laying the groundwork for adapting ecosystem models to the design of industrial systems. In more recent thinking, industrial ecology is being redefined and extended to include industrial symbiosis, design for the environment (DFE), industrial metabolism, cleaner production, eco-efficiency, and a host of other emerging terms describing properties of a so-called "eco-industrial system." Industrial symbiosis refers to the use of lessons learned from the observation of ecosystem behavior to make better use of resources by using existing industrial waste streams as resources for other industrial processes. An emerging discipline, DFE is altering the design process of human artifacts to enhance the reuse and recycling of material components of products. Industrial metabolism examines the inputs, processes, and outputs of industry to gain insights into resource utilization and waste production of industry, with an eye toward improving resource efficiency. Cleaner production is the systematic reduction in material use and the control and prevention of pollution throughout the chain of industrial processes from raw material use through product end of life (Business and the Environment 1998). Eco-efficiency calls on companies to reduce the material and energy output of goods and services, reduce toxic waste, make materials recyclable, maximize sustainable use of resources, increase product durability, and increase the service intensity of goods and services (Fiksel 1994).

Construction and operation of the built environment in the countries in the Organization for Economic Cooperation and Development (OECD), i.e. the major industrial countries, accounts for the greatest consumption of material and energy resources of all economic sectors and could benefit the most from employing natural systems models. Within the framework being defined by industrial ecology, construction industry would be well served by the definition of a subset, construction ecology, that spells out how this industry could achieve sustainability, both in the segment that manufactures the products that constitute the bulk of modern buildings and in the segment that demolishes existing buildings and assembles manufactured products into new or renovated buildings. As is the case with other industrial systems, construction would be aided in this effort by an examination of its throughput of resource, i.e. its "metabolism."

This chapter examines the potential for construction industry to incorporate the lessons learned from both natural systems and the emerging field of industrial ecology, primarily in its materials cycles, but also at larger scale for regional energy and materials flows. It also explores the issue of dematerialization and its relevance to the built environment. In many respects, the construction industry is no different from other industrial sectors. However, there are enough differences, especially the long lifetime and enormous diversity of products and components constituting the built environment, that it requires special attention and treatment. Consequently, attempts to apply ecology to this industry and to understand its metabolism present some unique problems not encountered in other industrial sectors.

## Construction industry compared with other industrial sectors

Buildings, the most significant components of the built environment, are complex systems that are perhaps the most significant embodiment of human culture, often lasting over time measured in centuries. Architecture can be a form of high art, and great buildings receive much the same attention and adoration as sculpture and painting. Their designers are revered and criticized in much the same manner as artists. This character of buildings as more than mere industrial products differentiates them from most other artifacts.

Their ecology and metabolism is marked by a long lifetime, with large quantities of resources expended in their creation and significant resources consumed over their operational lives.

The main purpose of the built environment is to separate humans from natural systems by providing space for human functions protected from the elements and from physical danger. Modern buildings have increased the sense of separation from the natural climatic processes and have made the underlying biological and chemical processes of Nature irrelevant for their occupants. Until humans achieved space travel, the extraction and conversion of materials for building construction was the most powerful expression of humankind's dominance over bioclimatic and material constraints. This has. in turn. created an ecological illiteracy and had profound psychological and human health impacts (Orr 1994). Concentrations of buildings affect microclimate (heat islands), hydrology (run-off), soils and plants (suffocation and compression), and create false natural habitats (nests on buildings). This increasing separation of ecological feedback loops inherent in the design, construction, and use of buildings since the Industrial Revolution has influenced many architects to reconsider this de-evolutionary and unsustainable path. The construction industry is extremely conservative and subject to slow rates of change because of regulatory and liability concerns as well as limited technology transfer from other sectors of society. The extended chain of responsibility and the separation of responsibilities for manufacturing materials, design and construction, operations and maintenance, and eventual adaptation or disposal have resulted in a breakdown of feedback loops among the parties involved in creating and operating the built environment.

Modern buildings, although products of industrial societies, are perhaps unique among modern technologies in terms of the diversity of components, unlimited forms and content, waste during the production process, land requirements, and long-term environmental impacts. In the USA, the construction industry, although representing only 8% of GDP, uses in excess of 40% of all extracted materials resources in creating buildings (Wernick and Ausubel 1995), which consume 30% of total US energy production in their operation. It is estimated that as much as 90% of the extracted stock of materials in the USA is contained in the built environment, making it a potential great resource or a future source of enormous waste.

The built environment interacts with the natural environment at a variety of levels. Individual structures may affect only their local environment, but cities can have an impact on the regional environment, by affecting the weather through changes in the Earth's albedo (Wernick and Ausubel 1995) and other surface characteristics, altering natural hydrologic cycles, and degrading air, water, and land via the emissions of their energy systems, as well as through the behavior of their inhabitants.

Buildings can be distinguished from other artifacts by their individuality and the wide variety of constituent parts. Buildings are assembled from a wide array of components that can be generally divided into five general categories:

1. manufactured, site-installed commodity products, systems, and components with little or no site processing (boilers, valves, electrical transformers, doors, windows, lighting, bricks);
2. engineered, off-site fabricated, site-assembled components (structural steel, precast concrete elements, glulam beams, engineered wood products, wood or metal trusses);
3. off-site processed, site-finished products (cast-in-place concrete, asphalt, aggregates, soil);

4 manufactured, site-processed products (dimensional lumber, drywall, plywood, electrical wiring, insulation, metal and plastic piping, ductwork);
5 manufactured, site-installed, low mass products (paints, sealers, varnishes, glues, mastics).

Each of these categories of building components has an influence on the potential for reuse or recycling at the end of the building's useful life and the quantity of waste generated during site assembly. Category 1 components, because they are manufactured as complete systems, can be more easily designed for remanufacturing, reuse, and disassembly, and thus have a excellent potential for being placed into a closed materials loop. Category 2 products also have this potential although engineered wood products, a relatively new technology, have not been scrutinized as to their fate. Concrete products fit into the first three categories, and the extraction of aggregates for further use is technically and, in many cases, economically feasible (Figure 1.1). Category 4 products are in some cases more difficult to reuse or recycle, although metals in general are recycled at a very high rate in most countries. Category 5 products are virtually impossible to recycle, and in many cases are sources of contamination for other categories of products, making their recycling more difficult.

Buildings as artifacts of human society are also distinguished to a large extent by their relatively large land requirements and the environmental effects of the co-option of this

*Figure 1.1* Mixed rubble including concrete aggregate, brick, masonry, and ceramic tile has substantial economic value when it can be reused in concrete mix design, as is the case with this material in The Netherlands.

valuable ecological resource. The built environment significantly modifies natural hydrologic cycles, contributes enormously to global environmental change, has tremendous effects on biodiversity, contributes to soil erosion, has major negative effects on water and air quality, and is the source of major quantities of solid waste. In the USA, construction and demolition waste accounts for the majority of industrial waste, amounting to perhaps 500 kg per capita or of the order of 136 million metric tons (MMT) annually. In the USA, the proportion of this waste that is reused or recycled is not known, but it is probably under 20% of the total mass and perhaps closer to 10%. Only concrete, recycled for its aggregates, and metals are recycled at high rates because of their relatively high economic value.

The construction industry also differs from other industrial sectors in that the end products, buildings, are not factory produced with high tolerances, but are generally one-off products designed to relatively low tolerances by widely varying teams of architects and engineers, and assembled at the site using significant quantities of labor from a wide array of subcontractors and craftspeople. The end products or buildings are generally not subject to extensive quality checks and testing and they are not generally identified with their producers, unlike, for example, automobiles or refrigerators. Unlike the implementation of extended producer responsibility (EPR) in the German automobile industry, which is resulting in near closed-loop behavior for that industry, buildings are far less likely to have their components returned to their original producers for take-back at the end of their life cycle. Arguably, EPR could be applied to components that are routinely replaced during the building life cycle and that are readily able to be decoupled from the building structure (chillers, plumbing fixtures, elevators). The bulk of a building's mass is not easily disassembled, and at present little thought is given in the design process to the fate of building materials at the end of the structure's useful life (Figures 1.2 and 1.3).

Most industrial products have an associated lifetime that is a function of their design, the materials constituting them, and the character of their service life. The design life of buildings in the developed world is typically specified in the range of 50–100 years. However, the service lives of buildings are unpredictable because the major component parts of the built environment wear out at different rates, complicating replacement and repair schedules. Brand (1994) describes these variable decay rates as "shearing layers of change," which create a constant temporal tension in buildings (Figure 1.4). Brand adapted O'Neill *et al.*'s (1986) hierarchical model of ecosystems to illustrate the issue of temporal hierarchy in buildings that can be related to the spatial decoupling of components. Faster cycling components such as space plan elements are in conflict with slower materials such as structure and site. Management of a building's temporal tension might be achieved with more efficient use of materials through spatial decoupling of slow and fast components. Components with faster replacement cycles would be more readily accessible. This hierarchy is also a hierarchy of control, i.e. the slower components will control the faster components. However, when the physical or technical degradation of faster components surpasses critical thresholds, the faster components begin to drive changes to the slower components such that dynamic structural change can occur. For example, in a typical office building electrical and electronic components wear out or become obsolete at a fairly high rate compared with the long-lived building structure. At some critical threshold the motivation to maintain the overall building ebbs and the building rapidly falls into disuse and disrepair simply because of the degradation of the faster, more technology-dependent components. Odum (1983) developed the concept of

*Figure 1.2* A 1960s era student residence hall at the University of Florida in the process of demolition. Although 12,000 bricks were recovered for reuse in new construction, in excess of 90% of the brick was unrecoverable because of the high-strength Portland cement mortar used to bind the bricks together and the lack of provision for disassembly of multistorey brick structures.

"emergy," the energy embodied in the creation and maintenance of a factor or process, as a means to quantify the relative contributions of different components to the operation of a hierarchy (see Chapter 2). Odum's theory predicts that the control of faster components by slower components is reflected in the latter's higher emergy transformity values. Transformity values are efficiency ratios of total emergy to actual energy, normalized in solar equivalent joules, that enumerate a process's relative capacity to influence system behavior. Using emergy to distinguish more carefully between slower and faster components and processes would allow designers to couple buildings to external processes of manufacture, reuse, and recycling more rationally. As such, this theory provides a quantitative framework for relating building design to its material components based on their relative contributions to the functions of an "ecosystem" that includes the built environment and the materials and processes that sustain it.

There are many similarities between construction industry and other industrial sectors when it comes to materials utilization. First, construction is in fact closely tied to industries that produce many of the products that ultimately constitute the built environment. Consequently, segments of these industries providing built environment products could be said, in a larger sense, to be a part of the construction industry. Thus, the environmental and resource impacts of their production systems are an integral part of the overall impacts of the built environment. Closing the loop for the manufactured systems that constitute

*Figure 1.3* Pallet of bricks recovered from the student residence hall at the University of Florida.

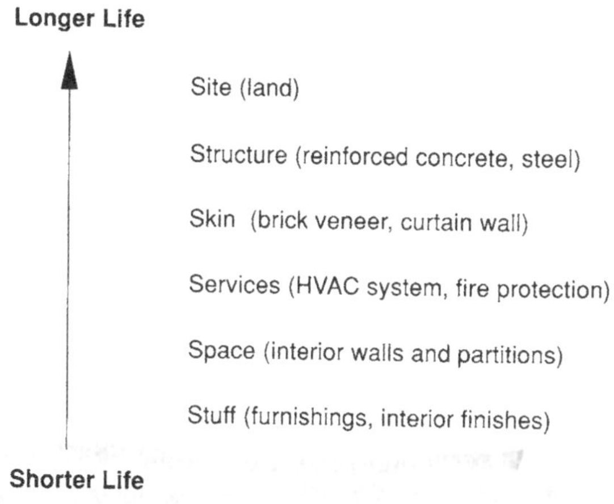


*Figure 1.4* Temporal hierarchy of building components (after Brand 1994).

much of the cost of construction is dependent on the originating industries applying DFE principles to allow disassembly and recycling or reuse of components as well as the establishment of reverse distribution systems in tandem with EPR legislation.

Placing construction materials into a closed-loop system is hampered by many of the same problems hampering other industrial sectors, such as the automotive and electronic industries. Components are often made of materials that are difficult if not impossible to recycle. They are, for the most part, not designed for disassembly to facilitate recycling. There are no requirements for manufacturers or suppliers to take back assembly waste or the worn-out products. The situation with regard to closing the materials loops varies from country to country, with, for example, US industry at one extreme and German industry at the other. American industry functions in an economy marked by a strong culture of almost pure market response, low levels of government intervention, and a history of cheap resources and low waste disposal costs. Consequently, recycled content or remanufactured products must compete with virgin resource-based products that are subject to only minor environmental cost internalization at best. The US federal government and other public sector organizations have a recent history of requiring the procurement of recycled content products, and consumer surveys have shown a favorable response to environmentally sensitive products. Several federal agencies, such as the National Park Service, the US Post Office, and several branches of the Department of Defense, are requiring "green" building design. This is having an impact on the building products industries, on the building design professions, and on the buying public. However, the purchasing cost of building products reflects little or no shifting of responsibility for environmental impacts addressing, for example, poor forestry practices, production emissions, and, in the case of construction, relatively large demolition and building assembly waste. German industry functions within a strong regulatory framework that constrains industry to a higher standard of materials use than is present in the USA. The Duales Deutschland System, EPR, and other regulatory systems are forcing German manufacturers to create products that are taken back by their producers to become raw materials for new products.

## Materials and sustainability

Sustainability is affected by anthropogenic materials use as a result of (1) environmental effects of mass materials movement during extraction, (2) depletion of high-quality mineral stocks for industrial use, and (3) dissipation of concentrated materials resulting from wear and emissions. Mass materials movements and their negative environmental impacts are a recently identified phenomenon. As humans deplete the relatively accessible and valuable stocks of minerals, there are fewer of these resources available for future generations, and the energy needed to extract more dilute stocks and the distances to them will both undoubtedly increase. The dissipation of artifacts is the thermodynamic equivalent of increasing entropy or conversion from useful to useless (Georgescu-Roegen 1971; Ayres 1993).

The Earth, along with its biosphere, is essentially a closed system with respect to materials and materials flux. Organizations studying materials cycles are producing convincing arguments that the environmental damage caused by extraction of primary materials is exceeding the capacity of natural systems to cope with the damage being caused by the mass material movements accompanying their extraction. The Wuppertal Institute estimates that the materials flux of human processes is twice the flux caused by all natural forces and systems combined, including hurricanes, earthquakes, tornadoes, and volcanoes, excluding sea floor spreading and continental subduction. Almost thirty years ago, Brown (1970) suggested that humankind had already become a major geologic force. He noted the need for increased recycling efficiency and a lowered demand on extraction as the source for metals to both protect the environment and address the worldwide disparity in resource availability between rich and poor nations. Accompanying the Wuppertal Institute scenario is the hypothesis that sustainability requires that the human-induced materials flux should be no greater than the natural flux. Parallel to the enormous quantities of matter being moved by humankind is the co-option of the order of 40% of all terrestrial and aquatic biomass by humans for their own use at the expense of all other species (Vitousek *et al.* 1986). Additionally, humans are also co-opting over 50% of all accessible water run-off worldwide, which is expected to increase to 70% in the next three decades (Postel *et al.* 1996). One-third to one-half of the Earth's surface has been transformed by human activities, and more nitrogen is fixed by humans than by all natural sources combined (Vitousek *et al.* 1997). The introduction of tens of thousands of synthetic chemicals, many of them hazardous, into the global environment is another factor that is causing documented illnesses and disturbances to the reproductive systems of animals, including humans, throughout the world. The net effect of all these human disturbances is not clearly understood, but the result can only be catastrophic if these trends continue, especially if synergism and positive feedback loops amplify these negative effects.

With regard to materials, the Wuppertal Institute suggests that the materials input per service unit (MIPS) must be reduced by a factor of 10 to move into a regime that could be considered sustainable (von Weizsäcker *et al.* 1997). Alternatively, it could be said that resource efficiency must be increased by a factor of 10 to achieve the same end. The Factor 10 Club is laying the groundwork for an international effort that originated with the Cournoules Statement in 1994, calling on industry and governments to transform their policies to effectively dematerialize their countries' economies. Dematerialization is the reduction of the quantities of materials needed to serve economic functions or the decline over time in the weight of materials used in industrial end products (Wernick *et al.* 1996). It should be noted that this proposal for dematerialization does not distinguish between virgin and recycled or reused resources. Closing materials loops could produce, in effect, a factor 10 reduction in human-induced materials flux from the Earth, with a far smaller reduction in aggregate materials throughput.

In addressing dematerialization, Bunker (1996) notes that, instead of being an environmental or sustainable development response, dematerialization is not much more than an attempt to increase profitability, that it is not a new idea because industry always strives to lower the unit costs of production. The intensity of use (IOU) index measures materials mass per unit of GDP, and for all industrialized countries IOU indices have been generally falling for many decades, indicating, by this metric, a steady dematerialization of their economies. In fact, industries compete to offer ever more lines of products, increase labor productivity, and in effect, increase demand and the consumption of materials. In housing, for example, over the past thirty years the average American home has steadily increased in size from 170 to 220 square meters while the average number of occupants has fallen from 3.5 to 2.5. Aggregate materials use or throughput, in contrast to IOU indices, is steadily increasing, and environmental damage is climbing proportionately. Also neglected in discussions of dematerialization are the toxic by-products associated with extraction and processing of, for example, metals such as copper, zinc, platinum, and titanium. Part of the problem with clearly assessing dematerialization is the substitution of lighter weight materials for heavier ones. In what is a classic scenario in materials use, high-technology polymers and carbon composites are rapidly replacing metals in many applications (Williams *et al.* 1989). Although dematerialization in an IOU sense is occurring by shifting to these alternatives, the environmental damage caused by the production of these materials and their general non-recyclability can make the benefits of dematerialization questionable.

True dematerialization must focus on virgin resource extraction rather than just dematerialization in the IOU sense, and the environmental impacts of the technologies and substitutions creating dematerialization need to be carefully scrutinized. Dematerialization must also focus on a shift to reuse, recycling, and remanufacturing, all important aspects of closing materials loops. Additionally, de-energization, decarbonization, and detoxification of the industrial system should accompany dematerialization if significant resource and ecological benefits are to be achieved. It must also be kept in mind that, although human ingenuity can perhaps effectively dematerialize the global economy, Ayres (1993) notes:

> There are no plausible technological substitutes for soil fertility, clean fresh water, unspoiled landscapes, climatic stability, biological diversity, biological nutrient recycling and environmental waste assimilative capacity. The irreversible loss of species and ecosystems, and the buildup of greenhouse gases in the atmosphere, and of toxic metals and chemicals in the topsoil, groundwater and in the silt of lake-bottoms and estuaries, are not reversible by any plausible technology that could appear in the next few decades. Finally, the great nutrient cycles of the natural world – carbon, oxygen, nitrogen, sulfur, and phosphorous – require constant stocks in each environmental compartment and balanced inflows and outflows. These conditions have already been violated by large-scale and unsustainable human intervention.

Finally, Hayes (1978) suggested that a sustainable world would be one in which "Material well-being would almost certainly be indexed by the quality of the existing inventory of goods, rather than by the rate of physical turnover. Planned obsolescence would be eliminated. Excessive consumption and waste would become causes of embarrassment, rather than symbols of prestige."

## Lessons from natural systems

Many authors have suggested that human industrial systems can and must use the metaphor of biological systems as guidance for their design. The field of ecological engineering emerged from Odum's (1983) pioneering work, which explored how functions and services could be optimized at much greater efficiencies by integrating human and natural systems through adept redesign. These lessons can be explored at a large or systems scale as well as at the small or microscopic scale in terms of the metabolism of natural systems versus industrial systems. At large scale this might mean that industry should recast itself as an industrial ecosystem comprising an interrelated network of producers and consumers that would function much as a natural ecosystem (Frosch and Gallopoulos 1989, 1992; Frosch 1997). Industrial processes would function much as biological organisms in that excess energy and waste from some systems would serve as inputs for industries requiring energy and which can use the waste in their production systems (Ausubel 1992). After their "birth, life, and death" at one scale, the products of industry would ultimately be metabolized and reutilized at another scale, mimicking the closed, waste-free cycles of natural systems. This proposition has profound implications for designers and builders, given the scale of energy and materials use in the built environment.

There are many questions to be answered in attempting to redesign industry to behave like Nature. Do natural systems in fact use resources optimally or can technology actually improve on the energy and matter utilization of Nature, perhaps through observing Nature itself? Are there limits to using the natural system metaphor for industrial systems, and, if there are, what are they? Can humankind really live off current solar income, as has been suggested, or is this impossible if quality of life for present and future populations is to be maintained? What is the human-carrying capacity of the Earth if adequate natural systems functions are to be maintained? Can natural systems perform many critical functions required by humankind and, in effect, substitute for the work of industry in some cases? This last question was at least partly answered recently by a team that estimated the mean value of a range of natural systems to be about $33 trillion or almost twice the total world economic output (Costanza *et al.* 1997). However, this estimate is at best controversial in the sense it depends to a great extent on people's "willingness to pay" for ecological services.

Ayres (1989) described some of the analogies between natural and economic systems by noting that natural systems themselves might not have always been sustainable. Alternatively, it can be said that no natural system is sustainable over long time scales. Changes in natural systems reflect experiments that shift the composition of processes, functions, and species, both independently and in response to novelty of system composition or of context (changing conditions). Evolutionary history is studded with unprecedented leaps of novelty that rendered unsustainable many systems that had endured for eons. Stage 1 of life on Earth consisted of fermentation-based life forms functioning and replicating by anabolism, generating carbon dioxide waste that accumulated in the atmosphere. This "waste" proved to be the resource for the next leap of evolution. The anerobic stage 1 organisms were followed in stage 2 by organisms employing photosynthesis to utilize the carbon and discharge oxygen as waste, thus killing most extant biota, for which oxygen was a toxic gas. Oxygen, initially a "toxic waste," created the conditions for novel stage 3 organisms that utilized oxygen to metabolize a larger range of molecules and allowed them to function with far greater energy, stamina, and diversity of shapes and sizes in an enormous variety of new environments. The emergence of oxygen was a radical shift in context that permitted an explosive increase in opportunities for biota that would have been unimaginable beforehand. In this manner, novelty periodically resets the standard for what is sustainable (Holling *et al.* 1995). Ayres (1989) suggests that the present industrial system, so dependent on fossil fuel-based energy systems, is analogous to the stage 1 fermentation cells that essentially convert stocks of carbon fuels to waste carbon dioxide. A similar catabolism–anabolism metabolic behavior is characteristic of industrial systems except that industrial systems, unlike ecosystems, metabolize their energy-matter throughput into largely useless waste.

Another related view is that the current industrial systems are the equivalent of type I or pioneer species, also known as r-strategists, which rapidly colonize areas laid bare by fire or other natural catastrophes. Their strategy of maximum mobility and reproduction involves investing all their energy in seeds and rapid growth and minimizes investments in structure. r-strategists are mobile, surviving by being the first at the scene of a disturbance and securing resources before they are eroded away (Begon *et al.* 1990; Holling *et al.* 1995). However, when the resource base has been expended, their populations will diminish to very low levels. They are not competitive in the long run, and only excel at outcompeting each other in a loose "scramble competition," eventually losing out to the K-strategists. In natural succession, type II species supplant type I species because they spend less energy on generating seeds and more on systems such as roots that will enable their survival during periods of lower available resources. Type III or K-strategists live in synergy with surrounding species and are far more complex than the other types. K-strategists, unlike r-strategists, are not mobile but survive longer at higher density by developing highly efficient resource and energy feedback loops. Both K- and r-strategists are present everywhere, with r-strategists surviving in subdued populations in "older" sites and exploding in population in "younger," disturbed sites. K-strategists invest more in structure than in mobility, and this is the template around which their complex interrelationships efficiently conserve the flow of energy and resources. In a similar manner, it could be said that industrial systems behave in a similar fashion (Graedel and Allenby 1995; Karamanos 1995). Type I industrial ecosystems are the typical industrial processes of today, linear systems with little or no recovery of materials from the waste stream. Type II are emerging industrial ecosystems that include reuse and recycling in their processes but also require significant primary material inputs for their functioning. Closed-loop type III industrial ecosystems with full materials recovery do not exist at present, partly because of a lack of technology and partly because of poor product design. Perhaps industrial ecology is simply another stage in a process of never-ending change in which human-designed systems "naturally" evolve in a manner similar to natural ecosystems (Erkman 1997). The question for humankind that may emerge from this observation of nature is how to move as rapidly as possible from our current type I global economy to a type III economy or from a r-strategy to a K-strategy (Benyus 1997; Shireman *et al.* 1997).

Both natural and industrial systems require energy to reproduce and maintain their functions. Natural systems, for the most part, use solar flux or stored solar energy in the form of biomass for their functioning whereas industrial systems use a wide variety of energy sources. The intensity of industrial operations requires energy sources that are refined to the highest quality by geological forces operating over millions or years. Natural systems are characterized by their use of renewable energy sources. In the present era, industrial systems operate largely by using stored solar energy in the form of fossil fuels, but these are being consumed at a pace on the order of 10,000 times their regeneration rate.

Natural systems are sustained by the emergence of surprise (Holling 1986) and novelty (Kauffman 1993) and by the diversity of information found in genetic codes, which instruct the fabrication and operation of organisms. This diversity is present at several levels: within each population of a species, across all populations of a species, and across all species in communities (Begon *et al.* 1990). Ecosystems are also sustained by a diversity of ecological functions that process energy, matter, and information in a shifting balance of competitive and cooperative relations. Functional diversity is maintained at several levels. Within each range of scales, different functions are partitioned among species that occupy separate, narrow niches. For example, at the scale of a stand of trees, a variety of different bird and mammal species occupy individual niches, each of which focuses on a different resource (insects, fruits, and seeds). Functional diversity is maintained across all scales in a system when specific functions are performed at different scales. For example, tiny birds may eat individuals of a species of insect found on tree branches, but flocks of birds will appear to eat the same insect when that insect's population explodes and it is evident as a swarm across an entire stand of trees (Peterson *et al.* 1998). Industrial systems tend to function similarly, with technological information being the equivalent of genetic codes (Rothschild 1990). They cooperate through strategic alliances and absorb one another through acquisitions. They struggle to occupy niches and fiercely compete to dominate their environment, their markets. It would be interesting to determine the degree to which industrial systems are more resilient because of the redundancy of function within scales and across scales. For example, corporate buy-outs or bankruptcies can reduce the diversity of companies performing the same function at the same scale. Cross-scale resilience, the performance of the same function at different scales, is lowered by the replacement of local manufacturers by enterprises distributing goods at larger scales (national or global).

Benyus (1997) coined the word "biomimicry" to describe the use of lessons from the natural world to develop a concept of sustainability for humankind. One example she provides is the powerful natural adhesives produced by mussels to anchor themselves to rocks in strong ocean currents and how scientists are studying the biological processes that are used in their synthesis. By learning about the chemistry of natural systems, the potential exists to create whole new classes of materials that are strong and lightweight yet can be decomposed into harmless substances when they have outlived their use. Benyus suggests that the questions that should be put to new innovations and the industrial systems that produce them, if they are to mimic nature are: Does it run on sunlight? Does it use only the energy it needs? Does it fit form to function? Does it recycle everything? Does it reward cooperation? Does it bank on diversity? Does it utilize local expertise? Does it curb excess from within? Does it tap the power of limits? Is it beautiful?

## Industrial ecology and metabolism

Industrial ecology can be defined as the application of ecological theory to industrial systems or the ecological restructuring of industry (Rejeski 1997). In its implementation it addresses materials, institutional barriers, and regional strategies and experiments (Box 1.1) (Wernick and Ausubel 1997). Industrial metabolism is the study of the flow of materials and energy from the natural environment, through the industrial system, and back into the environment. It is directed at understanding the flows of materials and energy from human activities and the interaction of these flows with local ecosystems, regions, and global biogeochemical cycles (Erkman 1997).

As noted by Graedel and Allenby (1995), the rejection of the concept of "waste" is one of the most important outcomes of global biogeochemical cycles. In an ideal industrial system, both renewable and non-renewable materials would be utilized in a closed loop to minimize the input of virgin resources. Products degraded by age or service would be designed to be reverse-distributed back to industry for recycling or remanufacturing.

*Box 1.1* Issues confronting the implementation of industrial ecology

1 The material basis
- choosing the material
- designing the product
- recovering the product

2 Institutional barriers and incentives
- market and institutional
- business and financial
- regulatory

3 Regional strategies and experiments
- geographic, economic, political
- industrial symbioses

The processes creating the loops would be designed for zero solid waste to include zero emissions to water and air. Renewable resources would also be used in a closed-loop manner to the maximum extent possible and follow the same zero waste rules as for non-renewables. Renewable resources, being biological in origin, could be recycled by natural processes as simple biomass that could serve as nourishment for biological growth.

According to Richards and Frosch (1997), "… industrial ecology views environmental quality in terms of the interactions among and between units of production and consumption and their economic and natural environments, and it does so with a special focus on materials flows and energy use." They also go on to note that the integration of environmental factors can occur at three scales:

- microlevel (the industrial plant);
- mesolevel (corporation or group operating as a system);
- macrolevel (nation, region, world).

It is interesting to note that these three levels are identical to the levels at which natural systems are studied for their function.

Industrial ecology has evolved in two major directions since it became well known in the late 1980s. The first direction is the evolution of the concept of eco-industrial parks (EIPs) from "industrial symbiosis" to a much broader range of sustainability benefits. Under the rubric of industrial symbiosis, companies exchange otherwise useless waste as resources. The cluster of companies exchanging energy, water, and material by-products at the Kalundborg EIP in Denmark is the most frequently cited success story of industrial symbiosis. The EIP concept has rapidly evolved beyond mere symbiosis to encompass broader ideas of sustainability and may include shared workforce training, daycare centers, business incubators, provision of cleaner production consulting services, and collaboration with community leaders to establish public–private partnerships that benefit the local populace. Extending the concept of benefit sharing to regional scale can hypothetically result in "islands of sustainability."

The second major direction of industrial ecology is the optimization of materials flows by increasing resource productivity or dematerialization. The notion of a service economy which sells services instead of the actual material products is considered the *sine qua non* of this strategy, alternatively referred to as "systemic dematerialization." One of the questions facing industrial ecology is whether corporations can profit more from closing materials loops and behaving environmentally responsibly or through built-in obsolescence and open materials cycles (Erkman 1997). Perhaps a more fundamental question for industrial ecology is how to achieve a transition to a clean, highly resource-efficient industry while maintaining the viability of the economy and the profitability of corporations.

Industrial ecology also embraces the emerging new discipline known as design for the environment (DFE), which has as its goal the creation of artifacts that are environmentally responsible. DFE can be defined as a practice by which environmental considerations are integrated into product and process engineering procedures and which considers the entire product life cycle (Keoleian and Menerey 1994). A fundamental goal is balancing environmental, business, and technical considerations in product and process design. The term "green design" is used interchangeably with DFE, and is defined as " … not a rigid set of product attributes, but rather a decision process whose objectives depend upon the specific environmental problems to be addressed" (Office of Technology Assessment 1992). There are several complementary terms frequently used to describe various aspects of DFE: design for disassembly, design for remanufacturing, design for recycling, design for reuse, and others. They have as their common denominator the consideration of environmental effects and resource efficiency in the design of artifacts. This proactive approach to creating objects that can be readily adapted, removed, reprocessed, recycled and reused, embodies the concept of "front-loaded" design (Wilson *et al.* 1998). Front-loaded design has implications related to the coding of information in the genetic structure of an organism that dictates its life stages from inception to maturation and decline, including the ability to adapt to changing environmental conditions. The practice of coding materials according to chemical composition so that they can be more readily recycled when the product is disassembled has utility for the construction industry in closing materials flows loops in a cost-effective manner. The fuller design for energy efficiency, materials efficiency, and human and environmental health can only be achieved by encoding a "natural" lifespan within the materials and design of buildings.

Another related endeavor, ecological design, is also becoming a part of the tools included in a broad vision of industrial ecology and is defined as " … any form of design that minimizes environmentally destructive impacts by integrating itself with living processes" (Van Der Ryn and Cowan 1996). Examples of ecological design are sewage treatment systems that use constructed wetlands to process wastewater, homes that use dimensional lumber from sustainably managed forests, and agricultural practices that mimic natural plant communities. According to Van Der Ryn and Cowan, ecological design uses three key strategies to protect critical natural capital: conservation, regeneration, and stewardship. Conservation acknowledges the finiteness of resources and is directed at reducing the rate of their consumption. Repairing the damage done to ecological systems is the strategy of regeneration and is evidenced on a large scale by the ongoing effort to restore the natural flow characteristics of the Florida Everglades. Stewardship or deliberate care of ecosystems is a long-term commitment and marks an attempt to change the fundamental attitudes of humanity to nature. Ecological design is based in place and knowledge of local ecosystems to include energy and materials flows. In terms of materials use, ecological design consists of "Restorative materials cycles in which the waste from on process becomes food for the next; designed-in reuse, recycling, flexibility, ease of repair, and durability" (Van Der Ryn and Cowan 1996).

## Ecologically sustainable architecture and construction

In the modern era of building design, several key figures emerge as the initiating forces whose ideas have coalesced into today's ecologically sustainable architecture. Among them are Frank Lloyd Wright, Richard Neutra, Lewis Mumford, Ian McHarg, and Malcolm Wells.

Frank Lloyd Wright's early education was under the tutelage of his mother, who employed Friedrich Froebel's Nature-based training. This early exposure to natural systems had a powerful influence on his life and architecture. The young Wright learned Nature's forms and geometries and " … as a result of his training [Wright] was far more interested in designing the world than in representing it; designing understood as discerning the underlying structure of nature and building on it" (McCarter 1997). As the prophet of "organic architecture," his goal was to create buildings that were " … integral to the site, to the environment, to the life of the inhabitants, integral with the nature of the materials …" (Wright 1954).

Richard Neutra, a pupil of Wright, starts off his seminal work, *Survival through Design* (1954), with the following statement on the contemporary condition of man's relationship with nature:

> Nature has too long been outraged by design of nose rings, corsets, and foul-aired subways. Perhaps our mass-fabricators of today have shown themselves out of touch with nature. But ever since Sodom and Gomorrah, organic normalcy has been raped again and again by man. That super-animal still struggling for its own balance.

Neutra points out how badly flawed human products are compared with Nature's offerings. Nature is dynamic while human artifacts are static and cannot self-regenerate or self-adjust. While nature evolves "naturally," progress in the human sense takes deliberate effort, energy, and considerable motivation. Human artifacts generally follow the precedents in Nature anyway; for example, a reinforced concrete structure bears more than glancing similarity to the skeletal structure of a vertebrate. Even coloration to stimulate interest in a building mimics the shades and tones of flora and fauna seeking to procreate. Nature' s form and function emerge simultaneously, while humans must first create a building's form and then allow it to function. Imitating nature is more than flattery on the part of humankind, it is also copying systems that function in an extraordinarily successful fashion. Neutra was also able to recognize what is now referred to as *biophilia*, by advocating close connections between living spaces and the "green world of the organic" (Neutra 1971).

Lewis Mumford articulated the problematic drift from preindustrial cities that were sensitive to Nature and geographic conditions to post-industrial versions that ignored Nature and sprawled, destroying compact urban forms as well as the countryside. He advocated the employment of *ecotechnics*, technologies that rely on local sources of energy and indigenous materials in which variety, craftsmanship, and vernacular are important, not only adding to our ecological consciousness, but also emphasizing beauty and aesthetics (Luccarelli 1995). An updated version of this thinking includes the concepts of *bioregionalism* and *biourbanism*, infrastructure design that maximizes the "free work" that natural systems can provide. Stormwater handling, clean-up of phosphate pollution from agriculture, and sanitary sewage treatment could all be handled by wetlands and other natural systems, eliminating much of the need for the typical infrastructure needed for these purposes (Williams 1999).

Until rather recently, the planning of the built environment existed in isolation, as has, to a great extent, been the case with many human endeavors. McHarg (1969) noted the glaring deficiencies in planning:

> The first was the absence of any knowledge of environment in planning – this was a totally applied socio-economic process. The next was the lack of integration within the environmental sciences. Geologist, meteorologists, hydrologists and soil scientists were informed in physical science, unknowing of life. Ecology and the biological sciences were only modestly aware of physical processes. Scientists in general had not revealed any interest in values nor in planning; and finally, there was no theory attempting to address the problem of human adaptations

Malcolm Wells (see Chapter 12) favors a fairly straightforward approach to ecological design, i.e. simply submeging buildings into the ground so that the surface of the planet is minimally affected by the buildings. He contends that the surface above the earth-covered buildings can still provide the same services as they did before the construction of the building. In his book *Gentle Architecture* (1981), Wells asks an important question: "Why is it that every architect can recognize and appreciate beauty in the natural world and yet so often fail to endow his own work with it?" He often criticized the greatest architects of that time of failing to be aware of or moved by the biological foundations of both life and art.

Since the beginning of the 1990s, many organizations worldwide have been articulating a concept, commonly known as sustainable construction, that seeks to change the nature of how the built environment is designed, built, operated, and disposed of. It is in some ways a return to the basics advocated by Wright, Neutra, Mumford, McHarg, and Wells, a return to organic, Nature-based design. Sustainable construction considers the life cycle of the built environment as a seamless continuity, from design through disposal. It can be defined as "the creation and maintenance of a healthy built environment using ecologically sound principles." Its goals are to maximize resource efficiency and minimize waste in the building assembly, operation, and disposal processes. Sustainable construction seeks to dovetail into the global sustainable development movement by moving construction industry onto a path where it adheres to principles that are able to provide a good quality of life for future generations. In doing so, the sustainable construction effort is considering how to alter construction materials cycles to reduce their environmental and resource impacts.

Executing the concept of sustainable construction poses numerous difficulties. For example, contemporary architecture's efforts relative to materials analysis and their sustainability aspects generally rely on vague criteria, resulting in a great deal of imprecision. Many of these criteria are intuitive and perhaps even sensible. A typical approach is to specify, as much as possible, materials that are natural and renewable, that are local and indigenous, and which have low embodied energy (Steele 1997). Specifying local materials has been cited as climatically appropriate, supportive of the local economy, and more economically viable (Zeiher 1996). An additional criterion that often appears in contemporary guides to "green" building materials is to avoid the use of synthetic materials. What is meant by "synthetic" is not clearly defined. Concrete could be said to be synthetic, yet variants of modern concrete have been used during all of recorded history. Metals, especially alloys, which constitute the bulk of metals in use, could also be considered synthetics. Plastics clearly fit this category in spite of their potential benefits and the fact that some varieties can be made from biomass.

In many countries, there are now references available to aid in the selection of low-impact, green building materials. The Environmental Policy of the Royal Australian Institute of Architects provides several principles for guiding materials selection. Principle 2 of the Environmental Policy calls for architects to minimize the consumption of resources and is accompanied by recommendations on how to implement this principle via use of renewable resources, recycling buildings and using recycled components, and designing for durability (Lawson 1996). *The Green Guide to Specifications* (UK Post Office Property Holdings) categorizes materials based on environmental issues: (1) toxic pollutants in manufacturing; (2) primary energy used in extraction, production, and transport; (3) emissions; (4) material, water, and oil resources; (5) reserves of raw materials; (6) wastes generated; and (7) various recycling aspects (Shiers *et al.* 1996). The resulting table of ratings (A, B, or C, with A being the best and C the worst rating) for each material or system contains sixteen ratings based on the environmental issues as well as cost, maintenance frequency, and replacement intervals. This depiction of materials performance, fairly typical of information being provided to building professionals, provides a large array of data and editorial opinions. Unfortunately, the various units of data (embodied energy, extracted materials mass, emissions rates) are incommensurable, i.e. they are unable to be reduced to a smaller number of factors that are tractable. Consequently, making decisions based on these data is quite difficult and subjective.

The American Institute of Architects' *Environmental Resource Guide* (ERG) provides excruciating detail on the life cycle effects of a wide range of materials and guidance on how to select the best product in terms of its overall environmental performance, but it does not address conventional criteria such as cost, performance, and availability (Demkin 1997). Many other current materials rating systems provide similar counsel but are far more subjective than objective. Again, they are all ultimately compromised by the inability to combine units of energy, mass, and toxicity into a single rating that is meaningful. Additionally, by focusing almost solely on environmental impacts, they do not provide a useful decision system for materials selection. This is not to say that these references lack worth as they do serve as a means of enlightening building industry professionals on the wide range of effects their work products are having both on Nature and their human occupants.

The result of these problems is that materials selection is easily the most difficult and contentious area of sustainable construction. Clearly, a method or system for selecting "green" building materials that extends beyond life cycle analysis is needed, one that is grounded in natural systems principles and focusing on the fate of the materials at the end of their life cycle. Ultimately, the problem is to move to a much deeper integration of ecological ideas into design.

## Defining construction ecology and metabolism

Efforts to change the materials cycle in construction are hampered by many of the same problems facing other industries. The individuality and long life of buildings poses some additional obstacles. Three fundamental difficulties arise when considering closed-loop materials cycles for buildings:

1. Buildings are not currently designed or built to be eventually disassembled.
2. Products constituting the built environment are not designed for disassembly.
3. The materials constituting building products are often composites that make recycling extremely difficult.

These difficulties also increase resource consumption by the construction industry, because building components must be frequently replaced, buildings may experience different uses during their lifetimes, and they periodically undergo renovations or modernization. In each case, the inability to readily remove and replace components results in significant energy inputs to alter building systems, and large quantities of waste are the result. For example, renovation of commercial structures in the USA produces of the order of 200 kg/m$^2$ of waste.

In terms of energy consumption, contemporary buildings designed to just meet energy codes in the USA use energy at a rate of about 30 kWh/m$^2$ per annum, although best practices indicate that as low as 3 kWh/m$^2$ annually is readily achievable. New building technologies reported in Germany suggest that homes can be designed to use less than

*Figure 1.5* A block of five houses being disassembled in Portland, Oregon, by Deconstruction Services, Inc.

15 $kWh/m^2$ /year for heating, while Swedish studies suggest that heating energy can be reduced to less than 1 $kWh/m^2$ /year.

Clearly, a new concept for materials and energy use in construction industry is needed if sustainability is to be achieved. As noted at the start of this chapter, industrial systems in general are beginning to take the first steps toward examining their resource utilization or metabolism, and beginning the process of defining and implementing industrial ecology. In this same spirit, a subset of these efforts for construction industry, construction ecology, would help accelerate the move toward integrating in with Nature and behaving in a "natural' manner. Construction ecology should consider the development and maintenance of a built environment (1) with a materials system that functions in a closed loop and is integrated with eco-industrial and natural systems; (2) that depends solely on renewable energy sources; and (3) that fosters preservation of natural system functions. Construction metabolism is resource utilization in the built environment that mimics natural system metabolism by recycling materials resources and by employing renewable energy systems. It would be a result of applying the general principles of industrial ecology and the specific dictates of construction ecology.

The outcomes of applying these natural system analogues to construction would be a built environment (1) that is readily deconstructable at the end of its useful life; (2) whose components are decoupled from the building for easy replacement; (3) composed of products that are themselves designed for recycling; (4) whose bulk structural materials are recyclable; (5) whose metabolism would be very slow because of its durability and adaptability; and (6) that promotes health for its human occupants.

The deconstruction or disassembly of buildings and material reuse is one area of endeavor in which there has been a great upswing in activity and interest in the past few years. For example, a non-profit corporation in Portland, Oregon, employs several crews on a full-time basis to take apart houses and recover materials for resale in the do-it-yourself (DIY) market (Figure 1.5). Similar efforts are under way in numerous countries around the world, to include building systems that are able to be disassembled and reused (Kibert and Chini 2000). The next stage must be to reconsider building component design, creating products that are easily disassembled and made from constituent materials able to be recycled.

## Summary and conclusions

The result of a shift toward construction ecology and its corresponding metabolism creates a host of issues and problems to be resolved. Can construction be readily dematerialized in the sense recommended by the Factor 10 Club? Can a construction ecology and metabolism be implemented without significant changes in national policy that alter national accounting systems and internalize environmental costs? What lessons from natural systems are feasible for application to the built environment? What are the roles of synthetic materials in construction ecology? How can construction materials production and recycling be integrated with the other components of the industrial production system? These are all difficult questions that must be answered to move forward into an era approximating sustainability in the built environment. Nonetheless, examining nature and ecological systems for patterns of energy and materials metabolism for their potential adoption into human systems can provide a substantial improvement on current methods of attempting to green the built environment.

## References

Ausubel, J.H. 1992. Industrial ecology: reflections on a colloquium. *Proceedings of the National Academy of Sciences* 89: 879–884.

Ayres, R.U. 1989. Industrial metabolism and global change. *International Social Science Journal* 121: 363–373.

Ayres, R.U. 1993. Cowboys, cornucopians and long-run sustainability. *Ecological Economics* 8: 189–207.

Business and the Environment (BATE) 1998. Cleaner production reduces costs for Chinese manufacturer. *Business and the Environment ISO 14000 Update* (November) 4: 3–4.

Begon, M., Harper, J.L. and Townsend, C.R. 1990. *Ecology: Individuals, Populations and Communities*, 2nd edn. Boston: Blackwell Scientific Publications.

Benyus, J.M. 1997. *Biomimicry: Innovation Inspired by Nature*. New York: William Morrow.

Brand, S. 1994. *How Buildings Learn: What Happens After They're Built*. New York: Penguin.

Brown, H. 1970. Human materials production as a process in the biosphere. *Scientific American* 223: 194–208.

Bunker, S.G. 1996. Raw material and the global economy: oversights and distortions in industrial ecology. *Society and Natural Resources* 9: 419–429.

Costanza, R., d'Arge, R. and De Groot, R. 1997. The value of the world's ecosystem services and natural capital. *Nature* 387: 253–260.

Demkin, J.A. (ed.) 1997. *Environmental Resource Guide*. New York: The American Institute of Architects/John Wiley.

Erkman, S. 1997. Industrial ecology: a historical view. *Journal of Cleaner Production* 5(1): 1–10.

Fiksel, J. 1994. *Design for Environment: Creating Eco-Efficient Products and Processes*. New York: McGraw-Hill.

Frosch, R.A. 1997. Closing the loop on waste materials. The industrial green game: overview and perspectives. In *The Industrial Green Game*. Richards, D.J. (ed.). Washington, DC: National Academy Press, pp. 37–47.

Frosch, R.A. and Gallopoulos, N.E. 1989. Strategies for manufacturing. *Scientific American* 260 (September): 144–151.

Frosch, R.A. and Gallopoulos, N.E. 1992. Toward an industrial ecology. In *Treatment and Handling of Wastes*. Bradshaw, A.D., Southwood R. and Warner, F. (eds). London: Chapman & Hall, pp. 269–292.

Georgescu-Roegen, N. 1971. *The Entropy Law and the Economic Process*. Cambridge, MA: Harvard University Press.

Graedel, T. and Allenby, B. 1995. *Industrial Ecology*. Englewood-Cliffs, NJ: Prentice Hall.

Hayes, D. 1978. *Repairs, Reuse, Recycling – First Steps to a Sustainable Society*. Worldwatch Paper 23. Washington, DC: The Worldwatch Institute.

Holling, C.S. 1986. The resilience of terrestrial ecosystems: local surprise and global change. In *Sustainable Development of the Biosphere*. Clark, W. C. and Munn, R. E. (eds). Cambridge: Cambridge University Press.

Holling, C.S., Schindler, D.W., Walker, B.W. and Roughgarden, J. 1995. Biodiversity in the functioning of ecosystems: an ecological synthesis. In *Biodiversity Loss: Economic and Ecological Issues*. Perrings, C., Maler, K.-G. Folke, C., Holling, C.S. and Jansson, B.-O. (eds). Cambridge: Cambridge University Press.

Karamanos, P. 1995. Industrial ecology: new opportunities for the private sector. Industry and Environment. *United Nations Environmental Program* (October–December) 18(4): 38–44.

Kauffman, S.A. (1993) *The Origins of Order: Self-organization and Selection in Evolution*. New York: Oxford University Press.

Keoleian, G.A. and Menerey, D. 1994. Sustainable development by design. *Air & Waste* 44 (May): 645–668.

Kibert, C.J. and Chini, A. 2000. *Overview of Deconstruction in Selected Countries*. CIB Report No. 252 available online at www.cce.ufl.edu

Lawson, B. 1996. *Building Materials, Energy and the Environment: Towards Ecologically Sustainable Development*. Red Hill ACT, Australia: The Royal Australian Institute of Architects.

Luccarelli, M. 1995. *Lewis Mumford and the Ecological Region: The Politics of Planning*. New York: The Guildford Press.

McCarter, R. 1997. *Frank Lloyd Wright*. London: Phaidon Press.

McHarg, I. 1969. *Design with Nature*. New York: Natural History Press.

Neutra, R. 1954. *Survival through Design*. New York: Oxford University Press.

Neutra, R. 1971. *Building with Nature*. New York: Universe Books.

Odum, H.T. 1983. *Systems Ecology: An Introduction*. New York: Wiley-Interscience.

O'Neill, R.U., DeAngelis, D.L., Wade, J.B. and Allen, T.F. 1986. *A Hierarchical Concept Ecosystems*. Princeton: Princeton University Press.

Orr, D.W. 1994. *Earth in Mind: On Education, Environment, and the Human Prospect*. Washington, DC: Island Press.

Office of Technology Assessment (OTA) 1992. *Green Products by Design: Choices for a Cleaner Development*. Office of Technology Assessment. US Congress. OTA-E-541.

Peterson, G., Allen, C. and Holling, C.S. 1998. Ecological resilience, biodiversity, and scale. *Ecosystems* 1: 6–18.

Postel, S.L., Dailey, G.C. and Ehrlich, P.R. 1996. Human appropriation of renewable fresh water. *Science* (6 February) 271: 785–788.

Rejeski, D. 1997. Metrics, systems, and choices. In *The Industrial Green Game*. Richards, D.J. (ed.). Washington, DC: National Academy Press, pp. 48–72.

Richards, D.J. and Frosch, R.A. 1997. The industrial green game: overview and perspectives. In *The Industrial Green Game*. Richards, D.J. (ed.). Washington, DC: National Academy Press, pp. 1–34.

Rothschild, M. 1990. *Bionomics: The Economy as Ecosystem*. New York: Henry Holt.

Shiers, D., Howard, N. and Sinclair, M. 1996. *The Green Guide to Specification*. Croydon, UK: Post Office Property Holdings.

Shireman, B., Kiuchi, T. and Blake, A. 1997. *An Exploration of Industrial Ecology and Natural Capitalism*. Industrial Ecology III, 24–26 April 1997, San Francisco, California.

Steele, J. 1997. *Sustainable Architecture*. New York: McGraw-Hill.

Van Der Ryn, S. and Cowan, S. 1996. *Ecological Design*. Washington, DC: Island Press.

Vitousek, P.M., Ehrlich, P.R., Ehrlich, A.H. and Marson, P.A. 1986. Human appropriation of the products of photosynthesis. *Bioscience* 36: 368–373.

Vitousek, P. M., Mooney, H.A., Lubchenco, J. and Melillo, J.M. 1997. Human domination of earth's ecosytems. *Science* (25 July) 277: 494–499.

von Weiszäcker, E., Lovins, A. and Lovins, H. 1997. *Factor Four: Doubling Wealth, Halving Resource Use*. London: Earthscan.

Wells, M. 1981. *Gentle Architecture*. New York: McGraw-Hill.

Wernick, I.K. and Ausubel, J.H. 1995. National materials flows and the environment. *Annual Review of Energy and Environment* 20: 463–492.

Wernick, I.K., Herman R., Govind, S. and Aususbel, J.H. 1996. Materialization and dematerialization: measures and trends. *Daedalus* 25: 171–198.

Wernick, I.K. and Ausubel, J.H. 1997. Industrial ecology: some directions for research. Pre-publication draft (May), Program for the Human Environment, The Rockefeller University, http://phe.rockefeller.edu/

Williams, D. 1999. Biourbanism and sustainable urban planning. In *Reshaping the Built Environment*. Kibert, C.J. (ed.). Washington, DC: Island Press

Williams, R.H., Larson, E.D. and Ross, M.H. 1989. Materials, affluence, and industrial energy use. *Annual Review of Energy* 12: 99–144.

Wilson, A., Uncapher, J., McManigal, L., Lovins, L.H., Cureton, M. and Browning, W. 1998. *Green Development: Integrating Ecology and Real Estate*. New York: John Wiley.

Wright, F.L. 1954. *The Natural House*. New York: Horizon Press.

Zeiher, L. 1996. *The Ecology of Architecture*. New York: Whitney Library of Design.

## Part 1

# The ecologists

*Jan Sendzimir*

Systems ecologists recognize the courage of builders who reach beyond their normal horizons to learn from the world surrounding the built environment. How might knowledge of ecological processes help us better harmonize the life cycles of buildings with the dynamic ways in which the environment builds and recycles structures? Ecologists have waged a similar battle for the past half-century to draw the attention of natural scientists from molecules to cells to organisms to populations to communities to landscapes to the biosphere. Specialists and generalists eternally implore one another to reset their sights and focus anew. The challenge of assessing how an activity fits within the environment is both easier and more difficult. It is easier because far more people genuinely want to understand the environment now that a series of crises (climate change, acid rain, ozone hole) have brought the consequences of human endeavor home to them. It is more difficult because closer scrutiny shows a far more complex picture than originally imagined, to the point where surprise and uncertainty appear unavoidable. Science may not deliver the certainty ("the smoking gun") that many feel is necessary to act on, but it can help us frame our questions and our activities in much more intelligent ways as we examine our dynamic and complex biosphere.

What are the key drivers of this age? The importance of industrial machinery seems to fade as we increasingly recognize the ascending role of biological complexity. Our popular appreciation of this emerges in the modern urban myth about a car mechanic scorning a surgeon, telling him that his high fees are unjustified because his work is no more complicated than a valve job. The surgeon replies that the mechanic will appreciate surgery better if he tries to do the valve job next time with the engine running. But this appreciation is stuck decades past at the level of the organism. How can we begin to understand the interplay between millions of organisms, species, and communities within the interweaving cycles of materials, nutrients, water and energy? Four ecologists take this challenge up from a systems perspective in the following chapters. They offer a variety of views on how systems (ecological and otherwise) persist through cycles of building and destruction before suggesting how to apply these ideas in managing the processes that create, recycle and resurrect human shelter.

Each of the authors brings unique experience to the application of systems science in understanding natural and human systems. Starting from meteorology and geochemistry, H.T. Odum (Chapter 2) built many of the key foundations of what is currently known as systems ecology. It would be an exacting task to determine what discipline H.T. Odum has *not* integrated into systems thinking over the past fifty years. The ease and clarity with which his chapter weaves together many fields into an elegantly simple working approach reflects a profound and comprehensive understanding that has been fundamental to the development of more than one generation of ecologists.

James Kay (Chapter 3) works to advance the theory of how systems develop structure as a function of flows and feedbacks of different quality energies and applies such ideas in the adaptive management of regional land development projects. Tim Allen (Chapter 4) experiments with plants as tests of how systems develop and change and has been a leading contributor to the development of hierarchy theory to explain the multiscale structure of systems.

Garry Peterson (Chapter 5) uses computer simulation to test ideas about how fire shapes the mosaic patterns of forest landscapes and how animals move in response to pattern. His simulations have been the testing grounds for different ideas of the adaptive management of long-leaf pine forests in north Florida.

The extent and depth of the systems revolution in ecology is evident from the start. Explanations begin here with fundamental laws of thermodynamics to trace patterns of cause and effect that eventually organize associations of materials and organisms, even humans. This is the return path of an inquiry that relentlessly pursued the underlying causes outward to "the next larger system" from biology to ecology to economy to meteorology, biogeochemistry and physics. Along the way several basic Nature myths (Nature in balance, evolution is continuous progress) died. As Allen and Kay argue, ecological systems do not emerge from some inherent state of balance, nor do they grow and persist in smooth, continuous ways that sustain this equilibrium. Structure starts to appear when things lose balance ("far from equilibrium"), as when a gradient of energy emerges. Like an earthquake launching a tidal wave, a new energy source sets up a moving slope that raises the potential to do new things. Many systems cannot make use of this new potential; they are too small or too big to match and engage the new energy, or *exergy* – the higher quality of energy that makes novelty possible (Kay). But some systems persist because they do match and can organize around that potential to ride that wave and develop new forms of organization or structure. Contrary to the common notion that change arises from introduction of novelty from "outside," new systems emerge by rearranging their own parts: they "self-organize." The new structure from that rearrangement is competitive in proportion to the increased capacity to use or amplify a diversity of new (and old) energy sources (Odum). The characteristics of these new systems – open to exchanges from above and below within a hierarchy of different systems that also self-organize – is summarized in the term SOHO (self-organizing hierarchical open system). As SOHOs develop into mature attractors they build structure to increase their feedbacks to make ever more effective use of resources and increase survivability (Kay). Peterson provides a clear introduction to terms, especially "ecological processes" and "scale," that allows us to visualize what and how interactions occur in SOHOs that cause dynamic changes over time and space.

To equate the dynamic complexity of human society with emerging patterns such as whirlpools in an energy gradient may startle and even offend those who ascribe society's sophistication to human ingenuity. Actually, the authors simply point out how ingenuity becomes competitive when it does not fight but rides energy gradients intelligently, and that this will become an imperative in an increasingly energy-scarce and $CO_2$-rich world. Ingenuity has not been discarded or snubbed; it has been set within the biophysical context of the world as the basis for a more nimble, adaptive, and competitive construction industry, as the chapter by Bisch confirms (see Chapter 11).

New systems may vary, even wildly, in their behavior as they develop feedback loops to reinforce the behavior and structure that emerges. The systems term, "attractor," describes the tendency for new structure to behave within certain bounds, as if its gyrations are loosely hemmed in by some hidden attraction or gravity that constrains the behavior of any system that starts within its orbit (Kay). Chaos theory explores how such seemingly wild behavior actually gives a system much more resilience to change (Peterson). As a counterintuitive example of "chaotic" resilience, a human heart has thousands of options in its behavior; that is why we can take so many different forms of exercise and body positions in our stride. The ways in which these systems persist by being open to exchanges and flexible in response to disturbance give a wider view of security than that which comes from closed, tight structures that resist any change at all. Design to leave buildings open to breathe as they adaptively ride energy gradients would avoid the "building sickness" of suffocating, overinsulated structures. Both Allen and Peterson also warn us not to be taken in by some illusion of evolutionary perfection, as if we should strive for a perfect end state, the "crown of creation," with the initial stroke. Mistakes and accidents are crucial sources of learning that mold design. The potential to learn increases with the recyclability of a building's materials and the flexibility of its original design, so they advise us to design the building to "use history" as it learns to adapt to changes in the environment and society.

History seems to be an irregular or syncopated series of episodes of new structure emerging around pulses of new energy or resources followed by longer intervals when resources slow to a trickle (Allen). Unprecedented forms of life sprang up to exploit novel resources (plants/sunlight, animals/biomass) in biological evolution. Similarly, the design and construction of the built environment have responded with waves of innovation to pulses of new energy and resources. Collective, permanent housing arose around the steady food supplies of domesticated plants and animals. Empires emerged when food surpluses permitted labor specialization, especially standing armies (Allen). Currently, our stupendous urban densities grow out of a capacity to use energy and materials at quality levels that are orders of magnitude larger than even a century ago. How then, Odum asks, can we innovate to ride "the prosperous way down' from the heights of the fossil fuel age to an energy-lean era? How best can we utilize the remaining resources of concentrated energy and materials to retain flexibility in our ways of life as climate changes? What can we learn from ecosystems about intelligently riding the waves *and* the troughs until the next wave arises? A number of tools are offered to visualize and integrate one's system of interest within the overall system such that they reinforce one another.

Self-organization may appear spontaneous when viewed in biological evolution, but Odum proposes that builders can deliberately use *amplification* and *energy matching* as design principles for selecting materials and designs that can best be matched to "mutually amplify" the potential of existing energy and materials resources. In the 1960s Odum pioneered ecological engineering, the science of integrating human and natural systems. This discipline views problems "from the next larger scale" to better fit human civilization to the environment so that they mutually reinforce one another. For example, ecologically engineered wetlands can efficiently convert treated wastewater to clean water that nurtures an attractive biodiversity, but they cannot handle intense concentrations of raw sewage or industrial wastes. When wetlands are overmatched, as in the latter case, they reorganize into degraded systems that provide little water treatment, wildlife habitat, or aesthetics (unless one has a particular affection for benthic worms).

Kay proposes four basic design principles to establish a built system within the hierarchy of other systems in the landscape: (1) interfacing, (2) bionics, (3) using appropriate biotechnology, and (4) use of non-renewables only to bring renewables on-line. *Interfacing* involves maintaining the survival potential of natural systems by factoring in their limited potential to provide energy or absorb wastes when interfaced with man-made systems. The principle of *bionics* is that the "the behavior and structure of large scale man-made systems should be as similar as possible to those exhibited by natural ecosystems [especially the capacity to learn and adapt]." *Using appropriate biotechnology* involves delegating functions to natural subsystems wherever possible. Regarding the last principle, Kay warns that no resource is inherently "renewable;" it becomes so only if our use is lower than the rate of replenishment.

Experiments in ecological re-engineering of the built environment can begin with design principles gained from the study of ecosystems. But how can we actually *see* the system and determine where a process or structure functions in the system (at what level or scale within the hierarchy does it operate?)? Odum presents several tools to view systems and measure the contributions of the whole and of the parts. His diagramming method allows one to clarify how one views the energy basis and organization of a system. What energy sources drive a system and in what order of importance? How are the different system components linked, and what new combinations are possible? Contributions to the system can be estimated by quantifying the diagram pathways. The total sum of energy (of one kind or quality) required to create or to run a component or process is a measure of its contribution to the overall system. This sum indexes the energy memory (emergy) of all inputs that converge in the web to make a component function. Odum proposes that rational indices (emergy per mass or per unit of energy flow) help to identify the intensity of a resource's contribution and where it fits within the whole system hierarchy. Is it a driving force or a less concentrated resource that requires amplification? This allows us to match it properly to other resources in the building cycle. From what distance can resources be sustainably imported or disposed of? What materials work best together? Odum's energy perspective allows us to reassess or complement current knowledge and practices that use various materials alone and in combination.

All systems eventually collapse and rebuild. Kay and Peterson point out that the same feedback loops that reinforce and sustain a system, despite shifts in the surrounding environment, can also realign and drastically change the system. Such catastrophic collapses are normal to systems dominated by feedback loops, and can be so swift, profound and unexpected, that, as Kay asserts, "Inherent uncertainty and limited predictability are inescapable consequences of these system phenomena." No builder could survive under such uncertain conditions; buildings attract people precisely because they deliver reliable shelter, water, utilities, and work space. Such daunting complexity might justify abandoning innovation in favor of the status quo or any design or practice no matter how unsustainable, for how would any effort make a difference in such chaos? But the lesson here is that denial of change and uncertainty precipitates bigger catastrophes. If we hide in larger fortresses that use technical sophistication to resist the environment we will miss the sea level rising around us. On the other hand, probing and embracing the uncertainty of a dynamic environment can mitigate the seriousness of change, and may enable construction of more flexible, more invigorating, and healthier interior environments. We may be able to learn to ride and even steer change in the way we integrate our buildings to the general environment. In this way, uncertainty becomes the inspiration that keeps the built environment adaptive and intriguing. Kay and Peterson offer a variety of suggestions for how we can design and steer the building cycle so it usefully engages our unpredictable environment.

Kay stresses that any initiative to apply construction ecology should begin by examining how socioeconomic factors fit within the environmental context. This means assessing what sustainable livelihoods are possible for society, what the potential for ecological integrity is across the wider region, and how the two interact. In the first case, livelihoods that do not compromise those of future generations are sustainable if people can survive shock or stress while improving or sustaining their material and spiritual conditions. People's capacity to adapt and thrive forms the core of the builder's understanding, and from there one looks outward to the environment. This means assessing three facets of ecosystem integrity: its current state of vigor, its resilience to disturbance, and its "potential to continue to self-organize: to develop, regenerate, and evolve in its normal environmental circumstances." These facets form the context that constrains and shapes the sustainable livelihoods of society, through delivery of exergy, materials, and information that allow society to self-organize. But society can alter that context, especially given the intense qualities of information, designs, and energy at our command today. Kay advises that we can better untangle these interactions by describing how they link various levels within the overall hierarchy, briefly describing the entire social/ecological situation as a SOHO.

While the science of describing a SOHO is quite young, much can be learned in the attempt. Fundamental to such efforts is the need to assess the levels of the system, how they are nested together, and how they are linked in the system based on both the quantity and quality of energy exchanges. This follows Allen's admonition not to confuse complexity with complicatedness. Complex systems compete and survive but are often no more complicated than the systems they replace, though they have a newer and higher quality of organization. So, insight must be sought in defining this new quality of organization, and raw quantitative (efficiency) measures may not capture the qualitative change (effectiveness) in what energies are flowing and how the system performs. Much remains to be learned about how to measure quality, in terms of both energy and information. But evidence continues to point toward these factors as pivotal to long-term sustainability, so we have to develop and test quality indices even as we apply them in experiments in construction ecology. This mandate to test and experiment as we engage uncertainty in society and environment comes from the dynamic nature of systems. It means that we can seek standards and norms but cannot rely on them forever. We must reformulate them as the world changes. Kay concludes that it is useless to build construction ecology as a "traditional normal science" when we constantly have to reassess criteria.

> Our normal way of applying ... criteria is based on an assumption that an incremental change in the context (that is the influence of the design on the natural system) will result in an incremental change in the natural system. But self-organizing systems do not work in this way. There can be substantial changes in context, which the system can buffer itself from, and hence there will be no change in the system's state. However, once this buffering capacity is used up, a very small change in context can cause dramatic change in the system's state.

So how can we move forward in understanding and design when, as Kay points out, "We know the questions to ask, but not how to answer them." Both Kay and Peterson recommend methods such as *adaptive management* (AM) to provide a framework to experiment with designs and flexibly modify the experiments as society and nature evolve. In contrast to the straight shunt (analysis – decision – implementation) of top-down linear management, AM increases the capacity to learn by making the process cyclic and self-correcting. AM completes the cycle with a monitoring phase to link the implementation and analysis phases, such that the initial hypotheses can be reformulated in the light of experience, and new hypotheses seed new design. This means that design responsibility does not stop with initial construction. It requires that at the outset a building's materials and design allow modification over time and that over the long run someone monitors the building's performance relative to the behavior of the complex of society and nature (SOHO). In this way a building and the people that design, build, monitor (live in it) and deconstruct it become a complex adaptive system that can self-organize using AM as a map of how to move through the phases of collaboration. Such processes integrate the construction enterprise vertically by broadening our vision (to all scales within the environment) and our ways of action (to all levels of society). Our capacity to adapt increases as we pursue design in a self-correcting cycle wherein we can see and draw on the resources and experience of more levels within society.

Peterson offers Holling's thought model, the *adaptive cycle*, as a map of how systems run through the phases of birth, death and rebirth. These phases, rapid growth (r), conservation (K), release (omega), and reorganization (alpha), allow one to assess where a system is in the cycle and how sensitive it is to disturbance and change. He illustrates the consequences of each phase by describing the strategies and structures that organisms have developed to be competitive at each stage in the cycle. Some species do well in early stages (r) by being mobile and flexible in capabilities. Others thrive in the build-up phase (K) by being very specific in function and resistant to change. Some species outcompete others by inviting catastrophe, proliferating burnable materials in an ecosystem prone to fire and then quickly resprouting after the fire. He summarizes the "active" approaches to change as learning, insurance, resistance, and management and shows where and how they are applied within the adaptive cycle.

Other chapters have described the various levels nested within a landscape hierarchy, but Peterson allows us to see interactions within and *between* these levels, sometimes spanning from the smallest to the largest levels. These *cross-scale interactions* are key to understanding how ecosystem *resilience*, the capacity to persist in the face of disturbance, collapses when the system becomes overdeveloped. With time, ecosystems can load up with biomass and become so overconnected with structure that the slightest effects add up, like sparks jumping between dust motes in a grain elevator, to an explosion. Catastrophic collapse appears as if from "nowhere" to surprise us with the collective action of tiny phenomena that we previously ignored as insignificant. For example, periodically, under the right combinations of weather and land cover development, little grasshoppers that most years are just garden pests transform into a plague of locusts that sweeps an entire region, denuding it of vegetation.

Peterson describes how our surprise at ecosystem collapse comes both from Nature and from society, from cross-scale complexity as well as the way our vision has been narrowed by our past successes in science and enterprise.

While fame and fortune come to those who compete best by mastering physical laws within a small range of scales (laboratory, factory, field, building), any view that looks beyond that domain appears to be a distraction. But Peterson points out that while the processes governing the narrow domains of human mastery are "homogenous, ahistorical, and unchanging, ecosystems are diverse, evolving, historical entities." Society can see and reward achievements within these narrow domains but is blind to the slow, diffuse changes that foreshadow an ecosystem slipping to the brink of collapse. Global and regional changes have awakened us to the need to see and manage change at many levels, and one of the first steps is to recognize how our past means of success controls how and what we see. The next step is to recognize that natural and human systems have similar and different drivers of change. Both have a historical component of self-organization, the rearrangements of what was as new resources appear. But, as Peterson notes, the fact that only human systems use *anticipation* to address or effect change deepens the challenge of integrating them with natural systems.

Peterson recommends new approaches to construction ecology based on the emerging concepts of "resilience, ecological dynamics, and scale in ecosystems." He applies the adaptive cycle, which he reconfigures as the design, construction, operation and maintenance, and deconstruction phases, as a model to see and manage the dynamics of construction. In contrast to various strategies that different species adapt to compete in different phases of the cycle, buildings need to balance a set of strategies that see them gracefully through all the phases. For each phase he poses different sets of design principles based on lessons learned from biological adaptations to a similar phase in the ecological model of the adaptive cycle. Consider the ecological collapse (release) versus building deconstruction phases for example. In fluctuating environments made turbulent by disturbances such as fire or flood, some species compete by not investing in defenses and yield easily to destruction. Instead, they invest in strategies that allow them to rebuild rapidly and cheaply following destruction. Peterson suggests that building design could follow similar strategies when social or ecological turbulence makes it hard to forecast what design can persist in the face of changing human preferences or environmental disturbance. Buildings that are cheap and easy to build, modify, deconstruct, and recycle might allow their occupants the best platform from which to adapt to shifting circumstances. Over the long term such strategies might economically outcompete current fast/cheap or slow/expensive approaches. In the former case, expenses saved in hasty design may hinder efforts to learn, redesign, and modify the structure. Similarly, the use of cheap, composite materials may complicate deconstruction and preclude recycle or reuse, thereby boosting the costs of disposal and reconstruction. In the latter case, attempts to build a fortress to insulate against change may unnecessarily increase construction (and eventual rebuilding) costs when the building cannot survive change in either human perceptions and needs or in environmental conditions.

In summary, ecologists offer a variety of ways to visualize how ecosystems are structured and how they function. Different sets of processes operate at different scales, thereby structuring the landscape into different levels in a hierarchy. Function occurs within and across the different layers (scales) within the landscape hierarchy as materials, nutrients, energy, and information alternatively flow and are stored within the system. Measuring these flows and storages can help to assess the significance of a process or structure within the system. These configurations of structure and function change over time, sometimes quite rapidly, in phases of birth, growth, destruction, and recycle (the adaptive cycle). The pervasiveness and power of this dynamism is stressed as a key factor to consider in making the built environment more adaptive to change. The ecologists describe how ideas and practices used to understand and manage change in ecological systems can be applied to make the built environment more adaptive and sustainable in the face of changes in Nature and society.

# 2 Material circulation, energy hierarchy, and building construction

*Howard T. Odum*

At this, the beginning of a new millennium, the buildings of civilization are aging, and construction is increasingly squeezed between shortages of cheap materials and energy and the scarcity of unpopulated environment for discarded buildings, wastes, and their impact. For its purposes this chapter uses the general principles of energy processing, biogeochemical cycling, and emergy evaluation (spelled with an "m") to explain the patterns of construction and recycle that emerge as sustainable in self-organization on any system and any scale. With the help of analogous ecosystem examples, this chapter suggests ways to improve building construction, economic management, and the handling of materials now and in the times of declining availability of fuels and virgin materials ahead.

Everyone can observe actions of human individuals with trial and error building, developing infrastructure, processing materials, and choosing according to prices. But at a larger scale, the smaller actions are selected for or eliminated according to what works in the aggregate. Thus, it is possible to determine what will be successful from basic principles of energy, matter, and information that can indicate the main features of successful self-organization that will emerge. In the give and take of public affairs, people are rarely aware that the future patterns follow universal principles that will emerge after trials and failures within the political process. Brand (1994) described self-organization of civilization with his book *How Buildings Learn*, in spite of what architects were trying to otherwise do. What would be ideal is to plan in the first place what is predictable.

First, let us review some concepts and definitions that can be used to measure and interpret the building cycle process, whether it is Nature's construction of a tree or human construction of a city. The world of environment and human economy is made of items of different scales all together (Figure 2.1a). By conceptually separating items of similar scale, as shown in Figure 2.1b, the small things can be seen to have small territories of support and influence and are made and replaced rapidly. Larger items in Figure 2.1b have larger territories and take longer to grow and be replaced. The small and large are connected by the flows of energy during self-organization of pattern and structure. According to energy systems concepts, self-organization of Nature and human society develops structures and processes on each scale of size and time, transforming energy of one scale to make the products for the next.

## The energy hierarchy

In brief, the processes of environment and economy can be arranged in a series according to the successive energy transformations required to make one quantity from another

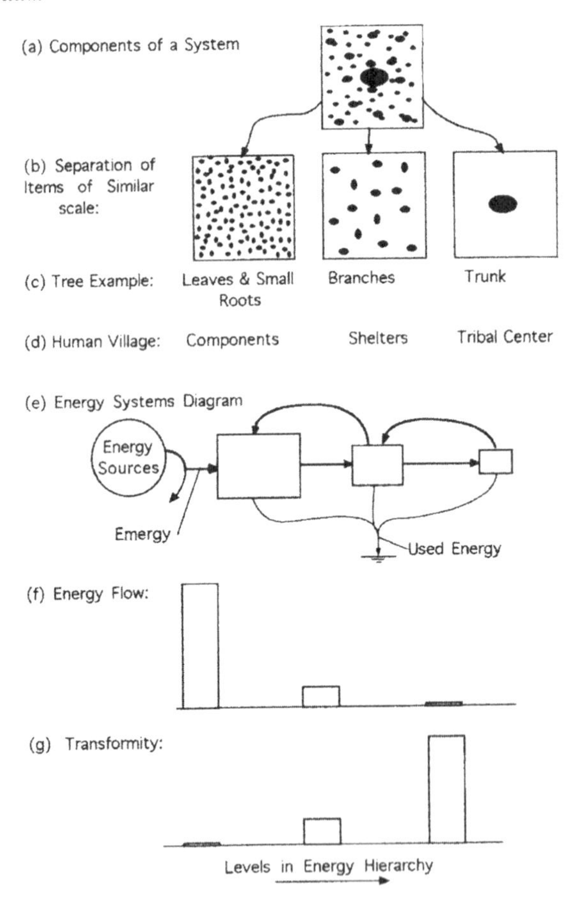


*Figure 2.1* Energy hierarchy concept illustrated with one source and three energy transformation steps. (a) Components of three scales viewed together. (b) Components of forest separated by scale. (c) Components of a tree separated by scale. (d) Components of a tribal village. (e) Energy flows in an energy systems diagram. (f) Flows of energy in the pathways from left to right. (g) Transformity of the energy flows.

(Figure 2.1e). Energy flowing in from the left builds units of structure and processing (the first box), with much of the degraded energy leaving the system as "used energy." Some of the transformed energy in the first box goes to support the next box, and so on.

## *Tree example*

An example of energy hierarchy is a tree in the forest. Many leaves and small roots in Figure 2.1c send organic substances and materials to branches. The branches send products to the tree trunk. The energy flows are represented in the energy systems diagram in Figure 2.1e. The energy driving the process comes from sunlight and wind (energy sources) on the left. At each step, work is done and much of the energy is degraded and is shown being dispersed (down pathways) as used energy that has lost its ability to do further work. At each step in the series, the quality of the energy transformed is increased. A joule of trunk wood energy is of higher quality than a joule of original sunlight. It is more concentrated, more usable by the forest system.

### *Feedback reinforcement*

In Figure 2.1e, note the feedback pathways by which the blocks downstream to the right feed back materials and services to help operate the upstream blocks. These are pathways that reinforce and select those processes that contribute. These feedback pathways often have high-quality materials or valuable control actions. For example, the branches support the leaves and the trunk supports the branches.

### *Human village*

A simple example involving human construction is a tribal village whose inhabitants use materials and foods entirely from the forest as the energy source for their construction work (Figure 2.1d). The first box on the left is the forest structures and processes. The second box to the right is for humans building their family shelters. The third box to the right is the tribal center, where people from the separate families use their materials and services to construct the tribal center's structure. The feedback by the tribal chief controls the family patterns, and the family work controls the processes in the forest that generate the materials and food.

### *Emergy flow*

In judging what is required for any pathway, it is incorrect to consider a calorie of chief's work as equivalent to a calorie of organic matter from the forest. Although energies of different kinds cannot be considered equal in regard to the work they do, there is an easy way to put them on a common basis. Each energy flow can be expressed in the units of energy of just one kind that would have been necessary in direct and indirect processes to make it. The energy of that kind used for the comparison is called *emergy*, spelled with an "m," and the unit of measure is the *emjoule*. In the examples in Figure 2.1, only one source is shown and all the other pathways receive its emergy flow. In other words, the flows through the blocks in this example all carry the same emergy, which is really a numerical "memory" of the source. In this chapter we express emergy in solar emjoules, abbreviated sej. The flow of emergy per unit time is called *empower*, with the units sej/time.

### *Transformity*

The energy that is transformed and remains beyond a conversion process decreases through each transformation, because of the nature of energy dispersal (the second energy law). In Figure 2.1 the items further to the right in the scale of successive work contributions have the same emergy but less energy (Figure 2.1f). The ratio of these two, emergy/energy, is defined as *transformity*. It increases from left to right and marks the position of anything in the energy hierarchy (Figure 2.1g).

### *Energy quality spectrum*

Energy of higher transformity is said to be of higher quality because more was required to develop it and because its uses have greater effects (for good or bad). For example, protein foods have higher transformity than vegetables. Genoni and Montague (1995) found that potential toxicity also increases with transformity.

Transformities range from 1 solar emjoule per joule for sunlight (by definition) to values 32 orders of magnitude larger for the genetic information of life. All energy flows and their transformations can be arranged in order of their transformity. As Alfred Lotka suggested the maximum power principle as the fourth law, perhaps this principle should be recognized as a fifth energy law. Transformity marks position in the universal energy hierarchy. In representing data on graphs it is useful to arrange energy flows in order of their transformity, thus forming a distribution according to energy quality. All the diagrams and graphs in this chapter are arranged from left to right in order of transformity.

## Materials and the energy hierarchy

Materials and their cycles in the organization of systems are dependent on the flows of energy. Here, several concepts are combined to explain the patterns of material distribution that result from the work of the energy hierarchy.

### *Material cycles*

Materials circulate through systems: water through the landscape, blood substances through the human body, and carbon through the biosphere. Estimating the rates of flow of a material and writing it on pathway diagrams is one of the standard ways of reporting data in a way that readers can visualize where flows are fast or slow. For example, biogeochemistry studies the circulation of the chemical substances, especially those that are incorporated into the production of organic matter by living organisms. For example, Figure 2.2 shows the circulation of material in a system which has three units of structure and processing. The numerical values on the pathways in this example are the flows of carbon into plant photosynthetic production of organic matter and transfers to consumers in a forest ecosystem.

## Material budgets

In many processes, a material may circulate without being changed into something else. In this case, the material is said to be conserved. Inflows equal what is added to storage plus outflows. For example, chemical elements such as carbon, phosphorus, and lead are usually conserved, being combined in different ways but not losing their identity as an element. Often, diagrams of material cycles are drawn with numerical values of storages and flows for the system's average condition, as if the web of material flows was in a steady state. The evaluated diagram shows the system's "budget" for that material in the same way that average purchases and expenditures in a household are called a monetary budget. The pattern of organic biomass materials used and recycled in a forest tribal village is similar. On a larger scale, Adriaanse *et al.* (1997) published quantitative diagrams of the materials budgets of nations.

### *Coupling of materials and energy*

In Figure 2.2b the circulation of material from Figure 2.2a is combined with the flows of energy from Figure 2.1. Both are necessary to the construction and operation of the system. Materials are said to be *coupled* to the energy transformations. Production processes can generate more useful products by incorporating materials into structural units that can use the materials, but work is required.

(a) Circulating Materials Through Structural Hierarchy

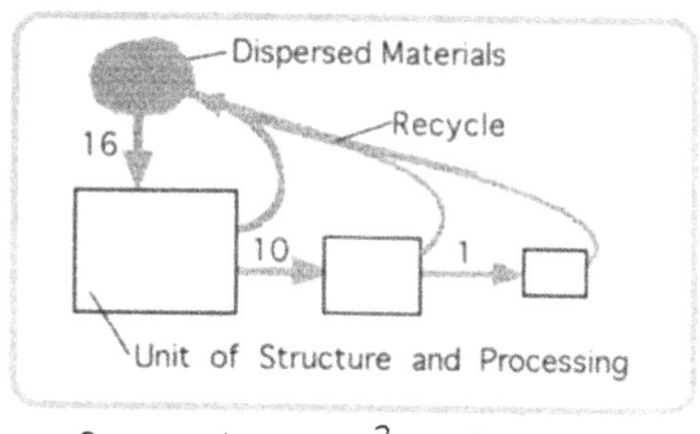


Grams carbon per $m^2$ per day

(b) Materials Combined with Energy Flows

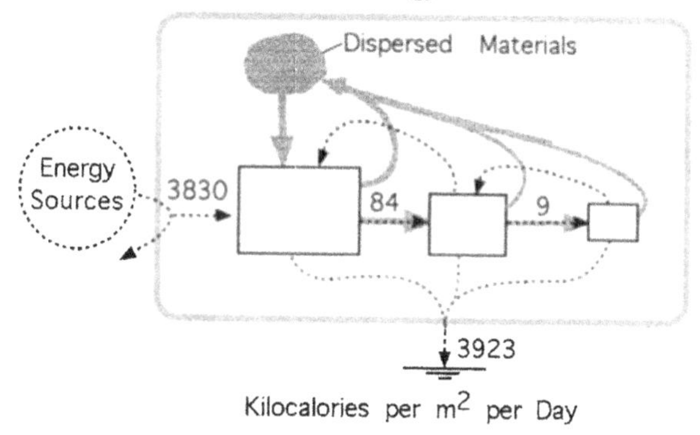


*Figure 2.2* Materials and energy flows with a hierarchy of structural units of different scales. Numerical values are for circulation of carbon through three levels of a rainforest (1, leaves; 2, branches and roots; and 3, tree trunks) (Odum 1970). (a) Circulation of material. (b) Combination of material and energy flows.

### *Energy required for concentrating materials*

Substances that are more concentrated than their environment tend to disperse and depreciate, a consequence of the second energy law. Conversely, to concentrate substances, available energy is required to do concentrating work. Most of the energy is degraded in the process and unable to do further work, while some energy is stored in the state of higher concentration (increase in Gibbs' free energy).

Figure 2.3a shows a gradient of increasing concentration from left to right. The spontaneous tendency is for the materials to disperse. moving to the left in the gradient. In Figure 2.3b, outside available energy is shown doing the work of pumping materials from lower to higher concentration on the right. The available energy adds the emergy to the product. As expected from the second energy law, more energy is used up than is stored. Genoni (1998) provides a number of examples.

As the available energy to the right of the energy transformation processes (production symbols) decreases, while the emergy increases, the transformity (emergy/energy) increases to the right. As emergy is added to the materials as they are processed to higher concentration, the mass emergy (emergy/mass) also increases along the energy hierarchy (Figure 2.3c).

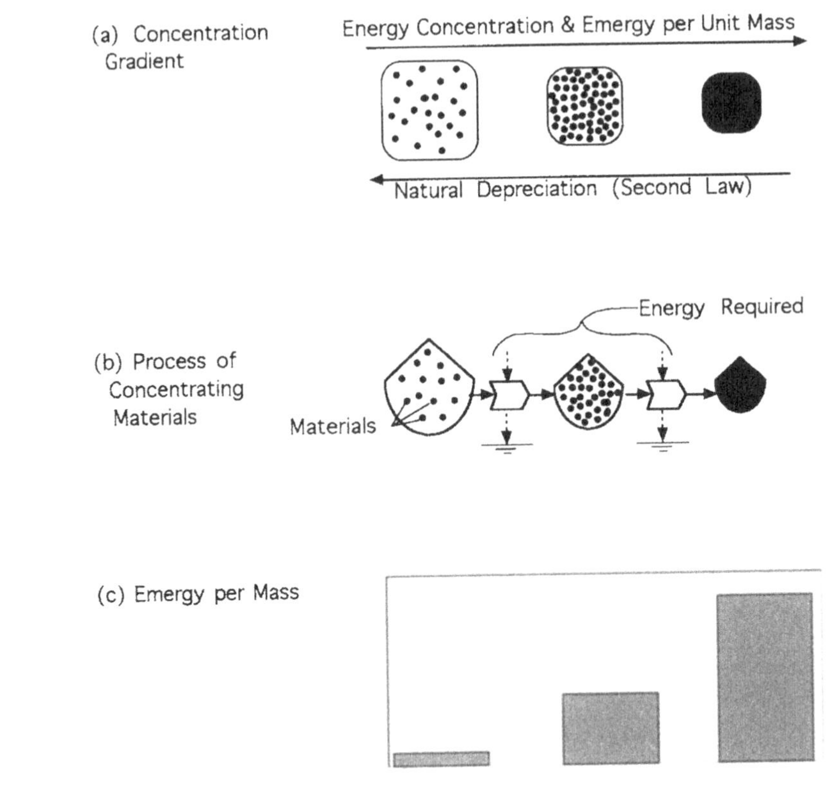


*Figure 2.3* Energy and the concentrations of materials. (a) Sketch of concentration increase from left to right. (b) Use of available energy to concentrate material. (c) Increase in mass emergy (emjoules per gram) with concentration.

### *Spatial convergence of materials with an energy hierarchy*

The self organization of systems of Nature and humanity converge into small centers the energy that is initially spread over the landscape. Even though energy flow diminishes through each successive transformation (often in structural blocks – Figure 2.4a), converging the energy maintains a high concentration, capable of feedbacks that reinforce the contributing processes. As the materials are part of the production and transformation process, they are also converged and concentrated into centers. Figure 2.4b illustrates some of the pathways of convergence and divergence as materials are processed to higher concentration in the centers and dispersed at lower concentration in the peripheral areas.

### Inverse relation of material flux and emergy per mass

Self-organization achieves additional performance by concentrating materials in ways in which the concentrations reinforce. However, the greater the concentration, the more emergy is required per unit weight and the less material can be upgraded. In Figure 2.5, for a system with constant empower, the quantity of material that can be processed in successive steps to a high emergy level is inversely related to the emergy per mass. Thus, high concentrations are scarce because of the nature of the energy hierarchy and its coupling to materials.

(a) Materials Combined with Energy Flows

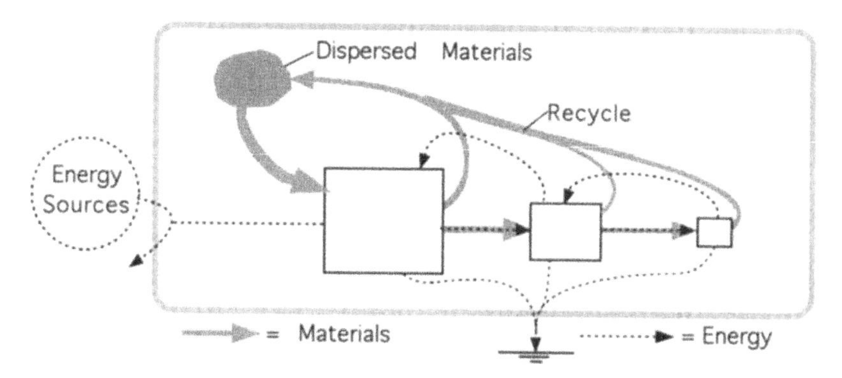


(b) Spatial Convergence of Materials

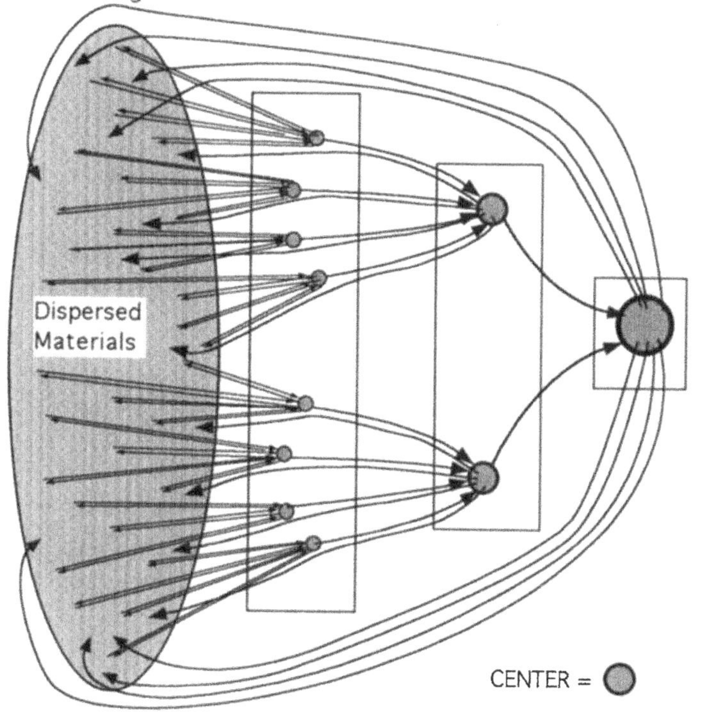


*Figure 2.4* The spatial convergence and divergence in the concentrating of materials. (a) Energy systems diagram of energy and materials. (b) Spatial pattern of material flows toward centers.

In Figure 2.4 there is one source of energy and emergy. In order for the centers to have an increase in ability to feed back and reinforce, energy and materials have to be more concentrated in centers in each step. Figure 2.5 shows that concentrating materials with the same empower leads to fewer materials being transformed to higher level. Thus, some of the materials are recycled at each stage. The materials recycled are less concentrated and lower in transformity. Values for water are less (Odum 1996a).

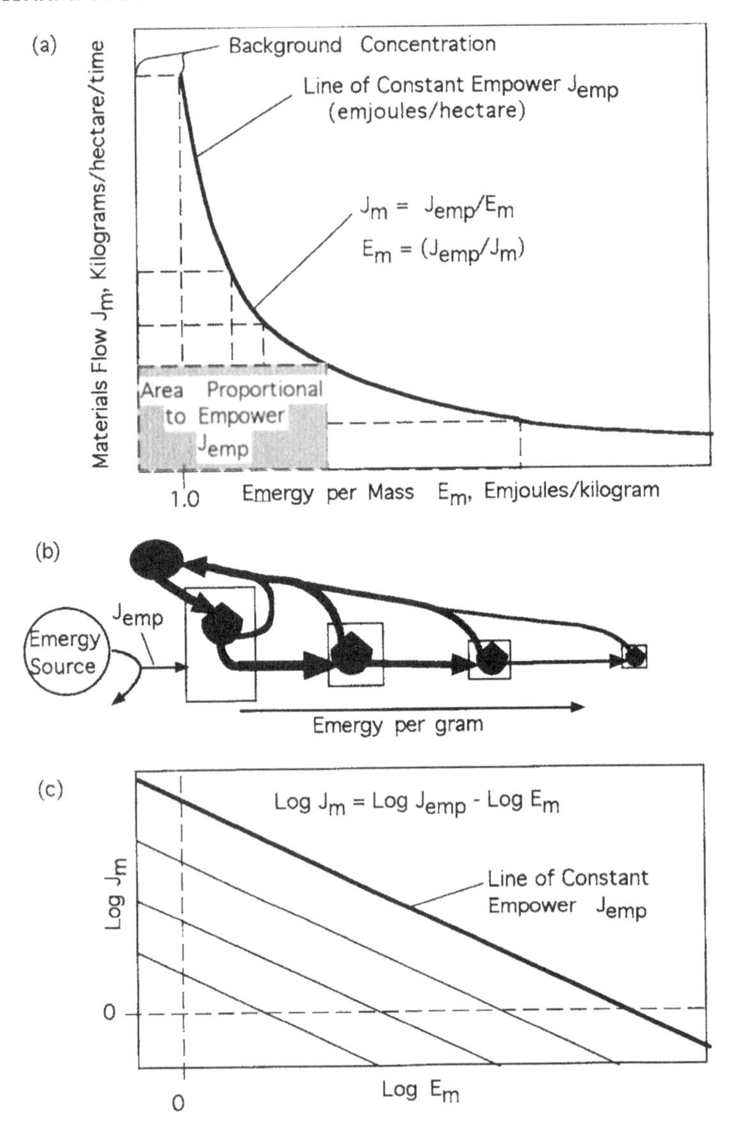


*Figure 2.5* Flux of material which can be concentrated by successive energy transformations where empower is constant. (a) Inverse relation of material flux and emergy per mass. (b) Energy system diagram with one emergy source. (c) Relationships on double logarithmic plot.

## *Coupling of materials with different zones of the energy hierarchy*

Materials of different kinds are found coupled with different levels of the energy hierarchy spectrum (Figure 2.5). Theoretically, more system performance results and a material contributes more when it interacts with energy that it can mutually amplify. By that concept, designs for material flows develop where energy flows interact with flows of somewhat higher or lower transformity. The kind of hierarchical pattern of material processing shown in Figures 2.3 and 2.4 may be at very different ranges of the energy hierarchy spectrum for different materials. Many heavy metals, for example, tend to go to the top of the biological range.

Figure 2.6 shows the way each material has its cycle and hierarchical skewed distribution occupying a zone in the universal scale of energy transformity. This plot is sometimes called an energy transformation spectrum.

### *Kinds of materials and their emergy per mass*

The mass emergy (emergy per mass) indicates the energy role of materials. Like transformity, emergy per mass can be used for the horizontal axis of graphs and diagrams. For example, Figure 2.7 has the mass emergy values for many materials over many orders

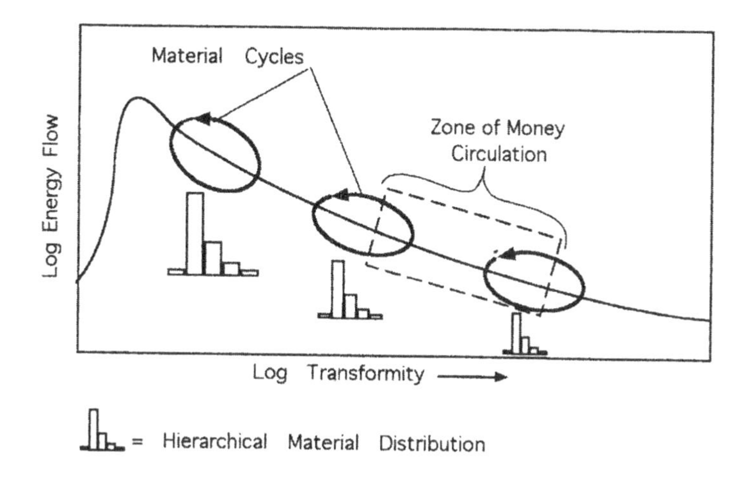


*Figure 2.6* Energy hierarchy spectrum showing the zones of productive interaction for different material cycles in different transformity ranges.

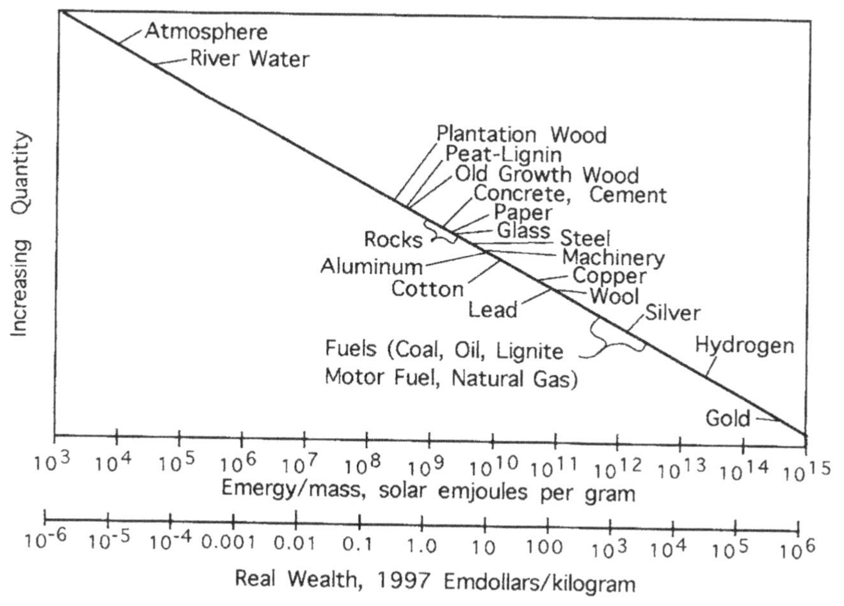


*Figure 2.7* Mass emergy (emergy per mass) of various materials used in construction of civilization and Nature. Included is a scale of equivalent emdollars per kilogram. See Appendix 1 for calculations.

of magnitude as calculated in Appendix 1. The diagram suggests the natural range of the energy hierarchy in which each kind of material tends to be bound and contribute to the system operation.

The more abundant a material is in the geobiosphere, the higher the concentrations of that material in the levels of the energy hierarchy (Laskey 1950). In general, it takes more emergy to concentrate materials that are initially scarce in the geobiosphere. In Figure 2.7, air and freshwater are on the left with low values, whereas scarce items, such as gold, when concentrated are on the right with high emergy value.

### *A sixth energy law for materials*

Classical thermodynamics (the science of energy considered near equilibrium) has three generally accepted laws:

1 Energy is conserved.
2 Energy concentrations are spontaneously dispersed.
3 The complexity of heat is zero at absolute zero (–273°C) when molecular motions cease.

For open systems, Lotka (1922a,b) proposed the law stated in emergy terms as the maximum empower principle (see 'Ecological engineering insights'). I proposed a fifth law (Odum 1996b, 2001) recognizing the universal energy hierarchy – all energy transformations form a series marked by transformity.

In this chapter, to summarize the relationships relating material distributions to the energy hierarchy, I propose as a sixth energy law:

> Materials are coupled to the energy transformation hierarchy and circulate toward centers of hierarchical concentration, recycling to dispersed background concentrations.

## Material valuation

Because money measures only what is paid to humans for their work, market values do not measure the contribution to the economy of real wealth arising from the environment's work. To measure all of the real wealth contributed when materials are used, emergy and empower valuation is necessary. These measures can be used to choose between alternative uses and processing of materials.

### *Emdollar equivalent of emergy*

In order to relate emergy to economic values, a monetary equivalent of emergy is calculated, called *emdollars*, the equivalent dollars of buying power. First, an emergy/money ratio is calculated for the economy for the year. The total emergy used by the state or nation (annual emergy budget) is calculated by adding up the emergy of everything used. Then the annual emergy budget is divided by the gross economic product for that year. For example, for the USA in 1997 the average emergy/money ratio is 1.1 trillion emjoules per dollar. To express the emergy values of materials or anything else in emdollars, divide by the emergy/money ratio.

### *Value of material concentrations*

As already explained in the section 'Materials and energy hierarchy', the higher the concentration of a material, the more emergy it contains. The higher the concentration, the more emergy and emdollars are contributed to construction when materials are used that have already been concentrated. The emergy per mass values plotted in Figure 2.7 are those for typical concentrations ready for use.

Because emdollar values include the services of both Nature and humans, they are usually higher than market values. Whereas humans and businesses have to use market values to keep their businesses economical, policies for the larger scale public benefit and its environment should use emdollars, otherwise monetary decisions will benefit a part of the system while ignoring losses to the whole system.

As documented in geochemical texts, the materials of the Earth are found in tiny quantities throughout the Earth's crust, oceans, and atmosphere. These low concentrations of materials are the "background concentrations." As they are the lowest concentrations on Earth, they cannot diffuse to any lower concentration. Relative to their environment they have no available energy storage and their emergy value is zero. Figures 2.2 and 2.4 showed the coupling of energy to materials starting with processes concentrating materials from the background.

## Emergy and economic geology of ores

The skewed patterns of useful materials developed by geologic cycles have been related to market values and energy for mining and processing, one of the principles of economic geology. Curves of the quantity of ores as a function of concentration are used to estimate which reserves are commercial. Page and Creasey (1975) published curves on the amount of rock and fuels required to produce refined metal from ores of different metal concentration (Figure 2.8). The emergy in the rocks, the emergy in the necessary fuels from these graphs, and emergy from human service were added to calculate the total emergy to refine a gram of metal, shown in Figure 2.8c. The rocks with lowest concentration of metal were given the emergy per mass of the average Earth cycle ($1 \times 10^9$ sej/g) that is carrying low concentrations. Emergy values in goods and services were calculated from 1975 costs given by the authors (70% for services and 30% for fuels) multiplied by emergy/money values for that year ($7.6 \times 10^9$ sej from Appendix D in Odum 1996b). Figure 2.8a and c have similar shapes. More resources (emergy) are required to concentrate low-concentration ores than those that were already more concentrated by the geologic work.

Back-calculations were made with the relationship in Figure 2.8c so as to plot emergy stored by geological process as a function of concentration of metal. Emergy/mass from geological work for each concentration of metal in rock equals the total emergy/mass to concentrate metal from background rock minus the emergy of human services and fuels used. The total emergy per mass of fuels and services to concentrate metal from rock with background concentration (without metal-specific geologic work) is $19.9 \times 10^9$ sej/g in Figure 2.8c.

For example, for ore with a copper concentration of 1.6% in Figure 2.8c:

$$\text{Emergy/mass} = 19.9 \times 10^9 \text{ sej/g} - 2.1 \times 10^9 \text{ sej/g} - 1.72 \times 10^9 \text{ sej/g}$$
$$= 16.1 \times 10^9 \text{ sej/g}$$

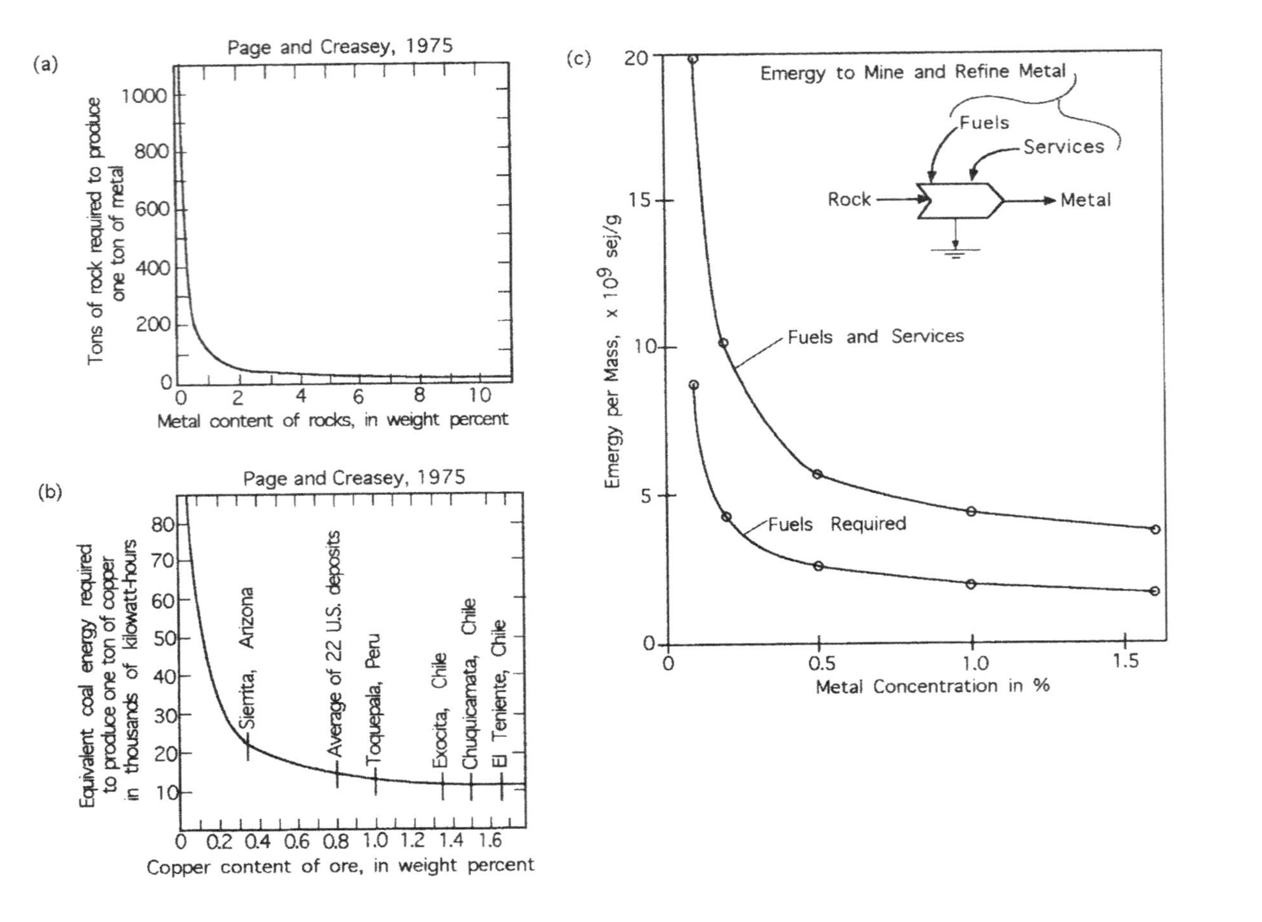


*Figure 2.8* Quantity of resources required to make concentrated metal, modified from Page and Creasey (1975). (a) Rock required. (b) Fuels required. (c) Emergy required to concentrate metal ores derived from (b).

The results of these calculations generate a graph of emergy per gram for concentrations of metal in ores in Figure 2.9a, which also shows the very low concentrations of copper as a dispersed element in the ocean and the Earth's crust. These are given a small emergy per mass value in proportion to their fraction of the Earth cycle. The lowest concentrations are in the sea, without available energy relative to the environment and thus no emergy. Equivalent 1999 emdollar values of these ores are also shown with the second scale for the vertical axis on the right.

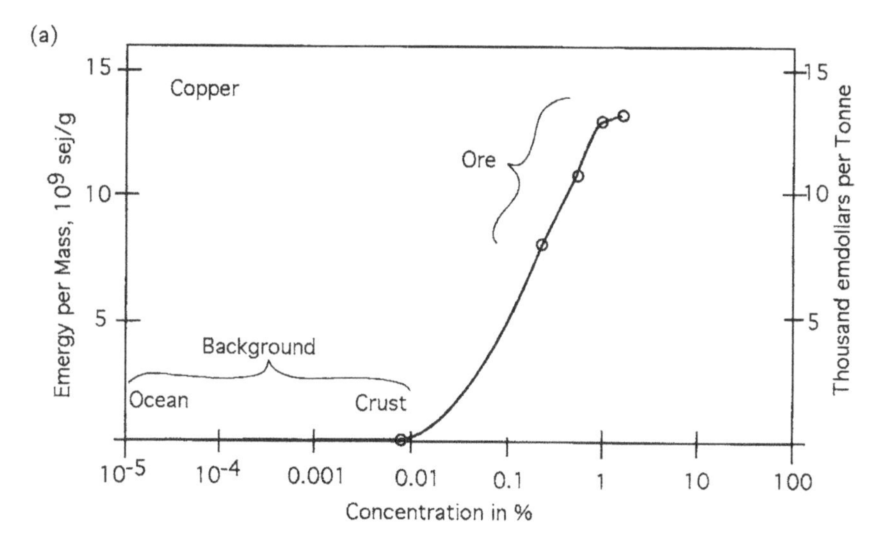


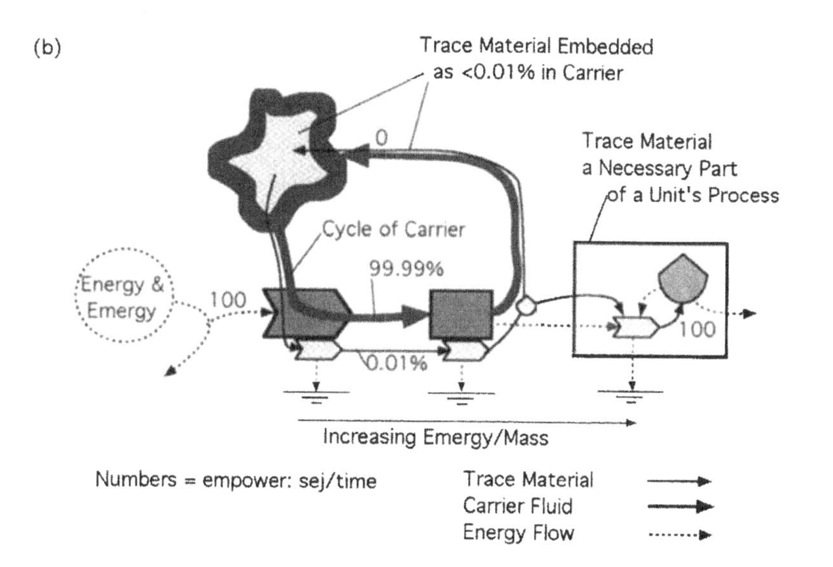


*Figure 2.9* Emergy per mass of dilute and concentrated materials. (a) Emergy per mass and metal concentration calculated from Figure 2.8c. (b) Energy systems diagram showing passive carrier and active production zones of material participation.

### *Critical concentration for material contribution to production*

As Figure 2.9a suggests, there are two zones in the distribution of elements in the geobiosphere. On the left, very dilute materials are carried along with flows of air, water, and earth cycles in proportion to their fraction of these circulating masses with that fraction of the empower. At a higher concentration, the materials participate in autocatalytic productive processes in which use of their special characteristics is a necessary part. The relationship of energy and emergy to the material cycles is shown diagrammatically in Figure 2.9b. On the left, materials are a small fraction of the carrying mass. But on the right the material is used to generate special products incorporating the emergy of the available energy used for the unit. The products carry the whole empower of the process with much higher emergy per mass. The critical concentration at which the autocatalytic process is using the material for its specific characteristics is analogous to the critical concentration of energy when flow switches from laminar flow to turbulence.

Figure 2.9b may help explain the division between materials that are not commercial and those of higher concentration that are economical, a property of ores noted by DeYoung and Singer (1961). Materials that are commercial occur at higher concentration than the critical point in each cycle where the material is used for its special properties and the processes store substantial emergy.

### *Emergy–concentration graphs*

To facilitate emergy–emdollar evaluations, graphs of emergy–emdollar value versus concentration, such as Figure 2.9a, need to be constructed for common materials and building components. See Figure 2.10 for another example, the graph for emergy and concentration for lead (Odum 2000).

### *Principles of biogeoeconomics of materials*

Money circulates in that zone of the universal energy hierarchy in which humans operate. Note the dashed line representing money circulation in the energy transformation spectrum in Figure 2.6. Boggess (1994), studying the circulation of phosphorus in a Florida watershed, showed how the transformity zone of money circulation and market prices overlapped only the higher transformities of the circulation of phosphorus. The range of transformities at which a material cycle overlaps identifies the transformity range over which material cycles are economical. In the case of cycles at lower transformities, either economic subsidies or the work of ecosystems is required. For very high-transformity materials, such as gold, the economic market values are also high, but they undervalue the real wealth and thus can result in inappropriate uses.

### *Transformity and toxicity*

Genoni and Montague (1995) and Genoni (1997, 1998) showed the increasing toxicity and impact of materials with increased transformities. In the case of a substance such as cocaine, with very high transformities and large impact, efforts to control availability can cause black market monetary values to be greater than open market values. It is possible that in the case of some very high-transformity materials, such as some found in chemistry laboratories, values may be outside the top of the range of the human economy.

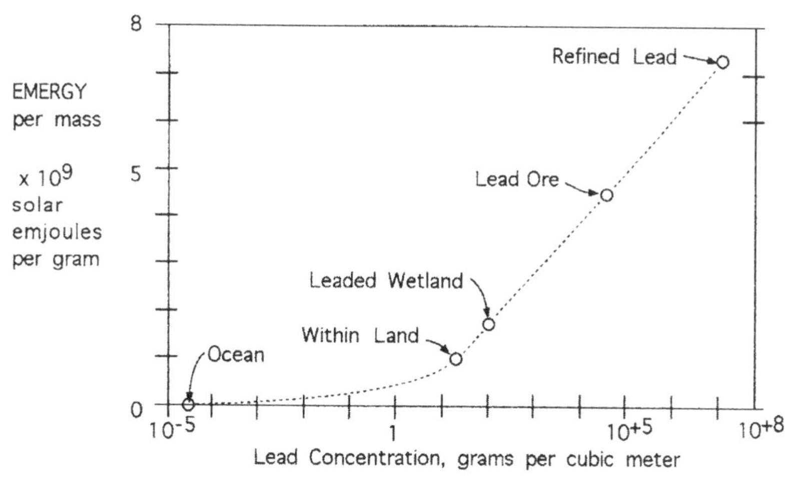


*Figure 2.10* Emergy per mass as a function of lead concentration (Odum 2000).

## *Total material accounting*

The Wuppertal Institute in Germany developed an interesting way of overviewing national economies by adding together all material flows on a weight basis, including fuels. The overview analysis is not primarily concerned with budgeting the cycles of each kind of material. The flow values in Figure 2.11a are an example of the material budget assembled for West Germany (Adriaanse *et al.* 1997; Bringezu 1997). The concentration of material flow is an index of environmental impact. Disturbances of the environmental materials, as in mining, are included as "hidden flows," a major part of each nation's annual total. The hidden flows in other countries supplying imports are regarded as a backpack ("rucksack") of embodied material responsibility, a materials footprint.

The relationship between materials and the economy has been studied using material/money ratios (Wernick *et al.* 1999). These ratios have fallen in developed countries, and this has been interpreted as a decoupling of matter and money. Wernick (Chapter 7) also uses ratios of material use (paper) to economic product as a measure of dematerialization. However, it is not that materials were used less, but that there was much accelerated circulation of money and money supply associated with urbanization and concentration of fiscal affairs.

Total material accounting aggregates materials with a range of real wealth values of more than ten orders of magnitude (emergy/mass range in Figure 2.7). Material per unit energy was used by Hinterberger and Stiller (1998) to represent the coupling with energy, although energies of different transformities were considered to be equal. Other than fuels, the embodied energy in the materials and environment were not included.

To relate the total material accounts to concepts in this chapter, emergy was calculated for the total material account for Germany redrawn in energy systems form in Figure 2.11b. Fuel use and imports in Germany in 1994 (from Adriaanse *et al.* 1997) were each multiplied by their transformities and added to main flows of environmental empower (from Bosch 1983). This total national empower was divided by the total material account to obtain a national emergy per mass index ($6.7 \times 10^7$ sej/g). The value is intermediate between that for fluids (air and water) and that for solids (earth and buildings) (Figure 2.7).

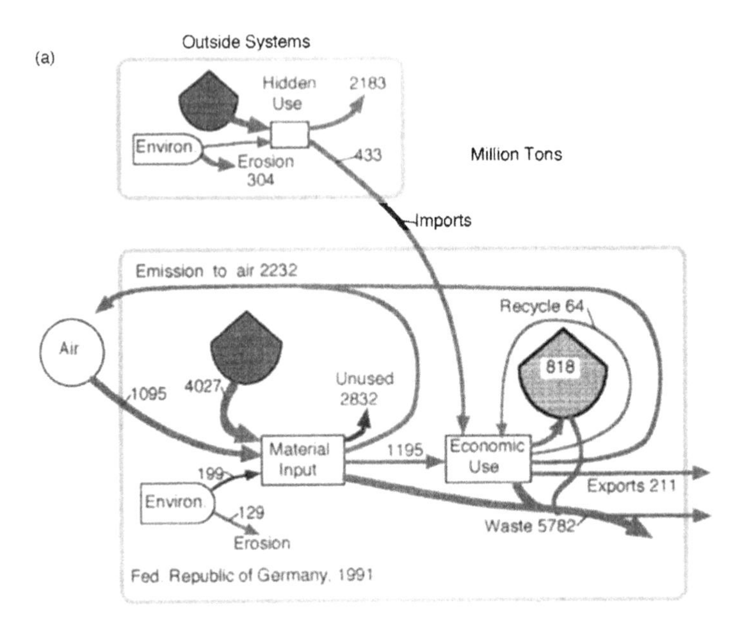


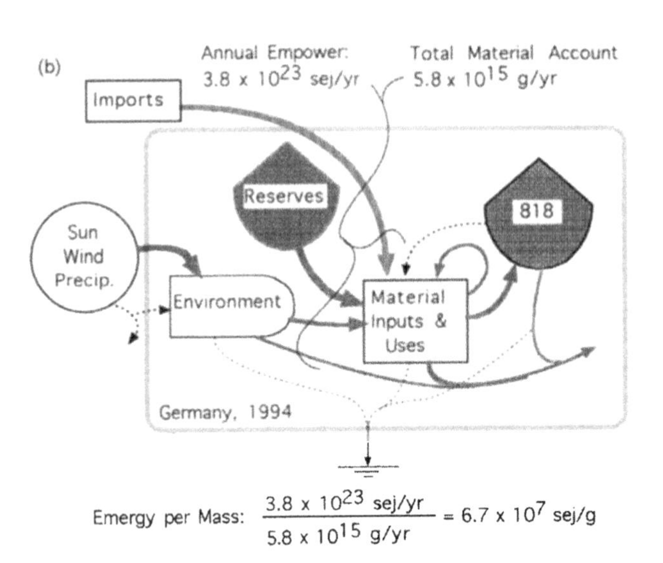


Emergy per Mass: $\frac{3.8 \times 10^{23} \text{ sej/yr}}{5.8 \times 10^{15} \text{ g/yr}} = 6.7 \times 10^{7} \text{ sej/g}$

*Figure 2.11* Aggregate material accounting of Germany from Wuppertal Institute (Adriaanse *et al.* 1997). (a) System diagram of the material flows given by S. Bringezu and H. Schutz. (b) Overview of material account of Germany with energy sources and emergy evaluation added.

## Metabolism and the structural unit

Each unit that develops in systems exhibits similar patterns and processes of materials and energy utilization. The processing of materials and energy is called *metabolism*, and either the rate of energy processing or material processing can be used to represent the metabolism quantitatively. In Figure 2.12 the essence of a unit of system structure and its processes is enlarged to show the details of inputs to produce structure and the processes that decompose the structure and release the materials.

Figure 2.12a shows the series of structural units of an energy chain and its energy hierarchy. The first structural unit is enlarged to show the component flows and storages of metabolic structure and process (Figure 2.12b). The figure shows inflowing materials being incorporated in constructive production and then returning to the pool of environmental materials, by depreciation, parts replacement, dispersion by the work the unit does for the outside, and by removal and consumptive destruction arranged by the

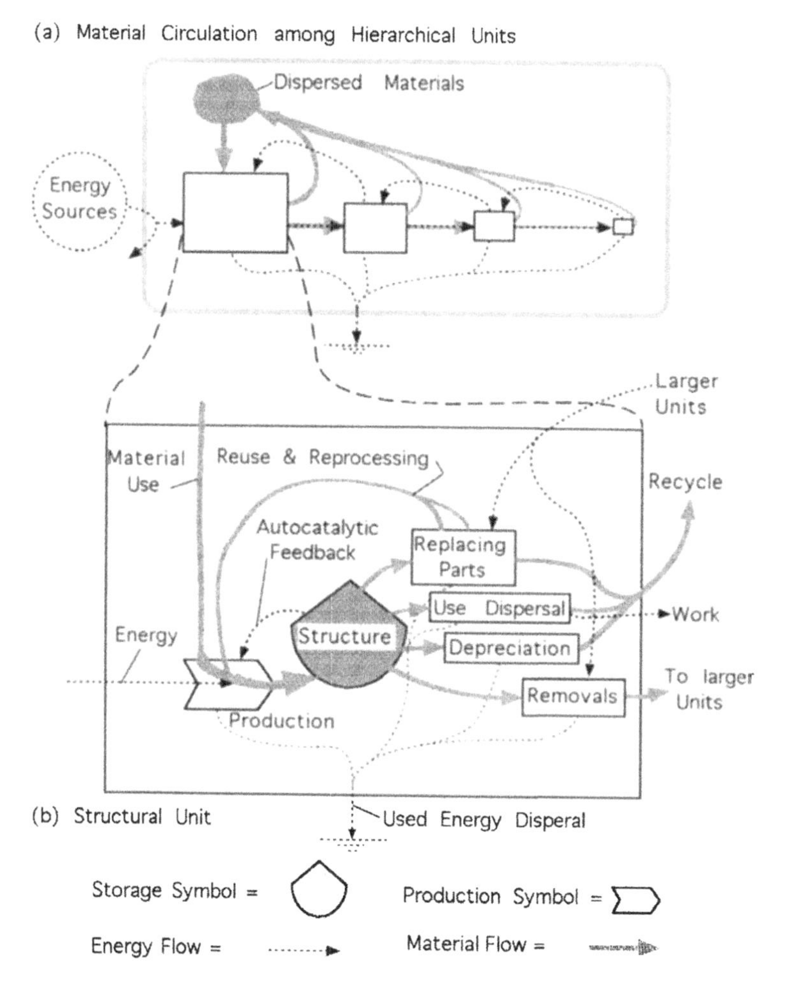


*Figure 2.12* Pathways and metabolism of a typical structural units like those found on many scales. (a) Material circulation coupled to an energy hierarchy. (b) Details of the pathways and storage of metabolism of one structural unit.

next level in the energy hierarchy. These processes are explained with Figure 2.12b as follows.

### *Construction (production)*

The incorporation of materials into a new product with the help of an energy source is called *production*, for example construction of a tree or a building. Note the pointed block symbol used to show the production process in Figure 2.12b. The rate of production is proportional to the local availability of the ingredients, and sometimes the shortage of a material or energy (limiting factor) slows down the process. Production also uses already-built structure to organize and control the addition or reuse of parts for growth and differentiation. For example, construction of initial foundations and frameworks in buildings guides the later finishing processes. As represented in Figure 2.12b the structure feeds back actions to implement the production process. Such feedbacks are called autocatalytic.

### *Depreciation and material release*

A store of materials in which the material is at greater concentration than in the environment will spontaneously disperse. Molecules of matter are dispersed by heat. Heat causes molecular motion in all materials, and the higher the temperature, the greater the motion. In other words, all matter tends to diffuse (a property described by the second energy law). Sometimes matter is dispersed by reacting with other substances. The spontaneous loss of structure is called *depreciation*. Depreciation releases materials that were previously bound into structure. Figure 2.12b shows the release of materials from depreciating structures being recycled outside the unit.

### *Replacing parts and wholes*

As a structure depreciates, a part may become non-functional and have to be replaced. Replacing parts of something larger to keep it operating is called *maintenance*. Eventually, the larger parts of the structure become non-functional and have to be removed and replaced. Replacing a whole unit is really maintenance at the next larger scale, where the unit being replaced is a part. As soon as a main structure is removed, function is interrupted until the replacement structure is reconstructed .

### *Material dispersal from use of structures in work*

Any unit is part of a larger system to which it makes a contribution in product or service. The process of doing that work causes losses of structure and dispersal of materials, also shown in Figure 2.12b. The total loss of structure is the sum of the loss which occurs without use plus the additional losses due to use of the structure in performing work for the surrounding system. The useful work carries emergy to the next system.

### *Destruction and recycle*

Depreciation through small-scale processes such as diffusion and erosion is too slow to return large units into the material cycle. Systems develop a destruction process that uses energy and work from larger scales to disperse the structure while utilizing some of the stored energy. In ecosystems, animals, microorganisms, and fire often supply this function. The equivalent operations in cities are the specialized business operations for removal and redistribution of parts, materials, and waste dispersal.

## Life cycle minimodel

One way of determining the similarities and differences among systems is to make overview models, using systems languages that are more rigorous than using words, which usually have many meanings. The energy systems language is used along with the equations for simulation that the diagrams define. Computer simulation of models is useful for finding the consequence of the structures and processes that have been included in a model. Simulation models are studied here to suggest answers to questions about structure and materials.

In ecosystems, the processes of depreciation, maintenance, decomposition, and destruction are sometimes aggregated as the process of consumption, represented by one block in Figure 2.13. In biological systems, the materials released by all these processes are sometimes collectively called respiration. For overview purposes, a simple model of structure is a balance of production and consumption, with materials recycling through the processes and structure (Figure 2.13a). This is a modification of the production–respiration procedure introduced for overall metabolism of ecosystems by the author in 1956. In ecosystems, respiration is the internal consumption that includes depreciation, maintenance, and work delivered to the outside of the unit.

### *Balance of construction and consumption*

Let us define $P$ as construction (production) and $C$ as consumption, with both measured in the same units of materials processed. A system, on average, may be in balance when $P = C$, and the ratio $P/C$ is 1. For example, in a forest the rate of carbon processing in production may be, on average, nearly equal to the rate of carbon recycle in all the consumption processes. In a forest village there could, on average, be balance between the organic matter used in the construction of shelters and the village center and the return of disorganized organic parts and wastes back to the forest.

### *Equations for a construction mini-model* RESTRUCT*.bas*

Diagramming a model with energy systems symbols and constraints generates a mathematical model that can be computer simulated. Figure 2.13b shows the equations that express the behavior of the simplified overview model shown in Figure 2.13a. The equations are incorporated in a BASIC language simulation program RESTRUCT.bas (shown in Appendix 2).

### *Overview simulation*

After assigning values based on growth of short-lived trees to calibrate the model, simulations were run, such as that in Figure 2.13c. In a typical run, the unit grows using outside energy to incorporate materials and develop structure. As structure develops, autocatalytic action facilitates further growth. The switching control in the model stops

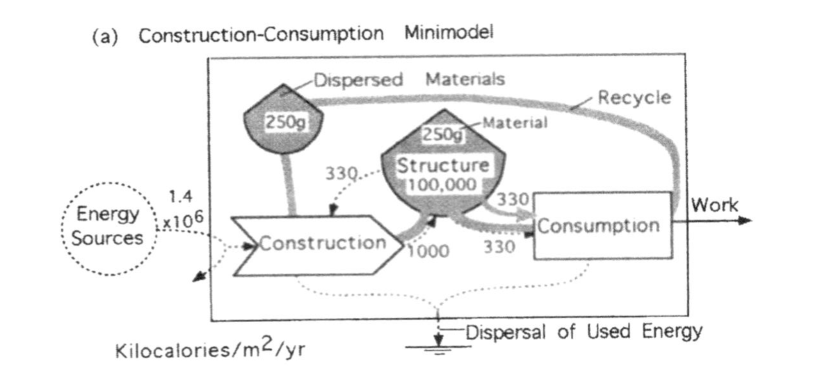


(b) Equations for Simulation

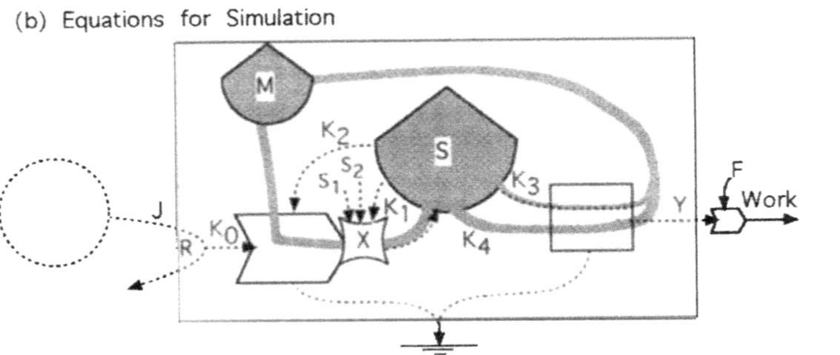


Unused Energy Inflow: $R = J - X*K_0*R*M*S$

and therefore: $R = J/(1 + X*K_0*M*S)$

If Total Material $T_m$ is Constant and fr is the fraction in S:

$$M = T_m - f_r * S$$

Rate of Change of Storage S is Inflow minus Outflow:

$$dS/dt = X*K_1*R*M*S - X*K_2*R*M*S - K_3*S - K_4*F*S$$

Switching Action [X]: If $S < S_1$ THEN $X = 1$; IF $S > S_2$ THEN $X = 0$

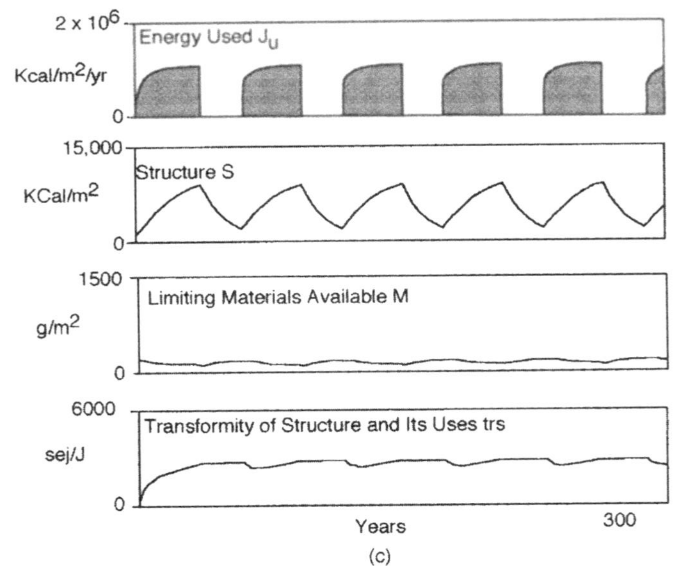


*Figure 2.13* Simplified overview model of construction (production) and consumption. (a) Systems diagram including energy and materials. (b) Equations derived from the diagram. (c) Typical simulation of the pulses of construction, depreciation, and replacement.

the growth when it reaches an upper threshold, after which the unit gradually loses structure owing to the various processes of depreciation and consumption. As loss of structure means loss of function, the switching action rebuilds the storage when the storage reaches a minimum threshold. At no point is the system in steady state, but if it is considered over a long period of time a repeating pattern is seen (Figure 2.13c), and average conditions may be calculated.

The model represents the growth of a short-lived tree, or the construction of a village dwelling from forest wood. The version shown here was calibrated for tree structure with the values in Figure 2.13, but the results are similar when values for a village dwelling but with more biomass and longer turnover time are substituted.

### *Emergy of construction and storage*

According to the definition of emergy as all of what is required to make something, the emergy of construction is the sum of the inputs to the construction as long as construction is in action. Inputs of energy and materials are multiplied by emergy/unit to evaluate emergy inflows (the empower of gross production). Part of the emergy produced is used by the larger system in which the structure functions (assumed as a third of the input). The rest is dispersed in depreciation and the necessary maintenance replacement of parts (feedback work). The model in Figure 2.13 builds structure in bursts, in the same way that humans build houses. It provides a disaggregated way of looking at ecosystems as the process of building and replacing organismic structures such as trees.

Emergy relationships are included in the minimodel RESTRUCT.bas in lines 300–350. With regard to storage in structure, emergy decreases if the stored structure suffers losses from depreciation, supporting maintenance, or removal to another system (Figure 2.12). The transformity of the structure is the emergy stored divided by the energy stored. The emdollars of real value stored in the structure is the emergy stored divided by the emergy per money for the designated year. The transformity of the structural storage is plotted on the lower panel of Figure 2.13c.

One term in the equation represents the use of the structure by the next larger system. The services or products (yield *Y*) carry the transformity of the storage. Replacing and repairing frequently keeps a high level of structure and helps maximize performance at the scale of the model and the larger scale to which its output is directed.

### *Useful destruction*

The removal and forced recycle of a structure's materials by an outside system is destruction from the point of view of the smaller unit. From the larger perspective the destruction is useful, contributing to overall empower production by preventing materials from becoming a limiting factor. The more frequent the withdrawals of structure by the larger scale, the less structure and stored emergy is maintained, but more empower is transferred to the next level. Areas that are frequently disturbed develop smaller units with faster turnover. The appropriate structure for maximizing performance depends on the climate of interruption.

## Structural stages and succession

The building of structure and replacing it in cycles was considered in the life cycle minimodel section, but there is more that happens when ecosystems and human settlements develop. There is a sequence of stages, each contributing for the time when its conditions are appropriate. In ecology, the sequence of stages is called *succession*, and stages in developments of human structures appear to be analogous. Availability of materials affects these stages and the diversity of units that develop.

### *Excess resource and low diversity*

When there is an initial excess of available resources of energy and/or materials, initial production (construction) is rapid, and the result is flimsy and short-lived structures. In ecosystems, for example, weedy growth specialists prevail at first because they can maximize growth and begin to function at an early stage. Because they grow rapidly, they overgrow other kinds of development, and the result is a low diversity. The analogous human settlements are those that develop are first to colonize an area with large material reserves, such as new colonies and mining towns. In such settlements, diversity of occupations is small. Excess resource specialists predominate the mining of rich earth reserves, shown on the left side of Figure 2.14.

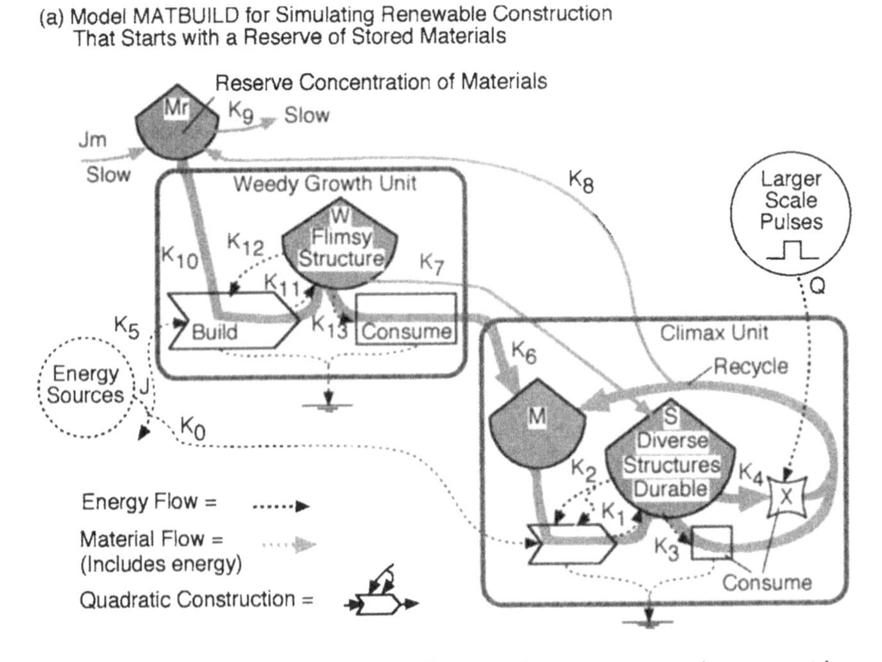


*Figure 2.14* Simulation model of succession, climax, and restart MATBUILD that starts with a reserve of materials. (a) Energy systems diagram. (b) Equations in the program. (c) Typical simulation run.

(b) Equations for the model MATBUILD simulating material in building settlements that include reserves and units for weedy growth and climax.

Remainder of Unused Energy: $R = J - K_0 * R * M * S * S - K_5 * R * M_r * W$

And $R = J/(1 + K_0 * M * S * S + K_5 * M_r * W)$

Material Available in Climax Unit: $M = T_m - f_r * S$

Rate of Change Equations:

Reserve of Stored Materials Mr: $dM_r/dt = J_m + L_8 * S - K_9 * M_r - K_{10} * R * M_r * W$

Weedy Structure W: $dW/dt = K_{11} * R * M_r * W - K_{12} * R * M_r * W - K_7 * W - K_{13} * W$

Climax S: $dS/dt = K_1 * R * M * S * S + K_7 * frw * W - K_2 * R * M * S * S - X * K_4 * Q - K_3 * S$

Total Materials in Climax $T_m$: $dT_m/dt = frw * K_6 * W + frw * K_7 * W - K_8 * fr * S$

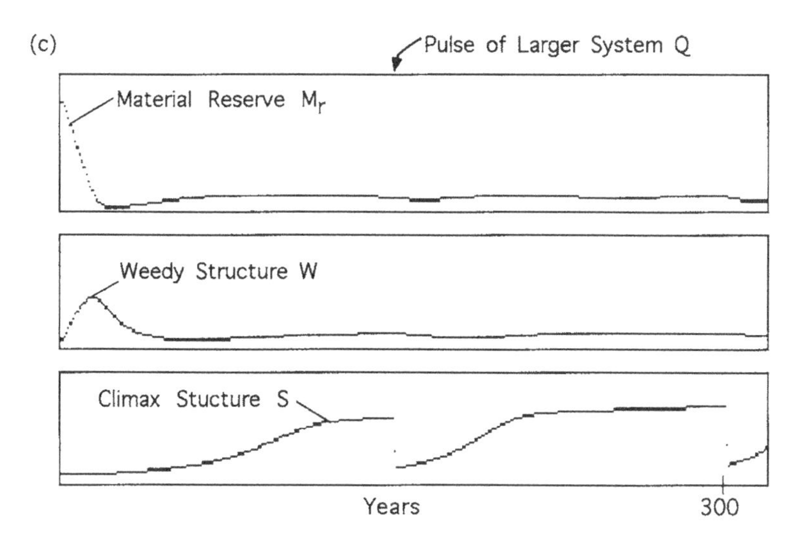


*Figure 2.14* Continued.

## *Efficient recycle and high diversity*

Once excess energy and materials have been incorporated into structures and storage, the weedy specialists are no longer ideally suited and are replaced by construction of longer-lasting units that are efficient in recycling the necessary materials for longer sustainability. Greater efficiency is achieved by a diversity of units. In ecosystems there is increase in biodiversity. In human settlements there is an increase in variety of occupations and an accompanying diversity of structures. The colonization of America by European immigrants resulted in the development of growth specialists, the capitalist economy, with simple structures that were predominant in the last century. In this century there has been increasing efficiency and diversity of structures and functions. The processing of materials is shifting from exploitative processing, shown on the left in Figure 2.14, to the reuse–reprocessing–recycle pattern on the right.

Complexity of materials also increases. Keitt (1991) found that the emergy of rainforest diversity increases with scale.

### *Simulation of life cycle and succession*

To understand better the interplay of growth and succession of buildings likely to occur in the future, the construction–consumption minimodel from Figure 2.13 was expanded as Figure 2.14a (program MATBUILD.bas in Appendix 3). The equations that are inherent in the systems diagram are listed in Figure 2.14b. For the initial run the model was calibrated for the building of woody structures using wood from forests. (Appendix 4 contains the numerical values used for calibration. Storages were normalized – expressed as 100 or 1000 at full development.)

In Figure 2.14a, the reserve of materials *M* shown in the upper left corner is exploited for accelerated, non-renewable use by the weedy rapid growth specialists *W* that prevail when there is resource excess. From these beginnings develop more sustainable climax units *S* (lower right), which exhibit lower depreciation, longer turnover time, and more efficient reuse of materials. Over a longer timescale, the central units of the next larger system outside the systems diagram cause a destructive consumption and removal when they pulse.

Figure 2.14c shows a typical run. The top graph shows the rapid depletion of the reserve as the weedy structures grow (second panel). The bottom panel shows the slower growth of the climax units that displace most of the successional weedy structure. At 150 years the system is partially restarted by the pulse of the surrounding system. The system repeats its pattern, except that the reserve available for growth is less.

Figure 2.14a is a general system model that relates structure and metabolism for any system. For example, the model applies to the smaller scale of growth of a forest that starts with an initial reserve of phosphate nutrients. With appropriate calibration, it also applies to the historical situation when an agrarian economy discovered the use of metals available from the reserves built by earth processes.

## Ecological engineering insight

*Ecological engineering* fits human civilization into environment so that they reinforce each other. Fitting the building enterprise into civilization and environment is part of the new field of *industrial ecology* and its subset, construction ecology. The principles relating materials to energy hierarchy provide suggestions for policy. It ought to be possible to use understanding more and trial and error less.

## Maximum empower principle

The *maximum empower principle* is a unifying concept that explains why there are material cycles, autocatalytic feedbacks, successional stages, spatial concentrations in centers, and pulsing over time. *Designs prevail that maximize empower*. The concept predicts that systems, such as forest production and building construction, that emerge in competitive trial and error develop patterns that maximize performance at all scales. The principle is a refinement of Lotka's (1922a,b) maximum power principle, which considered maximum energy use (power) at one scale. Systems prevail that utilize all available emergy sources, including stored concentrations of materials, wherever they are available. According to this principle, complex, high-quality, diverse construction should replace fast and flimsy structures when resources are no longer enough to support net growth (simulation example in Figure 2.14).

### *Principle of upscale overview*

Understanding and management of any process requires overview at the next largest scale. For example, building industries need policies for an area large enough to include the complete cycle of their materials. The overview needs to take into account the longer timescale of the life cycle of construction and reconstruction. An example of the wider view of construction cycles was provided at the workshop that serves as a basis for this volume by Connie Grenz of Collins Pine Company of Portland, Oregon. She explained how provision was made for longer sustained productivity rather than short-term exploitation profit. In Figure 2.15 her diagram of components and material flows was modified to include energy sources. Items are arranged left to right according to the energy hierarchy.

### *Characteristics for zones in the energy hierarchy*

Evaluating emergy per mass and transformities identifies the zone of a building enterprise in the energy hierarchy. The zone helps indicate the design properties needed for the system to be sustainable, such as the territory of support and replacement time; conversely, the territory and replacement times determine the emergy per mass of materials that are appropriate. On traveling from the center of a city to the countryside, the high emergy per mass and dense permanent buildings of the city gradually give way to the rural landscape of scattered small houses.

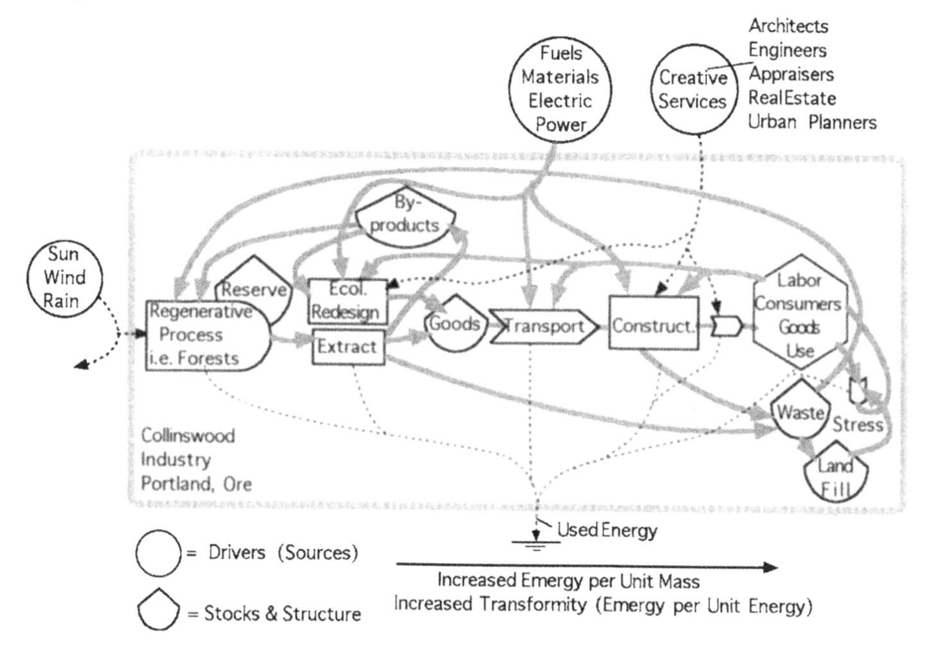


*Figure 2.15* Energy system diagram of the Collinswood pine industry of Portland, Oregon, showing the renewable regenerative forest processes on the left described by C. Grenz.

### *Adapting to pulses*

The large-scale pulses of the Earth include earthquakes, volcanic actions, catastrophic storms, disease epidemics, and economic storms. For building cycles to be most effective in the production that they support, their life cycles need to fit the frequency of the storms. In the case of smaller pulses, there are mechanisms of rapid repair. For example, tropical rainforests in hurricane belts develop mechanisms of rapid restoration, such as fast regrowth of foliage and conservation of mineral cycles. The shared information (genetic and learned) necessary for repair and replacement is the highest transformity of the building system for which continuity is required.

In the event of the less frequent, more catastrophic pulses, the forest is adapted to initiate rapid restart of the trees. In other words, the amount of structure that can be productively sustained is, on average, a function of the size and frequency of pulsing impacts received from the larger scale (a function of their position in the energy hierarchy). The adapted forest uses the energy of the pulses to accomplish the necessary replacement of structures. By analogy, hurricane destruction could result in urban renewal of coastal structures.

Odum *et al.* (1995) simulated a minimodel in which pulse energy, while causing destruction, also contributed to construction. There was an optimum pulse energy for maximum net contribution to structure. When pulse energy was large, less structure was maintained but the throughput of empower (gross production) increased.

Finding the optimum emergy in building structure means building patterns that minimize repair costs for small pulses but can be repeatedly rebuilt between the larger pulses. In 1967, in order to explain the small biological structure in oceanic ecosystems (mostly plankton), the author constructed a graph of the unit metabolism required for maintaining ecosystems (assuming 10% efficiency of construction) as a function of intervals between catastrophic pulses (Odum 1967). Fitting building construction to the Earth's climate of catastrophic pulses is well under way, adapting building codes and insurance premiums to various areas by trial and error.

### *The cement of forests*

A remarkable property of the forest building cycle is the cement of trees, lignin, a humic substance that undergoes various degrees of polymerization and aggregation. Lignin constitutes one-third of a tree, cementing the functional parts together. When a tree falls, the structure is shredded by animals and bacteria and the lignin is released as blackwaters or stored as peat. In pulp and paper mills, machines shred trees and lignin is removed in the wastewaters. The cement of forest construction has its own useful cycle. In human villages built of wood, recycle of materials is aided by the natural mechanisms of shredding and return of materials to the environment.

The massive permanent construction of civilization requires a cement that is useful throughout the building cycle. When it is time to replace a building, we blow up the structure and try to reuse the building materials, which is not easy because the cement of concrete construction does not disaggregate on demand. However, initiatives to use less dense cements were offered at the workshop that resulted in this volume.

### *Policy on returning wastes to the environment*

Waste materials that are too dilute to be reused or reprocessed economically require public management. If the emergy/mass ratio of a material is high enough that it is of public benefit to reprocess it (or, conversely, not to do so would affect adversely the environment), tax benefits or other incentives can be used to help pay for reprocessing. However, very low concentrations must be returned to the environment, but not to just any ecosystem. For example, peat filters, denatures, or holds potentially toxic compounds such as heavy metals and organic toxic compounds, and ecological engineering studies have revealed that the lignin chemistry of peat found in wetland ecosystems varies throughout the world (Odum *et al.* 1999). A complete life cycle for building construction may often include a wetland ecosystem in its loop. Many of the wastes of building construction are incompatible with the ecosystems of rivers, lakes, and oceans. The emergy per mass helps identify the appropriate zone of the Earth for material release.

## Global materials and construction

Building construction and material use by our civilization is limited by the energy hierarchy and the global cycles of materials. Figure 2.16 summarizes the necessary symbiosis between Earth processes and human construction in the use of the global cycle of materials. Figure 2.16a shows the series of transformations that provide the materials of our structures, starting with the slow work of the Earth in concentrating stored reserves on the left. Materials such as aluminum or iron are concentrated and increase in emergy values from left to right. Human uses start when materials are mined and processed into stocks available to the building industries. Next, construction incorporates materials into buildings, and they are eventually released to the global cycle at the end of the building life cycle. Human services are involved only on the right side of the global hierarchy. They are paid for with circulating money, as shown by the countercurrent of dashed lines in Figure 2.16b.

### *Emergy criteria for reuse, reprocessing, or environmental recycle*

When old structures are taken out of use, three pathways for materials are required (right end of Figure 2.16b). The highest quality components with some repair can be reused. Those remnants that are still concentrated can be fed back for reprocessing. The appropriate pathway for the least concentrated waste materials is recycling to the natural Earth processes capable of incorporating them in natural storages that benefit ecosystems.

The emergy per mass of a material can indicate which of these pathways will benefit the system by contributing more while using emergy of fuels and services least. For example, low concentrations of carbon have emergy per mass values like those of the environment, where there are processes that can process low concentrations. Materials with the highest mass emergy are economical, whereas those with lower emergy per mass may require some regulations and incentives from society. Otherwise, discarded materials will accumulate in slums and landfills, displacing lands and poisoning waters, an unsustainable condition.

### *Material basis of civilization*

As most people realize, the growth of our civilization in the last two centuries has been based on the rapid use of stored reserves of materials and energy. Such materials, which are being used faster than they are being replaced, are called non-renewable. (They are in

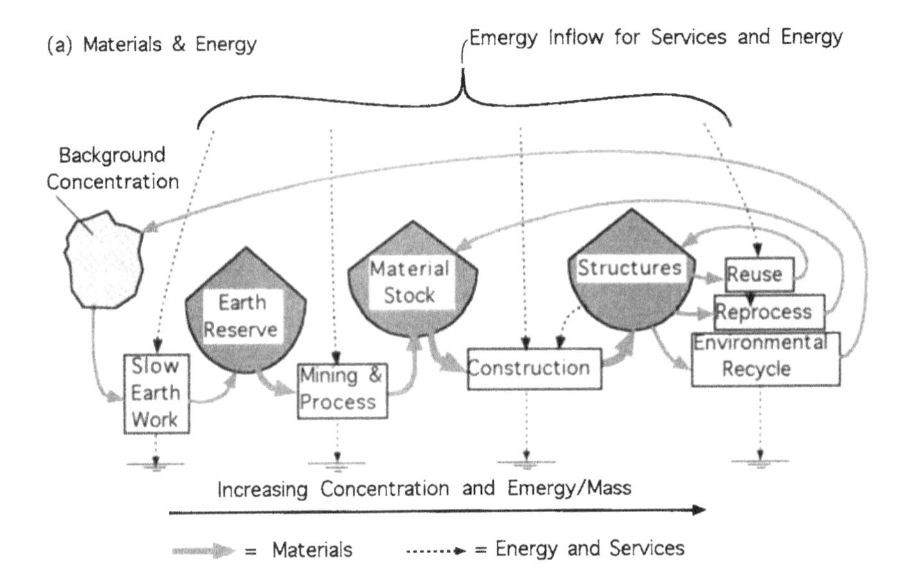


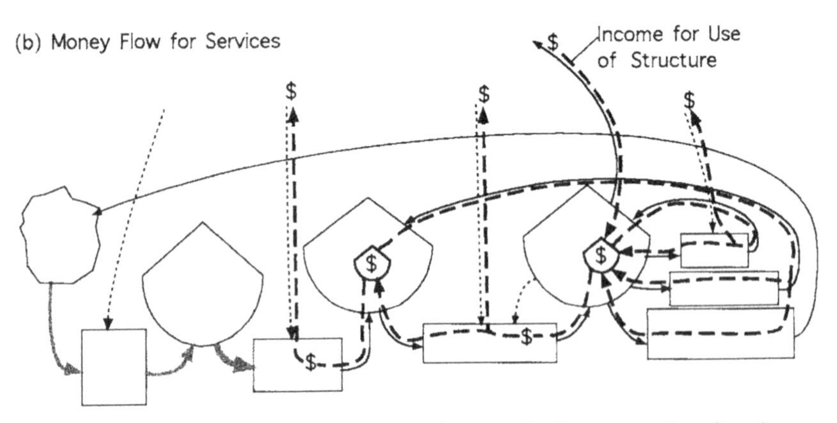


*Figure 2.16* Main pathways of material processing and storage in the system of earth and economy. (a) Material flows, reserves, stocks, and structures are shaded. Flows of energy are finely dotted lines. (b) Money circulation shown as a dashed line overlay of those pathways with human service.

fact renewed very slowly.) As the store of non-renewable reserves declines, construction will depend on the reuse and reprocessing type of recycle (on the right in Figure 2.16a), but these require fuels and services that will also become limited.

### *Pulsing paradigm*

In systems of all kinds and scales, observations show that the normal pattern is not at steady state for long. The usual sequence is growth to a maximum high-diversity state followed by removal and regrowth again. In other words, systems in the long run prevail by using a pulsing pattern to generate more. The simple cycle of building and replacement simulated in Figure 2.13c is an example.

Items on a small scale pulse often, whereas items on a larger scale pulse less frequently. Note the units and territories of different size in Figure 2.1. Smaller-scale items pulse many times before they are affected by the pulse of the larger-scale system of which they are a part. For example, the leaves of the forest grow and fall frequently, whereas the whole tree is replaced and restarted less frequently. In human settlements, the smaller structures are built and replaced frequently, whereas the larger structures and districts have longer lives.

### *Building life cycle and the pulse of the civilization*

On the global scale, the whole civilization is in a pulse based on use of the non-renewable reserves of energy and materials. We can develop more efficient life cycles for the buildings appropriate for a particular level of available energy and emergy. But the structure and diversity of the buildings that can be sustained is a moving target, rising in this century, declining again in the next.

### *Adapting construction for the prosperous way down*

In the future, when reserves (energy and materials) are less and the growth economy has ended, self-organization will be required to adapt buildings and material processing. To maintain prosperity, the population must fall at the same rate as the supporting empower. Reuse and reprocessing will replace most mining. Buildings will become more permanent and diverse (see 'Structural stages and succession'). The global excess of structure and infrastructure may not be sustainable, but the materials from excess structure that is no longer needed can be reprocessed to help form the more sustainable patterns of a lower-energy world. Values of emergy per mass can be used to select the materials that can be sustained in times of lower empower.

## Summary

In spite of, and because of, human creativity and purpose, civilization and ecosystems build structure and recycle according to the principles of energy hierarchy. Evaluation of the emergy of materials and buildings indicates where they belong in the landscape, being more concentrated at the center. Simulation of minimodels of building life cycles shows how the growth of biological structure and building construction follow similar principles of metabolism. Both operate in construction pulses. In the forthcoming era of non-renewable growth and the inevitable downturn to sustainable lower levels of structure and production, emergy–emdollar evaluation will provide a scale of value that will inform public policy on materials and human settlements with a view to achieving balance.

## Appendix 1: emergy of materials

| Material | Transformity (× 10³ sej/J) | Emergy/mass (× 10³ sej/J) | 1997 Em$ per kg[a] | Reference |
|---|---|---|---|---|
| *Environment* | | | | |
| Rain | 18 | $9 \times 10^{-5}$ | $8 \times 10^{-5}$ | Odum (1996b), Appendix C |
| Typical river water | 48.5 | 0.24 | 0.21 | Odum (1996b), Appendix C |
| Peat | 19.0 | 0.36 | 0.31 | Odum (1996b), Appendix C |
| Harvested pine logs | 8.0 | 0.24 | 0.21 | Odum (1996b), Appendix C |
| Wood chips | 15.6 | | | Doherty (1995) |
| Rainforest logs | 32.0 | 0.39 | 0.33 | Odum (1996b), Appendix C |
| Rainforest wood chips | 44.0 | 0.54 | 0.46 | Odum (1996b), Appendix C |
| *Rocks* | | | | |
| Granite | – | 0.50 | 0.43 | Odum (1996b), Appendix C |
| Global sediments | – | 1.0 | 0.85 | Odum (1996b), AppendixC |
| Mountain rocks | – | 1.12 | 0.96 | Odum (1996b), Appendix C |
| Metamorphic rock | – | 1.45 | 1.23 | Odum (1996b), Appendix C |
| Volcanic rocks | – | 4.5 | 3.9 | Odum (1996b), Appendix C |
| *Building* | | | | |
| Lumber | 42 | 0.88 | 0.75 | Buranakarn (1998) |
| Concrete block | – | 1.35 | 1.15 | Haukoos (1995) |
| Softwood plywood | 57 | 1.21 | 1.03 | Buranakarn (1998) |
| Hardwood plywood | 69 | 1.44 | 1.23 | Buranakarn (1998) |
| Particleboard | 158 | 2.4 | 2.1 | Haukoos (1995) |
| Ready-mixed concrete | – | 1.44 | 1.23 | Buranakarn (1998) |
| Cement | – | 1.98 | 1.69 | Buranakarn (1998) |
| Flat glass | – | 1.90 | 1.62 | Buranakarn (1998) |
| Glass | – | 4.7 | 4.0 | Haukoos (1995) |
| Float glass | – | 7.9 | 6.8 | Buranakarn (1998) |
| Ceramic tile | – | 3.1 | 2.7 | Buranakarn (1998) |
| Brick | – | 2.22 | 1.9 | Buranakarn (1998) |
| Slag | – | 7.0 | 6.0 | Buranakarn (1998) |
| Plastics | 96 | 3.3 | 2.8 | Buranakarn (1998) |
| Paper | 142 | 2.1 | 1.8 | Keller (1992) |
| High-density polyethylene | – | 5.3 | 4.5 | Buranakarn (1998) |
| Raw rubber | 393 | 5.5 | 4.7 | Odum *et al.* (1983) |
| Polyvinyl chloride (PVC) | – | 5.9 | 5.0 | Buranakarn (1998) |
| Vinyl floor | 194 | 6.32 | 5.4 | Buranakarn (1998) |
| Cotton | 865 | 14.4 | 12.3 | Odum *et al.* (1987) |

| Material | Transformity ($\times 10^3$ sej/J) | Emergy/mass ($\times 10^3$ sej/J) | 1997 Em$ per kg[a] | Reference |
|---|---|---|---|---|
| *Metals* | | | | |
| Pig iron | – | 2.8 | 2.4 | Buranakarn (1998) |
| Steel | – | 4.2 | 3.6 | Buranakarn (1998) |
| Steel | – | 10.7 | 9.2 | Sundberg *et al.* (1994) |
| Machinery | – | 6.7 | 5.7 | Odum (1996b), Appendix C |
| Aluminum ingots | – | 16.9 | 14.4 | Odum (1996b), AppendixC |
| Aluminum sheet | – | 12.7 | 10.9 | Buranakarn (1998) |
| Copper | – | 51.3 | 43.9 | Sundberg *et al.* (1994) |
| Lead | – | 73.4 | 62.7 | Pritchard (1992) |
| Wool | 3840 | 80 | 68 | Odum (1984) |
| Silver | – | 1,400 | 1,196 | Sundberg *et al.* (1994) |
| Gold | – | 589,000 | 503,418 | Bhatt (1986) |
| *Fuels* | | | | |
| Lignite | 37 | 444 | 380 | Odum (1996b), Appendix C |
| Coal | 40 | 1,160 | 991 | Odum (1996b), Appendix C |
| Crude oil | 54 | 2,322 | 1,984 | Odum (1996b), Appendix C |
| Natural gas | 48 | 2,640 | 2,256 | Odum (1996b), Appendix C |
| Motor fuel | 66 | 2,904 | 2,482 | Odum (1996b), Appendix C |
| Electric power | 170 | – | – | Odum (1996b), Appendix C |
| Hydrogen | 203 | 25,172 | 21,514 | Odum (1996b), Appendix C |

Notes

aEm$/kg in column 4 calculated as follows:

(Mass emergy in column 3)(1000 g/kg)/($1.17 \times 10^{12}$ sej per 1997 US$)

Abbreviations: sej, solar emjoule; J, joule; g, gram; kg, kilogram, $10^3$ g; Em$ (1997), 1997 US emdollars.

## Appendix 2: BASIC program RESTRUCT.bas for the construction minimodel in Figure 2.13

```
  4   REM RESTRUCT.bas (Tree Production & Consumption, H.T.Odum)
 10   LINE (0,0)-(320,60),,B
 20   LINE (0,70)-(320,120),,B
 30   LINE (0,130)-(320,180),,B
 35   LINE (0,190)-(320,250),,B
 40   REM Coefficients
 50   fr = .01
 60   K0= .0000036
 70   K1= 2.86E-09
 80   K2= 8.57E-10
 90   K3=.03
 95   k4=.03
100   REM Starting Values and Sources
105   S=1000
110   J=1400000!
115   F = 1
120   S1 = 2000
125   S2 = 9000!
130   Tm =200
135   T = .1
137   Trj = 1
140   REM Scaling factors
150   dt = 1
160   T0 = 1
170   S0 = 250
180   Mg = 25
190   trs0 = 100
200   REM Start of Loop
210   IF S<S1 THEN X = 1
220   IF S>S2 THEN X = 0
230   M=Tm-fr*S
235   IF M<.0001 THEN M= .0001
240   R=J/(1+X*K0*M*S):REM unused energy
250   ju=J-R: REM Energy used in Construction
260   Y = k4*F*S:REM Energy output to larger system
270   dS=X*K1*R*M*S-X*K2*R*M*S -K3*S-Y
280   S = S + dS*dt
300   REM Emergy and Transformity Equations
310   emj = ju * Trj:REM input emergy used
320   IF dS>0 THEN dEms = emj -Emy
325   IF dS<0 THEN dEms = dS*Trs
330   Ems = Ems + dEms*dt:REM emergy stored in S
340   Trs = Ems/S: REM Transformity of Storage S
350   Emy=Trs*Y:REM Emergy of output to larger sytem
360   T=T+dt
400   REM Plotting
410   PSET(T/T0,120-S/S0): REM Structure
420   IF X = 0 GOTO 440
430   LINE (T/T0,60)-(T/T0,60-.00003*ju): REM Energy used
440   PSET(T/T0,180-M/Mg): REM Available materials
450   PSET (T/T0,250-Trs/trs0):REM Yield to other systems
460   IF T/T0<320 GOTO 200
```

## Appendix 3: BASIC program MATBUILD.bas for simulating the minimodel of materials in succession in Figure 2.14 calibrated with Appendix 4

```
  4   REM MATBUILD.bas (Material in Building Succession & Climax),
      H.T.Odum Aug, 1999)
 10   LINE (0,0)-(320,60),,B
 20   LINE (0,70)-(320,120),,B
 30   LINE (0,130)-(320,180),,B
 40   REM Coefficients
 50   frw = .5
 55   fr = 1
 60   K0=9E-08
 70   K1=.0000005
 72   K2=.0000001
 74   K3=.01
 76   K4 = .5
 78   k5 = .0009
 80   K6 = .16
 82   k7 = .02
 84   k8 = .001
 86   k9 = .01:REM -?
 88   K10 = .01
 90   k11 = .01
 92   k12 = .001
 94   k13 = .08
100   REM Starting Values
105   S=100
110   J=1
115   Jm= 1
117   Q = 1
120   Mr=100
125   W = 5:REM Weedy, flimsy structure
130   Tm =500
140   REM Scaling factors
150   dt = 1
160   T0 = 1
170   S0 = 20
180   Mr0 = 2
190   W0 = 2
200   REM Start of Loop
205   Z = Z +1
210   IF Z > 150 THEN X = 1 :REM Introduces Large scale pulse
215   IF Z>150 THEN Z = 1
220   IF Z > 2/dt THEN X = 0
230   M=Tm-fr*S
235   IF M<.001 THEN M= .001
240   R=J/(1+K0*M*S+k5*Mr*W):REM unused energy flow
245   DMr = Jm -k9*Mr -K10*R*Mr*W +k8*fr*S
250   DS=K1*R*M*S*S +k7*frw*W-K2*R*M*S*S -K3*S-X*K4*S*Q
255   DTm = frw*K6*W +frw*k7*W -k8*fr*S
260   Dw = k11*R*Mr*W -k12*R*Mr*W -k7*W -k13*W
265   S = S + DS*dt
270   Mr = Mr +DMr*dt
275   Tm = Tm +DTm*dt
280   W = W +Dw*dt
290   T=T+dt
300   REM Plotting
310   PSET(T/T0,180-S/S0): REM ClimaxStructure
320   PSET(T/T0,120-W/W0):REM Weedy Structure
330   PSET(T/T0,60-Mr/Mr0): REM Material Reserve
350   IF T/T0<320 GOTO 200
```

## Appendix 4: spreadsheet for calibration of MATBUILD.bas

| | | |
|---|---|---|
| Inflows | Energy $J$ = | 1 |
| Material | $J_m$ = | 1 |
| Forcing pulses | $Q$ = | 1 |
| Starting states | Material reserve $M_r$ = | 100 |
| | Weedy structure $W$ = | 100 |
| | Climax structure $S$ = | 1,000 |
| | Climax material $T_m$ = | 1,100 |
| | Available material $M$ | 100 |
| Remainder | Energy $R$ = | 0.1 |
| Products | $R*M*S*S$ = | 10,000,000 |
| | $R*M_r*W$ = | 10,00 |

| *Coefficients* | | | | |
|---|---|---|---|---|
| Material fraction in weedy structure $frw$ = | | 0.5 | | |
| Material fraction in climax structure $fr$ = | | 1 | | |
| Energy use by weeds | $K_0*R*M*S*S$ = | 0.9 | and $K_0$ = | 0.00000009 |
| Climax structure production | $K_1*R*M*S*S$ = | 5 | and $K_2$ = | 0.0000005 |
| Use of structure in production | $K_2*R*M*S*S$ = | 1 | and $K_2$ = | 0.0000001 |
| Climax structure construction | $K_3*S$ = | 10 | and $K_3$ = | 0.01 |
| Pulsed consumption | $X*K_4*S*Q$ = | 500 | and $K_4$ = | 0.5 |
| Energy use by climax | $K_5*R*M_r*W$ = | 0.9 | and $K_5$ = | 0.0009 |
| Material from weed cons. | $K_6*frw*W$ = | 8 | and $K_6$ = | 0.16 |
| Weed to climax | $K_7*W$ = | 2 | and $K_7$ = | 0.02 |
| Climax to material reserve | $K_8*fr*S$ = | 1 | and $K_8$ = | 0.001 |
| Material reserve slow out | $K_9*M_r$ = | 1 | and $K_9$ = | 0.01 |
| Weed use of reserve | $K_{10}*R*M_r*W$ = | 10 | and $K_{10}$ = | 0.01 |
| Weed production | $K_{11}*R*M_r*W$ = | 10 | and $K_{11}$ = | 0.01 |
| Weed feedback | $K_{12}*R*M_r*W$ = | 1 | and $K_{12}$ = | 0.001 |
| Weed consumption | $K_{13}*W$ = | 8 | and $K_{13}$ = | 0.08 |

## References

Adriaanse, A., Bringezu, S., Hammond, A., Moriguchi, Y., Rodenburg, E., Rogich, D. and Schutz, H. 1997. *Resource Flows: The Material Basis of Industrial Economies*. Washington, DC: World Resources Institute.

Bhatt, R. 1986. Calculation table from unpublished report of policy research project, Lyndon B. Johnson School of Public Affairs, Austin, TX ("Policy Implication of Gold Emergy"), cited in Odum (1991).

Boggess, C.F. 1994. *The Biogeoeconomics of Phosphorus in the Kissimmee Valley*. PhD dissertation. University of Florida, Gainesville.

Bosch, G. 1983. Energy analysis overview of the Federal Republic of Germany. In *Energy Analysis: Overview of Nations*. Odum, H.T. and Odum, E.C. (eds). Working Paper WP-83-82. Laxenburg, Austria: International Institute for Applied Systems Analysis, pp. 197–223.

Brand, S. 1994. *How Buildings Learn: What Happens After They Are Built*. New York: Penguin.

Bringezu, S. 1997. From quantity to quality: material flow analysis. In *Regional and National Material Flow Accounting: From Paradigm to Practice of Sustainability*. Bringezu, S., Fischer-Kowalski, M., Kleijn, R. and Palm. V. (eds). *Proceedings of the ConAccount Workshop, 21–23 January 1997, Leiden, The Netherlands*. Wuppertal Special 4. Wuppertal, The Netherlands: Wuppertal Institute, pp. 43–57.

Buranakarn, V. 1998. *Evaluation of Recycling and Reuse of Building Materials Using the Emergy Analysis Method*. PhD dissertation, University of Florida, Gainesville.

DeYoung, J. and Singer, D.A. 1961. Physical factors that could restrict mineral supply. *Economic Geology*, 75th anniversary volume, 939–954.

Genoni, G.P. and Montague, C.L. 1995. Influence of the energy relationships of trophic levels and of elements on bioaccumulation. *Ecotoxicology and Environmental Safety* 30: 203–218.

Genoni, G.P. 1997. Towards a conceptual synthesis in ecotoxicology. *Oikos* 80: 96–106.

Genoni, G.P. 1998. The energy dose makes the poison. *EAWAG News* 45 (November): 13–15.

Haukoos, D.S. 1995. *Sustainable Architecture and its Relationship to Industrialized Building*. MS thesis, University of Florida, Gainesville.

Hinterberger, F. and Stiller, H. 1998. Energy and material flows. In *Advances in Energy Studies: Energy Flows in Ecology and Economy*. Ulgiati, S. (ed.). Rome: Museum of Science, pp. 275–286.

Keitt, T.H. 1991. *Hierarchical Organization of Energy and Information in a Tropical Rain Forest Ecosystem*. MS thesis, University of Florida, Gainesville.

Keller, P.A. 1992. *Perspectives on Interfacing Paper Mill Wastewaters and Wetlands*. MS thesis. University of Florida, Gainseville.

Laskey, S.G. 1950. Mineral-resource appraisal by the US Geological Survey. *Colorado School Mines Quarterly* 43(1A): 1–27.

Lotka, A.J. 1922a. Contribution to the energetics of evolution. *Proceedings of the National Academy of Sciences of the USA* 8: 147–151.

Lotka, A.J. 1922b. Natural selection as a physical principle. *Proceedings of the National Academy of Sciences of the USA* 8: 151–154.

Odum, H.T. 1967. Biological circuits and the marine systems of Texas. In *Pollution and Marine Ecology*. Olson, T.A. and Burgess, F.J. (eds). New York: Interscience/John Wiley, pp. 99–157.

Odum, H.T. 1970. Summary: an emerging view of the ecological system at El Verde, Puerto Rico. In *A Tropical Rainforest*. Odum, H.T. and Pigeon, R.F. (eds). Oak Ridge, TN: Division of Technical Information, Atomic Energy Commission, pp. I-191–I-277.

Odum, H.T. 1984. Energy analysis of the environmental role in agriculture. In *Energy and Agriculture*. Stanhill, G. (ed.). New York: Springer-Verlag, pp. 24–51.

Odum, H.T. 1991. Emergy and biogeochemical cycles. In *Ecological Physical Chemistry: Proceedings of an International Workshop, November 1990, Sienna, Italy*. Rossi, C. and Tiezzi, E. (eds). Amsterdam : Elsevier Science, pp. 25–65.

Odum, H.T. 1996a. Economic impacts brought about by alterations to freshwater flow. In *Improving Interactions Between Coastal Science and Policy: Proceedings of the Gulf of Mexico Symposium*. Washington, DC: National Research Council, National Academy Press, pp. 239–254.

Odum, H.T. 1996b. *Environmental Accounting, Emergy and Decision Making*. New York: John Wiley.

Odum H.T. 2000 Biogeochemical cycle of lead and the energy hierarchy. In *Heavy Metals in the Environment: Using Wetlands for Their Removal*. Odum, H.T. with Wojcik, W., Pritchard Jr., L., Ton, S., Delfino, J.J., Wojcik, M., Leszczynski, S., Patel, J.D., Doherty, S.J. and Stasik, J. (eds). CRC Press, Boca Raton, FL, Chapter 4.

Odum, H.T. 2001. An energy hierarchy law for biogeochemical cycles. In *Emergy Synthesis, Theory and Applications of the Emergy Methodology*. Brown, M.T. (ed.). Gainesville, FL: Center for Environmental Policy, pp. 235–248.

Odum, H.T., Odum, E.C. and Blissett, M. 1987. *Ecology and Economy: "Emergy" Analysis and Public Policy in Texas*. Policy Research Project Report #78. Lyndon B. Johnson School of Public Affairs, Austin, TX.

Odum, H.T., Wojcik, W., Pritchard, Jr., L., Ton, S., Delfino, J.J., Wojcik, M., Patel, J.D., Doherty, S.J. and Stasik, J. 1999. *Wetlands for Heavy Metals and Society*. Boca Raton, FL: CRC Press.

Odum, W.E., Odum, E.P. and Odum, H.T. 1995. Nature's pulsing paradigm. *Estuaries* 18: 547–555.

Page, N.J. and Creasey, S.C. 1975. Ore grade, metal production and energy. *Journal of Research, US Geological Survey* 3: 9–13.

Pritchard, L. 1992. *The Ecological Economics of Natural Wetland Retention of Lead*. MS thesis. University of Florida, Gainesville.

Sundberg, U., Lindegren, J., Odum, H.T. and Dohergy, S. 1994. Forest emergy basis for Swedish power in the 17th century. *Scandinavian Journal of Forest Research*, Supplement 1.

Wernick, I.K., Herman, R. Govind, S. and Ausbel, J.H. 1999. *Materialization and Dematerialization Measures and Trends*. Workshop Reader for the 10th Rinker International Conference. Gainesville: School of Building Construction, University of Florida.

# 3 On complexity theory, exergy, and industrial ecology

## Some implications for construction ecology

*James J. Kay*

## Introduction

Industrial ecology has been defined by Graedel and Allenby (1995) as "... the means by which humanity can deliberately and rationally approach and maintain a desirable carrying capacity, given continued economic, cultural and technological evolution. The concept requires that an industrial system be viewed not in isolation from its surrounding systems, but in concert with them." The Institute of Electrical and Electronic Engineers (IEEE) Electronics and the Environment Committee (1995) has defined industrial ecology as "... the objective, multidisciplinary study of industrial and economic systems, and their linkages with fundamental natural systems." Tibbs (1992) describes industrial ecology as follows: "... industrial ecology involves designing industrial infrastructures as if they were a series of interlocking manmade ecosystems interfacing with the natural global ecosystem". And O'Rourke *et al.* (1996) characterize industrial ecology as "... bringing systems thinking in ecology together with systems engineering (for design of products and processes) and economics"

While much of the literature on industrial ecology suggests that it originated in the early 1990s, Erkman (1997) and O'Rourke *et al.* (1996) trace its roots back to the early 1970s. One thread of development, which has been largely overlooked, can be traced to the Systems, Man, and Cybernetics Society of the Institute of Electrical and Electronic Engineers. In the early 1970s, Koenig, and later his son and their students at Michigan State, in collaboration with Chandrashekar and his students at the University of Waterloo, Canada, developed analytical tools, based on systems theory. These tools are used to analyze and design engineering systems so that they are as efficient and effective as possible, while minimizing their environmental impact. These tools are more sophisticated than any this author has seen in the industrial ecology literature. (Koenig and Tummala 1972; Koenig *et al.* 1972, 1975; Wong 1979; Wong and Chandrashekar 1982; Chinneck 1983; Chinneck and Chandrashekar 1984; Koenig and Tummala 1991; Tummala and Koenig 1993; Saama *et al.* 1994; Koenig and Cantlon 1998, 1999).

As part of Chandrashekar's team, this author undertook a master's thesis in systems design engineering entitled *An Investigation into the Design Principles for a Conserver Society* (Kay 1977). (Today it would be called *Engineering Design Principles for Industrial Ecology*.) The basic premise of the thesis was that all man-made systems must contribute to the survival potential of natural ecosystems. This thesis proposed that:

> a new branch of engineering to investigate and implement design strategies that are in line with this premise should be started. It would seem appropriate to call this branch ecosystems engineering, where ecosystems is used in the broad sense of H. T.

Odum to include industrial systems. This branch will bring together the disciplines of ecology, economics, engineering design, systems theory, and thermodynamics. This branch would be responsible for providing engineers in the field with the tables, rules of thumb and models (in other words the methodology) necessary for designing, implementing, and maintaining eco-compatible systems.

(Kay 1977)

In this regard, it was proposed in this thesis that hierarchical production–consumption models of all engineering systems be constructed and that their design as production-consumption systems should adhere to the following principles:

- The interface between man-made systems and natural ecosystems should reflect the limited ability of natural ecosystems to provide energy and absorb waste before their survival potential is significantly altered and the survival potential of natural ecosystems must be maintained. This is referred to as the problem of interfacing.
- The behavior and structure of large-scale (i.e. involving several different mass–energy transformation processes) man-made systems should be as similar as possible to those exhibited by natural ecosystems. This is referred to, after Papanek (1970), as the principle of bionics.
- Whenever feasible, the function of a component of a man-made system should be carried out by a subsystem of the natural biosphere. This is referred to as using appropriate biotechnology.
- Non-renewable resources should be used only as capital expenditures to bring renewable resources on line.

All of this suggests a root definition of industrial ecology. It is fundamentally about dealing with human transformations of mass and energy (i.e. industrial activities) from an ecosystem perspective. This begs the question: What is an ecosystem approach, and to what end? An ecosystem approach is about the application of systems thinking to the analysis and design of biophysical mass and energy transformation systems. The point of an ecosystem approach is to maintain a situation which is ecologically sound, i.e. has integrity, while providing humans with a sustainable livelihood (Kay 1991; UNDP 1998; Kay and Regier 2000). This chapter is about exploring an ecosystem approach for industrial ecology.

## *The challenges*

O'Rourke *et al.* (1996) observe that "many of the concepts of IE are not new. None the less, the IE literature fails to examine the lessons of the 1970s and 1980s attempts to reform industry." During the late 1970s and early 1980s, the efforts of others, besides this author, were applied to the advancement of design principles and methodologies for what we now call industrial ecology. [In addition to the work of Koenig and his colleagues, see, for example, O'Callaghan (1981) and Edgerton (1982).] So why do our designs not reflect these ideas?

There are three reasons. First, our society simply does not see the need for these design principles, at least not in a wholesale way. It is for this reason that I think that development of ecological economics is critical. Second, we do not grasp how to analyze mass–energy flow systems. Our ability to build production–consumption models is quite limited. Thus, our competence in dealing with the interfacing problem is wanting. Third, we really do not have a good understanding of how ecological systems work. Thus, it is quite difficult for us to understand our effects on ecological systems or what properties of ecological systems we should mimic. So, in the absence of clear direction as to how to implement the four design principles mentioned above and other similar ones, and given that lack of perceived economic incentives to do so, it is convenient simply to ignore them.

So, how to we rectify the situation? The first challenge is the recasting of economics such that it reflects the biophysical reality that humans are enmeshed in and dependent on a biosphere of natural ecosystems. This is the business of ecological economics and will not be dwelt on any further herein. Second is the need to develop thermodynamics so that we can adequately describe mass–energy flow systems. While we are quite competent in first law analysis (the analysis of the quantity of flow and efficiency), second law analysis (the analysis of the quality of flow and effectiveness) eludes us. Some very useful strides have been made in exergy (quality of energy) analysis (Moran 1982; Gaggioli 1983; Wall 1986; Szargut *et al.* 1988; Brodyansky *et al.* 1994) particularly network thermodynamics (Wong and Chandrashekar 1982; Chinneck 1983; Chinneck and Chandrashekar 1984; Peusner 1986), but the development of similar analysis techniques for material flow quality still remains as a challenge. The third and massive challenge, the one that is the crucial to our survival, is to understand ecosystems.

It is this last challenge that I took up for my doctoral work (Kay 1984). The key to this challenge is to understand ecosystems as complex, adaptive, self-organizing hierarchical open systems (SOHO systems for short) (Schneider and Kay 1994a; Kay *et al.* 1999). This is an area of theory and practice that is in its infancy.

Most readers will be quite unfamiliar with the notions and the language of complexity and self-organization theory. Yet in this chapter it is possible to provide only a brief overview summary of this theory. It must be left to the reader to pursue other works that elaborate on this topic. In particular, the following are written in a style suitable for a general audience and are recommended for further reading: Holling (1986), Casti (1994), Kay and Schneider (1994), Kay (1997) and Kay *et al.* (1999). Hopefully, the reader will be motivated to wade through the brief theoretical synopsis that follows. Valuable knowledge emerges from these theoretical considerations, as well as many important insights with direct practical implications for industrial ecology. These insights, with examples, are delved into in the discussion of the design principles and their application to construction ecology to be found later in this chapter. This chapter closes with an overview of an ecosystem approach design methodology for construction ecology.

## Ecosystems, sustainability, and complexity

The issue of complexity has attracted much attention in the past decade. This issue emerged in the wake of the new sciences that became prominent in the 1970s: catastrophe theory, chaos theory, non-equilibrium thermodynamics and self-organization theory, Jaynesian information theory, complexity theory, etc. A number of authors have focused specifically on self-organizing systems (Jantsch 1980; Peacocke 1983; Kay 1984; di Castri 1987; Wicken 1987; Nicolis and Prigogine 1977, 1989; Casti 1994). The term "complex systems thinking" is being used to refer to the body of knowledge that deals with complexity. Complex systems thinking has its origins in von Bertalanffy's (1968, 1975) general systems theory.

### *Complex adaptive self-organizing hierarchical open (SOHO) systems*

Spontaneous coherent behavior and organization occurs in open systems (such as natural ecosystems and human systems). Central to understanding such phenomena is the realization that open systems are processing an enduring flow of high-quality energy (exergy). In these circumstances, coherent behavior appears in systems for varying periods of time. However, such behavior can change suddenly whenever the system reaches a "catastrophe threshold" and "flips" into a new coherent behavioral state (Nicolis and Prigogine 1977). [A catastrophe threshold is a point of discontinuity at which continuous change in some variables generates sudden discontinuous responses. A simple example is the vortex that spontaneously appears in water draining from a bathtub or, more dramatically, the appearance of tornadoes "from nowhere". See Appendix 1 in Kay (1991).]

Schneider and Kay (1993, 1994a,b) have examined the energetics of open systems and taken Prigogine's work one step further An open system with exergy (high-quality energy) pumped into it is moved away from equilibrium, but Nature resists movement away from equilibrium. This is the second law of thermodynamics restated for non-equilibrium situations. When the input of exergy and material pushes the system beyond a critical distance from equilibrium, the open system responds with the spontaneous emergence of new, reconfigured organized behavior that uses the exergy to build, organize, and maintain its new structure. This reduces the ability of the exergy to move the system further away from equilibrium. As more exergy is pumped into a system, more organization emerges, in a stepwise way, to degrade the exergy. Furthermore, these systems tend to get better and better at "grabbing" resources and utilizing them to build more structure, thus enhancing their dissipating capability. There is, however, in principle, an upper limit to this organizational response. Beyond a critical distance from equilibrium, the organizational capacity of the system is overwhelmed and the system's behavior leaves the domain of self-organization and becomes chaotic. As noted by Ulanowicz (1997a) there is a "window of vitality", i.e. a minimum and maximum level between which self-organization can occur.

Self-organizing dissipative processes emerge whenever sufficient exergy is available to support them. Once a dissipative process emerges and becomes established, it manifests itself as a structure. These structures provide a new context, nested within which new processes can emerge, which in turn beget new structures, nested within which ... Thus emerges a SOHO system, a nested constellation of self-organizing dissipative process/structures organized about a particular set of sources of exergy, materials, and information, embedded in a physical environment, that give rise to coherent self-perpetuating behaviors.

A common example is the emergence of a vortex in bathtub water as it drains. The exergy is the potential energy of the water (due to the height of water in the bathtub), the raw material is the water, the dissipative process is water draining, and the dissipative structure is the vortex. A vortex does not form until the height of water in the bathtub reaches a certain level, and if the water height is too great, laminar flow occurs instead.

The theory of non-equilibrium thermodynamics suggests that the self-organization process in SOHO systems proceeds in a way that captures increasing resources (exergy and material); makes ever more effective use of the resources; builds more structure; and enhances survivability (Kay 1984; Kay and Schneider 1992; Schneider and Kay 1994a). These seem to be the kernel of the propensities of self-organization. This conception of self-organization, as a dissipative system, is presented in Figure 3.1.

How these propensities manifest themselves as morphogenetic causal loops and

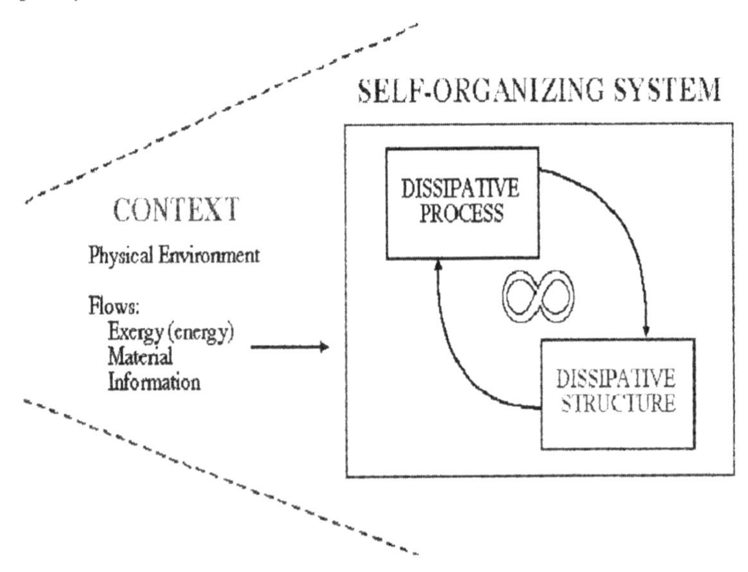


*Figure 3.1* A conceptual model for self-organizing systems as dissipative structures.

dissipative processes is a function of the given environment (context) in which the system is embedded, as well as the available materials, exergy, and "information", the last defined as factors embedded internally within the system that constrain and guide the self-organization. The interplay of these factors defines the context and associated constraints on the set of processes that may emerge. Generally speaking, which specific processes emerge from the potential set is uncertain.

This brief overview of a framework for discussing the dynamics of complex adaptive SOHO systems is intended to illustrate the very different kinds of dynamics associated with these systems and hence the very different kinds of considerations that need to be taken into account when examining them. Flips between attractors, organization about attractors, the spontaneous emergence of behaviors, their nested nature, and the importance of the second law of thermodynamics *vis-à-vis* exergy and non-equilibrium, requires a very different mindset for understanding dynamics (Caley and Sawada 1994). A central tenet of our work is that natural ecosystems and societal systems cannot be understood without understanding them as SOHO systems. Industrial ecology must take into account these considerations of complexity and self-organization.

### *Ecosystems as self-organizing systems*

Ecosystems can be viewed as the biotic, physical, and chemical components of Nature acting together as non-equilibrium self-organizing dissipative systems. As ecosystems develop or mature they should develop more complex structures and processes with greater diversity, more cycling, and more hierarchical levels to aid exergy degradation. Species that survive in ecosystems are those that funnel energy into their own production and reproduction and contribute to autocatalytic processes which increase the total exergy degradation of the ecosystem. In short, ecosystems develop in a way that systematically increases their ability to degrade the incoming solar exergy (Kay 1984; Kay and Schneider 1992; Schneider and Kay 1994a,b).

Keeping in mind that the more processes or reactions of material and energy that there are within a system (i.e. metabolism, cycling, building higher trophic levels), the greater the possibility for exergy degradation, Schneider and Kay showed that most, if not all, of Odum's phenomenological attributes of maturing ecosystems can be explained by ecosystems behaving in such a manner as to degrade as much as possible of the incoming exergy (Box 3.1) (Odum 1969; Schneider 1988; Kay and Schneider 1992). (See also the discussion of maximum empower in Chapter 2.)

*Box 3.1* Some expected changes in ecosystems as they develop.

The rationale behind these expectations is that ecosystems will develop so as to utilize more fully the exergy in the energy available to them. (Exergy utilization results in degradation of the exergy content of the energy.)

1 More exergy capture: Because the more exergy that flows into a system the more there is to utilize.
2 More energy flow activity within the system: Again, the more energy that flows within and through a system, the greater the potential for the use of its exergy.
3 More cycling of energy and material:
   - Numbers of cycles: More pathways for energy to be recycled in the system results in further utilization of the incoming exergy.
   - The length of cycles: More mature systems will have cycles of greater length, i.e. more nodes in the cycle. Each chemical reaction at or within a node results in further exergy utilization; the longer the cycle, the more the reactions and the more complete the exergy utilization.
   - The amount of material flowing in cycles (as versus straight through flow) increases. The ecosystem becomes less leaky, thus maintaining a supply of raw material for exergy utilization processes.
   - Turnover time of cycles or cycling rate decreases: More nodes or cycles in a system will result in nutrients or energy being stored at nodes in the system, resulting in longer residence time in the system.
4 Higher average trophic structure:
   - Longer trophic food chains: Exergy is utilized at each step of the trophic food chain, therefore longer chains will result in more thorough utilization.
   - Species will occupy higher average trophic levels: This will result in more exergy utilization as energy at higher trophic levels has a higher exergy content.
   - Greater trophic efficiencies: The exergy content of energy that is passed higher up the food chain will be more thoroughly utilized than that of energy that is shunted immediately into the detrital food chain.
5 Higher respiration and transpiration: Transpiration and respiration results in exergy utilization.
6 Larger ecosystem biomass: More biomass means more pathways for exergy utilization.
7 More types of organisms (higher diversity): More types of organisms will provide diverse and different pathways for utilizing exergy.

In terrestrial ecosystems, surface temperature measurements can be used to demonstrate that ecosystems develop so as to degrade exergy more effectively. Exergy degradation in a terrestrial ecosystem is a function of the difference in black body temperature between the captured solar energy and the energy reradiated by the ecosystem. [This is discussed in detail in Fraser and Kay (2002).] Thus, if a group of ecosystems are bathed by the same incoming energy, the most mature ecosystem should reradiate its energy at the lowest exergy level, i.e. the ecosystem would have the lowest black body temperature. The black body temperature is determined by the surface temperature of the canopy of the ecosystem.

Consider the fate of solar energy impinging on five different surfaces: a mirror, a flat black surface, a piece of false grass carpet (e.g. Astroturf), a natural grass lawn, and a rainforest. The perfect mirror will reflect all the incoming energy back toward space with the same exergy content as the incoming radiation. The black surface will reradiate the energy outward at a lower quality than the incoming energy, because much of the exergy is converted to lower-quality infrared radiation and sensible heat. The green carpet will reradiate its energy in a similar to the black surface but will differ because of its surface quality and different emissivity. The natural grass surface will degrade the incoming radiation more completely than the green carpet surface, because processes associated with life (i.e. growth, metabolism and transpiration) degrade exergy (Ulanowicz and Hannon 1987). Its surface temperature will be lower than the black surface or the Astroturf. The rainforest should degrade the incoming exergy most effectively because of the many pathways (i.e. more species, canopy construction) available for degradation. It will be even colder than the grass.

In previous papers, Kay and Schneider have discussed Luvall *et al.*'s experiments in which they overflew terrestrial ecosystems and measured surface temperatures (Luvall and Holbo 1989, 1991; Luvall *et al.* 1990; Kay and Schneider 1994; Schneider and Kay 1994a, Quattrochi and Luvall 1999). Luvall and his co-workers have documented ecosystem energy budgets, including tropical forests, mid-latitude varied ecosystems, and semiarid ecosystems. Their data show one unmistakable trend: when other variables are constant, the more developed the ecosystem, the lower its surface temperature and the more degraded its reradiated energy.

Work by Akbari (1995) on agricultural plots showed a similar trend. A lawn (single species of grass) had the warmest surface temperature, an undisturbed hay field was cooler, and a field that had been naturally regenerating for 20 years was coldest. These trends were confirmed over three years of observation. In addition, another field that had been regenerating for 20 years was disturbed by mowing. Its surface temperature immediately rose significantly, but very quickly it returned to its cooler predisturbance value. Very recently, Allen and Norman (personal communication) performed a set of experiments to explore the relationship between development and surface temperature in plant communities. Their experimental results so far demonstrate that the surface temperature of plant communities tends to increase when the plants are removed from their normal conditions. In other words, plant communities are coldest (degrading the most exergy) when they are in the normal conditions to which they are adapted. All this is evidence that exergy degradation is the name of the game in ecosystem development.

Recently Luvall and his colleagues at NASA have used these observations about natural ecosystems to develop a "green cities" strategy (Lo *et al.* 1997). This has been applied in several urban centers in the USA. The green cities initiative begins by using the same surface temperature measurements described above to generate a thermodynamic description of a city. A core tactic in this approach is to use the thermodynamic description to focus on roofing materials and the presence of flora, particularly trees. Both of these factors can dramatically alter the thermodynamic budget in a city. Using the analysis of surface temperature, areas of the city that could benefit from changes in roofing materials, more flora, etc. are identified. This initiative is discussed further later in the chapter.

There is much to be gained from examining ecosystems through the lens of exergy degradation. A number of ecosystem phenomena can be explained, and hypotheses concerning ecosystem development can be generated and tested. But there is more to the story. Most ecosystems will have many different options for exergy degradation available to them. Some will have different sources of exergy available. Different combinations of exergy sources and degradation possibilities may be equivalent from an exergy degradation perspective. So the number of possible variations on ecosystem organization, which are thermodynamically equivalent, may be significant. This quickly leads to a complicated set of possible organizational pathways. What is actually manifested may very well be a reflection of a collection of accidents of history.

The imperative of thermodynamics and exergy degradation is not the only one acting on living systems. Of equal importance is survival, an imperative that may not be consistent with maximum exergy degradation. Inevitably trade-offs will have to be made, and ecosystems, as they exist on the ground, will reflect these trade-offs (Kay 1984). There will not be single best solutions to the imperatives of exergy degradation and survival, just solutions that work longer than others. Furthermore, to add to the complexity and uncertainty, Dempster and Kay (personal communication) have shown that such systems must, by necessity, be recursively nested autopoietic (organizationally closed and self-duplicating, i.e. a cell) and synpoietic (organizationally open and evolving, i.e. a species) systems (Dempster 1998).

Box 3.2 summarizes the characteristics of ecosystems as self-organizing hierarchical open systems. These self-organizing characteristics require the consideration of very different issues, and the use of analytical tools that are very different from those which traditional ecological approaches would suggest are pertinent. In particular, the issues of complexity and uncertainty must be confronted head on.

### *Sustainability and complexity theory: some lessons*

Our partial understanding of ecosystems as complex systems suggests several lessons that need to be kept in mind when discussing sustainability (Kay and Schneider 1994; Schneider and Kay 1994a; Kay and Regier 1999).

There is growing comprehension that sustainability issues cannot be discussed in isolation. They must always be examined within their broader context. Every system is a component of another system and is, itself, made up of systems. Thus, a wetland must be understood in the context of the sub-watershed of which it is a part, and in terms of the processes and species that make it up. The body of thinking that deals with these issues is called hierarchy theory (Allen and Starr 1982; Allen and Hoekstra 1992; Allen *et al.* 1993). Its central tenet is that sustainability issues can only be understood in terms of systems embedded in systems which are also embedded in systems or, in the vernacular of hierarchy theorists, as *nested holons* (Koestler and Smythies 1969; Koestler 1978).

The hierarchical nature of complex systems requires that they be studied from different types of perspectives and at different scales. There is no one correct perspective. Rather, a diversity of perspectives is required for understanding.

*Box 3.2* Properties of self-organizing hierarchical open systems to consider when thinking about ecosystems (remember that SOHO systems are complex, adaptive, dissipative systems)

1 Open to material and energy flows.
2 Non-equilibrium: Exist in quasi-steady states some distance from equilibrium.
3 Thermodynamics: Maintained by energy *gradients* (exergy) across their boundaries. The gradients are *irreversibly* degraded (the exergy is used) in order to build and maintain organization. These systems maintain their organized state by exporting entropy to other hierarchical levels.
4 Propensities: As *dissipative* systems are moved away from equilibrium they become organized:
   - They use more exergy.
   - They build more structure
   - This happens in spurts as new attractors become accessible.
   - It becomes harder to move them further away from equilibrium
5 Feedback loops: Exhibit material or energy *cycling* – cycling, and especially autocatalytic cycling, is intrinsic to the nature of dissipative systems. The very process of cycling leads to organization. *Autocatalysis* (positive feedback) is a powerful organizational and selective process.
6 Hierarchical: Are *holarchically nested.* The system is nested within a system and is made up of systems. Such nestings cannot be understood by focusing on one hierarchical level (holon) alone. Understanding comes from the multiple perspectives of different *types* and *scale*.
7 Multiple steady states: There is not necessarily a unique preferred system state in a given situation. *Multiple attractors* can be possible in a given situation, and the current system state may be as much a function of historical accidents as anything else.
8 Exhibit chaotic and catastrophic behavior: Will undergo dramatic and sudden changes in discontinuous and unpredictable ways.
   - Catastrophic behaviour. The norm
     – *bifurcations*: moments of unpredictable behaviour
     – *flips*: sudden discontinuities, rapid change
     – *Holling four-box cycle*: Shifting steady-state mosaic
   - Chaotic behaviour: our ability to forecast and predict is always limited, for example between five and ten days for weather forecasts, regardless of how sophisticated our computers are and how much information we have.
9 Dynamically stable?: Equilibrium points for the system may not exist.
10 Non-linear: Behave as a whole, *a system*. Cannot be understood by simply decomposing into pieces which are added or multiplied together.
11 Internal causality: Non-Newtonian, not a mechanism, but rather is *self-organizing*. Characterized by goals, positive and negative feedback, autocatalysis, emergent properties, and surprise.
12 Window of vitality: Must have enough complexity but not too much. There is a range within which self-organization can occur. Complex systems strive for *optimum*, not minimum or maximum.

By their nature, complex systems are self-organizing. This means that their dynamics are largely a function of positive and negative feedback loops. Linear, causal mechanical explanations of their dynamics are precluded. In addition, emergence and surprise are normal phenomena in systems dominated by feedback loops. Inherent uncertainty and limited predictability are inescapable consequences of these system phenomena.

Complex systems organize about attractors (Figure 3.2). Complex systems have multiple possible operating states or attractors, and may shift or diverge suddenly from any one of them (Holling 1986; Kay 1991, 1997; Ludwig *et al.* 1997). Even when the environmental situation changes, the system's feedback loops tend to maintain its current state. However, when system change does occur, it tends to be very rapid and even catastrophic. When precisely the change will occur, and what state the system will change to, are generally not predictable. In a given situation, there are often several possible system states (attractors) that are equivalent. Which state is currently occupied is a function of its

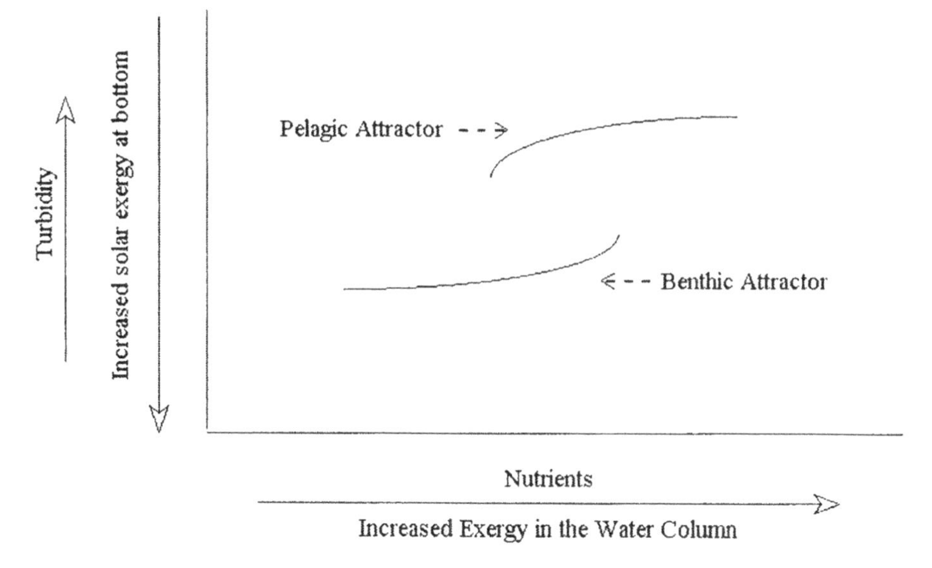


*Figure 3.2* Benthic and pelagic attractors in shallow lakes. Two different attractors for shallow lakes have been identified. In the *benthic* state, a high-water-clarity bottom vegetation ecosystem exists. As nutrient loading increases the turbidity in the water, the ecosystem hits a catastrophe threshold and flips into a hypertrophic, turbid, phytoplankton *pelagic* ecosystem. The relationship of these two attractors, from a thermodynamic perspective, is as follows. Let us assume that the benthic attractor is dominant and that the rate at which phosphorus is being added to the water is increasing. The benthic system has means of deactivating phosphorus. However, the amount of active phosphorus will increase, albeit slowly, effectively increasing the exergy in the water column. As this exergy increases, a critical threshold is passed which allows the pelagic system to self-organize to coherence. Once this occurs the exergy at bottom decreases rapidly as result of shading (turbidity), thus catastrophically de-energizing the benthic system. This results in the eventual reactivation of the phosphorus in the bottom muds which the benthic system had previously deactivated, thus strengthening the pelagic attractor even more. Assuming that the pelagic attractor is dominant, and if the level of active phosphorus in the water column decreases, a critical threshold is again reached below which it is no longer possible to capture enough solar energy to energize the pelagic system. In effect, the exergy in the water column decreases below the minimum level for the window of vitality of the pelagic system. As this occurs, the exergy at the bottom increases, thus re-energizing the benthic system. And so the aquatic system flips back and forth between the pelagic and the benthic regime depending on where in the water column the sunlight's exergy is available to energize the system.

history. There is not a "correct" state for the system, although there may be a state that is preferred by humans.

Thus, categorical statements about the "correct" way to proceed, that is the correct ecosystem for a given circumstance, cannot be deduced from scientific arguments. Furthermore, which response comes to pass may be a function of history or just the moment. Thus, there is an element of irreducible uncertainty about self-organizing behavior, uncertainty about what may come to pass as well as uncertainty about what ought to come to pass. These properties of inherent uncertainty and emergence limit the capacity to predict how an ecological situation will unfold.

Our premise is that resolving sustainability issues for ecological–economic systems entails understanding these systems as "self-organizing hierarchical open" (SOHO) systems. Such an understanding comes from thinking through the hierarchical nature of these systems by considering issues of type and scale, bounding, and nesting of the system. This "hierarchical systems description" of the important processes and structures, and their relationships and context, is essential. In addition, the self-organizing behaviors of the system need to be identified, described, and understood as far as possible. This involves identifying the attractors accessible to the system, the feedbacks which maintain the system at the attractors, the external influences which define the context for a specific attractor, and the conditions under which flips between attractors are likely. The overall understanding of a system's behavior that comes from studying it as a SOHO system is summarized in the form of a narrative of its dynamics. [A more detailed discussion of these notions can be found in Kay (1997) and Kay *et al.* (1999).]

The understanding of ecological–economic systems as SOHO systems requires a major change in some of the ways in which science and decision making are conducted. Traditional reductionistic disciplinary science and expert predictions, the basis for much of the advice given to decision makers, have limited applicability. Narratives about possible futures for given SOHO systems are better able to capture the richness of possibilities. Other epistemological "mindsets" or causal metatypes must be brought to bear, notably explanations based on morphogenetic causal loops that involve both positive and negative feedback processes and autocatalysis (Maruyama 1980; Caley and Sawada 1994; Ulanowicz 1997b). Expectations that decision makers can carefully control or manage changes in societal or ecological systems must be relinquished. Rather, adaptive learning and management, guided by a much wider range of human experience and understanding than disciplinary science, must form the basis for decision making in a sustainable society.

## Industrial ecology: the design of ecological–economic systems

The design principles mentioned in the introduction, in combination with the insights of complex systems theory, especially with reference to ecosystems, provide a theoretical basis for an ecosystem approach to industrial ecology. *Industrial ecology is taken to be the activity of designing and managing human production–consumption systems, so that they interact with natural systems, to form an integrated (eco)system which has ecological integrity and provides humans with a sustainable livelihood.* In essence, industrial ecology is about designing human ecological–economic systems that fit in with natural ecological systems.

### *The normative foundation*

This definition of industrial ecology establishes the *raison d'être* for industrial ecology: ecological integrity and sustainable livelihoods. Together, these two notions are the normative basis for the practice of industrial ecology. The first step in any industrial ecology enterprise must be to establish what constitutes a sustainable livelihood and ecological integrity in the given circumstances. Only when this has been done is there a basis for evaluating the outcome of an industrial ecology enterprise.

### *Sustainable livelihoods*

According to the United Nations sustainable livelihoods (SL) program:

> A livelihood system is an aggregate yet dynamic environment of human activity that integrates both the opportunities and assets available to men and women as means for achieving their goals and aspirations as well as interactions with, and exposure to, a range of beneficial or harmful ecological, social, economic and political perturbations that change their capacity to make a living ...
>
> Sustainable livelihoods are derived from people's capacity to make a living by surviving shocks and stress and improve their material condition without jeopardizing the livelihood options of other peoples, either now or in the future. This requires reliance on both capabilities and assets (i.e., stores, resources, claims and accesses) for a means of living.
>
> The sustainability of livelihoods becomes a function of how men and women utilize asset portfolios on both a short and long-term basis. Sustainability should be defined in a broad manner and implies:
>
> - The ability to cope with and recover from shocks and stresses;
> - Economic efficiency, or the use of minimal inputs to generate a given amount of outputs;
> - Ecological integrity, ensuring that livelihood activities do not irreversibly degrade natural resources within a given ecosystem; and
> - Social equity which suggests that promotion of livelihood opportunities for one group should not foreclose options for other groups, either now or in the future.
>
> In other words, SL is the capability of people to make a living and improve their quality of life without jeopardizing the livelihood options of others, either now or in the future.
>
> UNDP (1998)

Sustainable livelihoods is the socioeconomic impetus behind industrial ecology. The UNDP program on sustainable livelihoods provides a set of tools for evaluating, designing, and implementing sustainable livelihoods. It is left to the reader to investigate this further.

### *Ecological integrity*

Ecological integrity is the biophysical purpose of industrial ecology.

Ecological integrity is about three facets of the self-organization of ecological systems: (a) current well-being, (b) resiliency, and (c) capacity to develop, regenerate, and evolve (Kay and Regier 2000).

The first of these is about the ecological health of the system, about its vigor, its well-being and how well it is flourishing in the current circumstances. It concerns the current state of the ecosystem.

The second aspect of integrity is about the stress response capability of the ecosystem, something that is often referred to as its resiliency. It is about what happens when the system's state is disturbed by outside influences, i.e. its ability to reorganize in the face of change. Stress response is about how the system deals with change that disturbs it from its current attractor and which possibly flips it into the domain of another attractor. This has been discussed in detail in Kay (1991).

The third aspect of integrity concerns the ecological system's potential to continue to self-organize. This pertains to the system's ability to develop, regenerate, and evolve in its normal environmental circumstances. This is about its capacity to:

1 continue to develop, i.e. increase its organization relative to an attractor;
2 regenerate, to deal with birth–growth–death–renewal cycle [i.e. the Holling four-box model; see Chapter 5 and Holling (1986, 1992)], i.e. to deal with the multiple nested dual attractor problem; and to
3 continue to evolve, that is switch attractors spontaneously (emergent complexity).

Put in the parlance of complex systems, ecological integrity is about maintaining the integrity of the process of self-organization. This has three facets:

1 the current organizational state of the system;
2 the ability of the system to reorganize in the face of environmental change;
3 the system's capacity to continue to *self*-organize in its normal environment.

These three facets must be considered when evaluating the integrity of the ecological systems that emanate from the practice of industrial ecology.

We are only beginning to understand how to investigate and evaluate ecological integrity (Woodley *et al.* 1993). Much work on understanding ecosystems as complex self-organizing systems still remains to be done. In particular, the notion of attractors, and flips between attractors, has only been considered in the literature in the last fifteen years. Much remains to be learned about these complex behaviors. In the meantime it seems prudent, given our ignorance, to adopt the precautionary principle.

Together the notions of ecological integrity and sustainable livelihoods form the normative basis for industrial ecology.

### *A conceptual model for industrial ecology*

We have developed an integrated SOHO system model that portrays ecological–societal systems as dissipative complex systems (Regier and Kay 1996; Kay and Regier 1999; Kay *et al.* 1999; Kay 2000; Boyle *et al.* 2002). This SOHO system model provides a conceptual basis for discussing ecological integrity and human sustainability (refer to Figures 3.3–3.5). It furnishes us with an integrated, nested ecosystem description of the relationship between natural and human systems. As such, it can serve as a basis in industrial ecology for scrutinizing these relationships.

In this model, the elements of the landscape (e.g. woodlots, wetlands, farms, neighborhoods) that make up the societal and ecological systems are seen as self-organizing entities set in an environmental context. Self-organizing entities are understood through consideration of their constituent processes and structures, and the relationships between these. (For example, in a woodlot processes would be evapotranspiration and growth of biomass; structure would be the species that make up the woodlot; and a description of

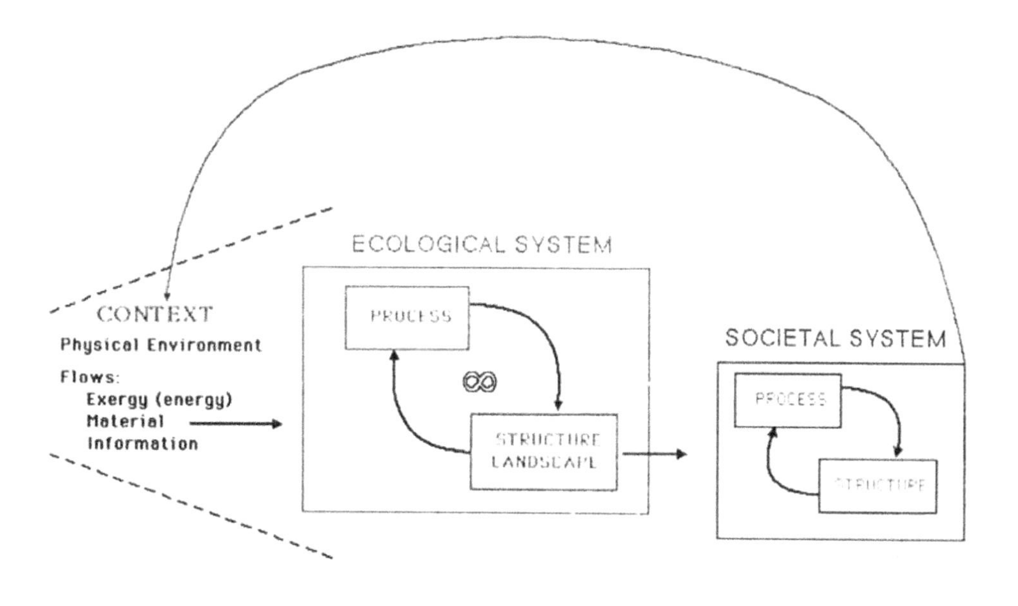


*Figure 3.3* A conceptual model of the ecological–societal system interface – single horizontal holarchical level.

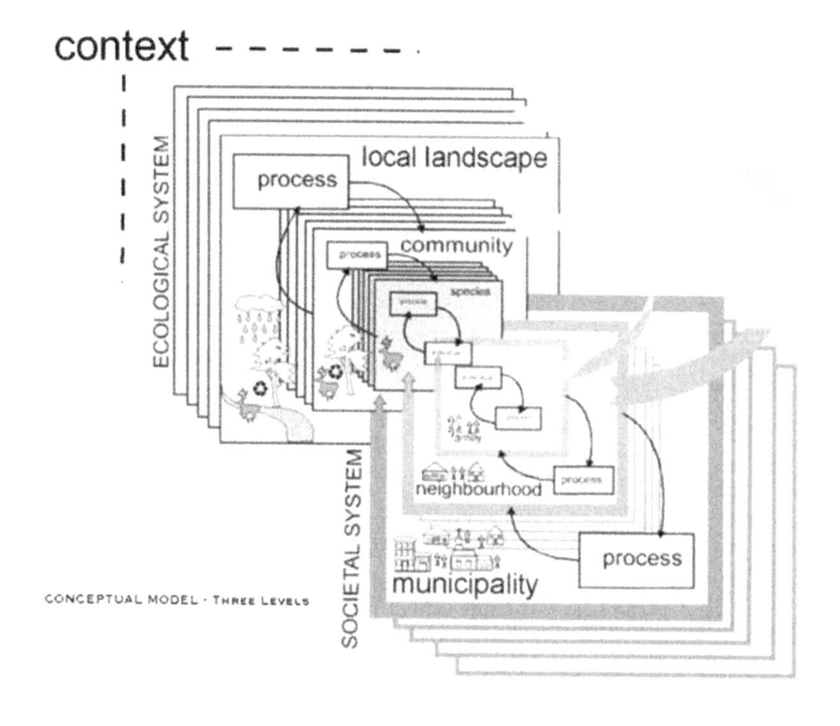


*Figure 3.4* Example of a nested model of the ecological–societal system

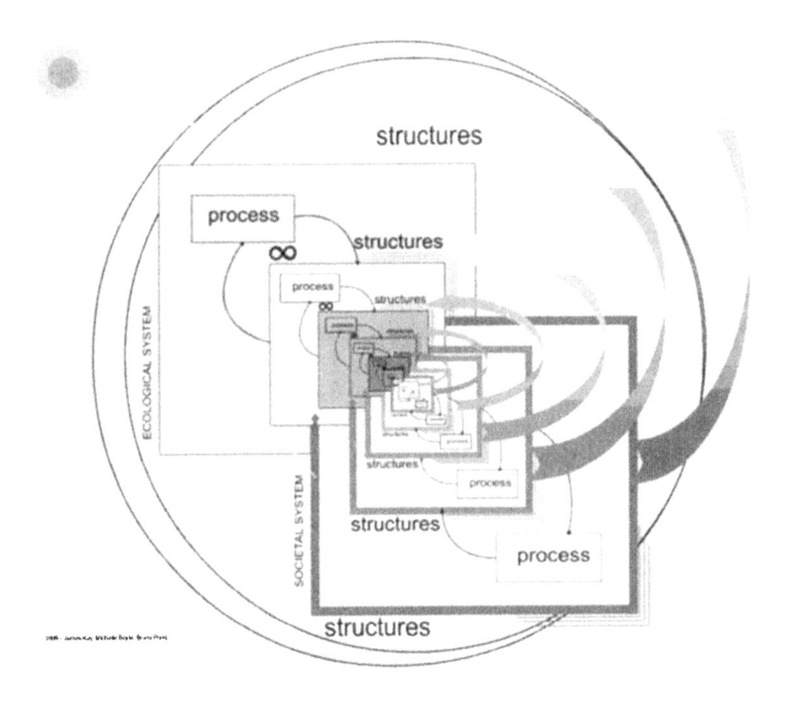


*Figure 3.5* The nested hierarchical conceptual model of the ecological–societal system

the relationship between these processes and structure would be Holling's four-box model. See Chapter 5 for an extensive description of the adaptive cycle or four-box model.) The processes involve the flows of material, energy, and information. The structures are the objects (i.e. trees) we see on the landscape. The processes allow for the emergence and support of structures, which in turn allow for the emergence of new processes, and so on. The recognition of this recursive relationship between process and structure separates this hierarchical conceptual model from more traditional ones.

Our conception of self-organization, as a dissipative system, was presented in Figure 3.1. It is a description of how a mass–energy transformation system emerges. This formulation is the kernel for the SOHO system model. Self-organizing dissipative processes emerge whenever sufficient exergy is available to support them. The details of the processes depend on the raw materials available to operate them, the information present to catalyze the processes, and the physical environment. The interplay of these factors defines the context for (i.e. constrains) the set of processes that may emerge. (Generally speaking, which specific processes emerge from the available set is uncertain.) Once a dissipative process emerges and becomes established, it manifests itself as a structure.

This basic description characterizes mass–energy transformation systems in terms of the exergy, materials, and information they consume and how these are used in dissipative processes. It focuses on the consumption side of the production–consumption duality of systems. However, when more than one such element is connected to form a larger system, the production aspects of the system become clear. Each element not only consumes exergy, materials, and information but also produces exergy, materials, and information for the next element in the concatenation. Each element provides the context for another element. So, horizontally, each element in the SOHO system model has, like Janus, two faces: its consumption face and its production face (Figure 3.6).

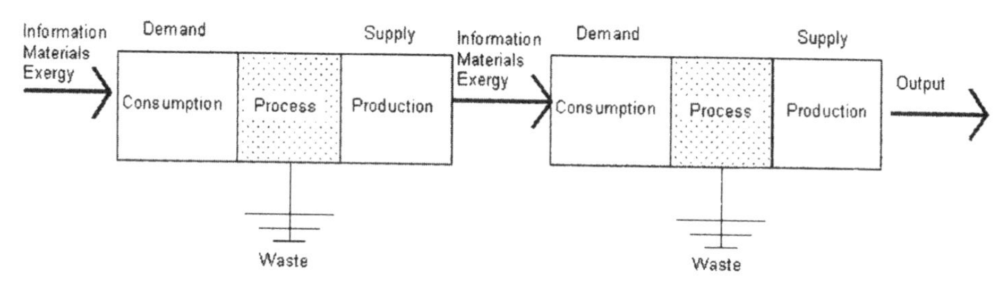


*Figure 3.6* Each component of a SOHO system consumes exergy, materials, and information.

## *The ecological–societal system interface*

Ecological communities provide the exergy, materials, and information required for human societies to sustain themselves. This is depicted in Figure 3.3. The societal system depends on the flow of exergy, materials, and information from the ecological system to support its processes and structures. These flows, along with the biophysical environment provided by the ecological systems, are the context for societal systems. The context constrains the possible societal processes and structures in a specific location. While Figure 3.3 illustrates a single ecological system providing the context for the societal system, the reality is that it is a suite of adjacent ecological systems (for example woodlots, fields, and wetlands adjacent to a farm) that provide this context. Alterations in these adjacent systems will alter the context and thus the possibilities for the system in question.

However, the societal system can also influence the ecological system in two ways. The first influence is through changes in the structure of the ecological system (for example cutting trees down in a woodlot, filling in wetlands, and all the human activities that involve removing or dismantling ecological structures on the landscape). Such actions, of course, alter the flows from the ecological systems to the societal systems and thus create a feedback structure on the landscape. This is represented in Figure 3.3 by the lower arrow back from the societal system to the structure in the ecological system. The feedback to the societal system occurs because changes in the ecological structure change the context for the societal system.

The second influence occurs when the context of the ecological system is altered by the societal system. For example, the run-off into a wetland or stream may be altered by human activities on adjacent properties. It is depicted in Figure 3.3 by the upper arrow from the societal system back up to the context of the ecological system. This influence is qualitatively different than the structural influence just discussed.

The resulting feedback loop has more steps and accordingly is more indirect. By changing the context of the ecological system, the societal system affects the ecological processes and, in turn, the ecological structure, and ultimately affects the societal system's own context. For example, modifying the run-off into a waterway can dramatically alter the character of the waterway, and hence the type of fish found in it, and therefore the sport fishery and associated economic system.

To summarize this discussion, each self-organizing entity resides in an environment that provides: (a) the biophysical surroundings in which the entity exists; and (b) flows of exergy, materials, and information that the entity depends upon for the continuation of the self-organizing processes that maintain its structure. The biophysical surroundings, in conjunction with the flows into the system, constitute the context for the self-organizing entity.

Referring to Figure 3.3, the relationship between societal systems and ecological systems is threefold:

1 *Ecological systems provide the context for societal systems.* In other words, they provide the biophysical surroundings and flows of exergy, materials, and information that are required by the self-organizing processes of the societal systems.
2 *Societal systems can alter the structures in ecological systems.* (For example, cutting down a woodlot, removing beaver from a watershed.) Changes in the ecological structure can then, of course, alter the context for the societal systems themselves.
3 *Societal systems can alter the context for the self-organizing processes of ecological systems.* (For example, a change in the drainage patterns into a wetland, a change in the local microclimate, such as a heat island effect, for a woodlot.) Changes in ecological process can alter ecological structure and consequently the context for societal systems.

### *The nested structure of the model*

Figure 3.3 applies to one hierarchical level, but, as observed earlier, sustainability and integrity issues can only be understood in terms of "nested holons". Figure 3.4 illustrates this idea of nesting. On the ecological side, "local landscape" can be thought of as a subwatershed, for example. The hydrologic cycle is an example of a process in the subwatershed. The structures that make up the subwatershed are the ecological communities (woodlots, wetlands, open fields, etc.). The communities are in turn made up of species. On the societal side, municipalities rest on the local landscape. These, in turn, are made up of neighborhoods, which are made up of families and businesses.

In many cases, the local subwatershed defines the context for the local municipality. However, the municipality can, and does, directly modify the ecological communities in the subwatershed and thus its own context. Similarly, the context for local neighborhoods is determined by the adjacent ecological communities. Nonetheless, the local neighborhood is quite capable of influencing ecological communities, through direct structural change (such as harvesting wood from a woodlot) or by changing the context of an ecological community (for example changing drainage patterns into a wetland).

Figure 3.4 and the examples are meant to be illustrative and not exhaustive, although they do demonstrate that such changes can cascade through the nested holons ultimately to affect individual families and businesses. Figure 3.5 shows the full conceptual model that would be used as a template to develop a situation-specific conceptual model in which the important levels, processes, structures, contexts, and influence/feedback considerations are specifically identified.

### *The challenge of constructing a SOHO system description*

Sustainability is about maintaining the integrity of the combined ecological–societal system. This means maintaining their self-organizing processes and structures. *This will happen naturally if we maintain the context for self-organization in ecological systems, which in turn will maintain the context for the continued well-being of the societal systems. It is this relationship that must be thought through if the promise of industrial ecology is to come to fruition.*

This will require building production–consumption models of each element in the combined ecological–societal system. These models will need to tell us about the relationship between the organizational state (attractor) of the element and its context.

This, in turn, requires a description of the context. The contextual description has four aspects the flow of energy, material, and information and the physical environment. Given this description, one then needs to link specific contextual states to organizational attractors. This turns out to be the fundamental challenge to the successful development of a program for industrial ecology. We are only beginning to explore the relationship between context and self-organization.

The most progress has been made in discussing energy flow. If one considers only the energy aspect of the flows in a SOHO system description, then network thermodynamics, using graph theoretic techniques, allows a complete system description. At the core of this description are measures of the quantity and quality of the flow. (Most readers will know that energy is measured in calories or joules. However the problem is that all joules are not equivalent. I can do less with a joule of crude oil than I can with a joule of household heating oil, and I can do less with a joule of both of these than I can with a joule of electricity. The quality of each of these joules is different and is measured by exergy.) Each component is described by the change in quality and quantity of flow between its inputs and outputs. This effectively involves measuring the gradient drop across the component and the associated flow through the component. Overall, the components together, that is the system, must conform to two rules, known variously as Kirchkoff's laws; the cutset and circuit equation; or the first and second laws of thermodynamics. In electrical systems, the measure of quantity is current and the measure of quality is voltage. In hydrodynamic systems, the measure of quantity is volume flow and the measure of quality is pressure. In general, it has been shown that, for any energy flow system, the quantity measure is flow of energy and the quality measure is exergy density (exergy per unit energy) (Ford *et al.* 1975; Wong 1979; Ahern 1980; Gaggioli 1980, 1983; Hevert and Hevert 1980; Edgerton 1982; Moran 1982; Wall 1986; Szargut *et al.* 1988; Brodyansky *et al.* 1994).

The details of how to measure these and how to do the analysis are not important here and are left for the reader. What is important is that this body of work has demonstrated unequivocally that any description of a real physical flow system must take into account *both* a quality and quantity measure if meaningful results are to arise from the analysis. Only with both types of measures can the full effect of the first and second laws of thermodynamics be taken into account. Most energy analysis has traditionally looked only at energy flow, and it has been demonstrated in a number of works that this has led to poor decisions, at both the micro (plant or building) and macro (describing the economy) level. In spite of the power of this form of analysis, it has begun to work its way into the engineering curriculum and textbooks only in the past decade.

Unfortunately, such a body of knowledge does not exist for the other aspects of the contextual descriptions necessary for a self-organizing hierarchical open systems description of ecological–societal systems. We can guess that similar measures of the quality and quantity of material flow are needed. While quantity measures of material flow are self-evident, quality measures elude us. Yet these are critical if we are to evaluate the implications of such strategies as material recycling. Even more unclear is the calculus of information. The central role of information in directing the emergence of self-organizing processes and structures has only recently been put forward and how to describe this role remains quite unclear.

This gap in our knowledge presents a major challenge to the development of industrial ecology. There are profound and fundamental theoretical issues related to complexity and self-organization that we must resolve before we will have a robust theoretical basis

*Box 3.3* Some design principles for industrial ecology

The design of production consumption systems should be such that:

1. The interface between societal systems and natural ecosystems reflects the limited ability of natural ecosystems to provide energy and absorb waste before their survival potential is significantly altered and the fact that the survival potential natural ecosystems must be maintained. This is referred to as the problem of *interfacing*.
2. The behavior and structure of large-scale societal systems should be as similar as possible to those exhibited by natural ecosystems. This is referred to, after Papanek, as the *principle of bionics*. (In the industrial ecology literature it is often referred to as *mimicry*.)
3. Whenever feasible, the function of a component of a societal system should be carried out by a subsystem of the natural biosphere. This is referred to as *using appropriate biotechnology*.
4. Non-renewable resources are used only as capital expenditures to bring renewable resources on line.

for discussing sustainability, ecological integrity, and industrial ecology. These issues revolve around the question of system description and quality. Even with this profound gap in our knowledge, there is still much that complex systems thinking can say about design and industrial ecology. This is explored in the next section.

### *Design principles*

Design principles and tools for industrial ecology have been proposed elsewhere (Ehrenfeld 1997; van Berkel and Lafeur 1997; van Berkel *et al.* 1997; Allenby 1999). There is much overlap between these principles and the four introduced at the beginning of this chapter (Box 3.3). These four are somewhat different in that they are explicitly derived from a systematic application of systems theory. They deal explicitly with the implications of the second law of thermodynamics, hierarchy, and attractors and, finally, although this is a curiosity more than anything, they were first published in 1977.

#### *Interfacing human natural (eco)systems*

The first of the design principles is: *the interface between man-made systems and natural ecosystems must reflect the limited ability of natural ecosystems to provide energy and absorb waste before their survival potential is significantly altered, and that the survival potential of natural ecosystems must be maintained.* This is referred to as the problem of *interfacing*.

In an ideal situation, efforts to address the interfacing problem would be based on an analysis of the situation using the SOHO system description discussed above. This description would deal with each component as a self-organizing production–consumption system. All the relevant flows of exergy, materials, and information and the effect of changes in these on self-organization would be accounted for. The nested nature of the system requires that design implications be considered at different spatial and temporal scales. In particular, the effects of the two different forms of feedback (from the societal to the natural system) would need to be thought through in detail.

The criteria for evaluating the implications of the design are the normative principles of sustainable livelihoods and ecological integrity. The nature of self-organizing systems requires that these criteria be applied in quite a different way from what we are used to. Our normal way of applying such criteria is based on an assumption that an incremental change in the context (i.e. the influence of the design on the natural system) will result in an incremental change in the natural system. But self-organizing systems do not work this way. There can be substantial changes in context, from which the system can buffer itself, and hence there will be no change in the system's state. However, once this buffering capacity is used up, a very small change in context can cause dramatic change in the system's state. For example, as documented elsewhere (see Figure 3.2), the incremental addition of phosphorus to a shallow lake will have little effect until a threshold is reached. After the threshold, a small change in phosphorus loading will trigger a massive and dramatic reorganization in the lake, a flip between attractors (Scheffer *et al.* 1993; Scheffer 1998; Kay and Regier 1999). Once a flip is precipitated, it requires a massive change in context to return the system to its original state. In the case of Lake Erie, billions of dollars and decades of phosphorus remediation programs were necessary to "clean up the lake," i.e. trigger a flip back to its earlier "clean" state.

The point is that determining how a particular design is going to affect ecological integrity and sustainable livelihoods cannot be achieved using a linear incremental approach to the relationship between cause (the design) and effect (the ecosystem's reorganization). Rather, a non-linear hierarchical mindset, which takes into account thresholds, cumulative effects, buffering, flips between attractors, and cross-scale dynamics, must be used. Put another way, an industrial plant could increase its discharge by 50% (an arbitrary figure) with little noticeable change in the surrounding natural ecosystems, but an increase of 55% could trigger a dramatic and irreversible change in the surrounding natural ecosystems. A linear incremental change approach to evaluating the implications of a design would not suggest such a possibility.

If there is one lesson to be learned from the past thirty years of dealing with environmental issues it is this: the complexity of the relationship between societal and natural ecosystems requires a significantly more sophisticated approach than that of normal scientific methods of analysis and evaluation.

Unfortunately, our current state of knowledge is not up to the task. The type of information needed to build up a SOHO system description is generally not collected. [For details on a monitoring program to accomplish this see Boyle *et al.* (1996, 2002).] In addition, the study of ecological attractors and flips between them is in its infancy. Our understanding of these phenomena is at best qualitative and ambiguous. Thus, we do not have an understanding of the relationship between context and self-organization of a system. This ignorance is a major stumbling block to discussing ecological integrity. We certainly do not know what the thresholds for flips are.

In terms of dealing with the interfacing of societal and natural ecosystems we are left in a quandary. While we understand how to frame the discussion of the implications of a particular design, we are not, at this time, in a position to make specific statements about the implications of a specific design that are sufficiently robust to allow us to proceed with confidence. We know the questions to ask, but not how to answer them. We do not have sufficient understanding of cause and effect relationships in these situations. This is a fundamental challenge for the practice of industrial ecology. How do we proceed in the face of such profound uncertainty? These are two strategies: adaptive management and the precautionary principle.

Adaptive management (see Chapter 5 for further discussion) involves assuming that one's design is at best a temporary transient solution to a situation. Then one must build into one's design the ability to change and adapt to changing circumstances. This requires that a design be inherently flexible. It also requires that comprehensive monitoring be carried out so that change in the environment can be detected sufficiently early to allow for appropriate change in the design. In effect, our design process must change so that the resulting systems have the capacity to reorganize. They should be constructed as self-organizing systems. This involves a profound shift in paradigm. We can no longer treat our designs as mechanical clockwork edifices designed to withstand the test of time.

Given our ignorance about how our interactions with natural systems will affect them, it behooves us to minimize these interactions. This is the precautionary principle. Whenever possible, we should limit the effluent from societal systems (both waste materials and energy) flowing across the interface into natural systems. We should minimize the displacement on the landscape of natural systems by societal systems. Given that human society is appropriating more than half the photosynthetic capacity of the biosphere, human systems must decrease their use of energy. In short, adoption of the precautionary principle mandates that our designs minimize their ecological footprint.

This is not inconsistent with some of the core design principles espoused in the industrial ecology literature. Closing the material flow loops, dematerialization, and life cycle efficiency are all strategies to decrease the ecological footprint of a design. However, curiously enough, the rationale for these strategies is very different in the industrial ecology literature from the one presented herein. The normal justification in the literature is based on these being properties of natural ecological systems that our designs should mimic. This line of logic is incorrect, as we shall see in the next section. But this does not negate the validity of these efficiency-related design principles. However, some words of caution from systems theory concerning efficiency are in order.

## ABOUT EFFICIENCY

As mentioned earlier, physical flows must be analyzed in terms of quantity and quality. Efficiency is about how well the quantity of flow is used. Effectiveness is how well the quality of the flow is used. When the flow is energy, quality is measured by exergy density. Effectiveness measures how much exergy is used in the system relative to a theoretical best-case scenario. [The theoretical best case, according to the second law of thermodynamics, is when all processes are performed reversibly and all the available work (exergy) is extracted from the energy.]

Second law analysis is the activity of studying how effectively the quality of a flow is utilized in a system. Only a handful of authors have suggested that quality and effectiveness (i.e. exergy analysis) are important considerations for design in industrial ecology (Brodyansky *et al.* 1994; O'Rourke *et al.* 1996; Connelly and Koshland 1997; Ayres *et al.* 1998). Yet it has been well demonstrated that focusing on efficiency alone will lead to poor decisions. Take, for example, electric radiant heat versus a natural gas forced air furnace. While the efficiency (first law) of an electric heater is essentially 100% (all of the electricity is converted to heat), its second law effectiveness is much lower than that of a natural gas furnace. This is because the quality of the electricity is wasted. The electricity could have been used to run a heat pump or other devices that, in turn, would have generated heat while doing other tasks as well. The point here is that focusing only on efficiency, as much industrial ecology work does, will lead to designs that use more exergy and produce more waste than they need to. Focusing only on efficiency leads to a design with a larger ecological footprint than necessary.

Another problem with the focus on efficiency is suboptimization. There is an underlying assumption that, if individual processes and subsystems are made efficient, then the overall system will be efficient. This assumption is only valid when the interconnections between elements of the system are strictly linear. This is rarely true in real physical systems. For example, we undertook to change a student residence cafeteria, so that less waste would be produced. Observation of the food which remained on plates after students had finished their meals revealed a large amount of untouched food and unopened packages that ended up in the garbage. Surveys of students revealed that this was because food was served on a fixed-price, "all you can eat" basis. A redesign of the cafeteria was undertaken so that students paid for what they put on their plates. This reduced the waste from plates after meals by 72%. The redesign seemed to be a big success. However, we monitored all the waste generated, from the time the food entered the university, through all the processing steps, until it was disposed of or consumed. The changes in the overall food system required by the redesign actually increased the waste generated in some subsystems. When all the waste generated was taken into account, a 45% decrease resulted.

The point is that any time one part of a system is optimized in isolation, another part will be moved further from its optimum in order to accommodate the change. Generally, when a system is optimal, its components are themselves run in a suboptimal way. One cannot assume that imposing an efficiency criterion on every component in a system will lead to the most efficient system overall. Generally, it will not. Instead, a nested approach must be taken when dealing with efficiency.

Allenby (1999) takes this observation one step further. He notes that one cannot talk about a sustainable process or plant, but only about a sustainable biosphere. Sustainability, like efficiency, must be a property of the overall nested system, not of each of the subsystems and components.

The exploration of the issue of interfacing natural and societal systems illustrates our profound ignorance of how the natural biosphere works. Dealing with this ignorance is the single most important scientific challenge facing our species. In lieu of the knowledge necessary to evaluate the implications of our designs for natural systems, the only currently viable way of dealing with interfacing, other than ignoring it in the design process, is to design for adaptability and to minimize the ecological footprint of our designs.

### *Mimicry of natural ecosystems*

The second design principle is: ***the behavior and structure of large scale man-made systems should be as similar as possible to those exhibited by natural ecosystems.***

This seems to be the cornerstone of industrial ecology and has been proposed by many authors (Kay 1977; Frosch and Gallopoulos 1989; Tibbs 1992; O'Rourke *et al.* 1996; Boons and Bass 1997; Erkman 1997). The rationale for this principle is typically in the form:

> The purpose of having man-made production–consumption systems mimic natural ones is to benefit from the "learning" that is embedded in the structure and behavior of natural systems. As long as man ultimately depends on the sun for energy and the earth for material resources he would be foolish to ignore the teachings of several billion years of evolution.
>
> (Kay 1977)

or

> Nature is the undisputed master of complex systems, and in our design of a global industrial system we could learn much from the way the natural global ecosystem functions. In doing so, we could not only improve the efficiency of industry but also find more acceptable ways of interfacing it with nature. Indeed, the most effective way of doing this is probably to model the systemic design of industry on the systemic design of the natural system. This insight is at the heart of the closely related concepts of industrial ecology, industrial ecosystems, industrial metabolism, and industrial symbiosis, all of which have been emerging in recent years.
>
> (Tibbs 1992)

Unfortunately, the application and discussion of this principle is often flawed by a romantic turn of the last-century Clementsian, Odumesque (E.P. not H.T.) view of ecosystems as superefficient, closed-loop, highly tuned systems. As Boons and Bass (1997) point out, this is simply not the case. (Nor are ecosystems random accidents of history, as suggested by the Gleasonian school of ecology.) Rather, as I have summarized briefly at the beginning of this chapter and have written about at length elsewhere, ecosystems are complex, adaptive, self-organizing hierarchical systems. There are two broad themes to self-organizations of ecosystems:

- coping with a changing environment;
- making good use of available resources.

At any time, the state of development of an ecosystem reflects a historical balancing act between these sometimes contradictory themes. An ecosystem that is superefficient and which has a highly articulated mass–energy flow network is usually quite brittle, i.e. unable to cope with change. So, to apply this principle successfully, we need to alter significantly our notion of how ecosystems develop. Ecosystems are not necessarily about "closing the loop," but rather about making effective (in a second law sense) use of the resources available while maintaining adaptability. Striking this balance is what maintaining ecological integrity is about.

In some instances, with respect to some resources, ecosystems are very leaky. For example, a shallow lake in a benthic regime will extract incoming phosphorus from the water column and bury it in the bottom muck, where it is deactivated. In effect, all the phosphorus is removed from the system. Ironically when the ecosystem can no longer accomplish this, a flip to a different regime (the pelagic) is precipitated, and, in the process, the phosphorus in the bottom muck is rereleased into the water column, where it fuels the new regime. (I cannot help but think of our practice of landfilling our garbage and how this buried waste may become a valuable resource when our "ecosystem" flips.)

From an efficiency point of view, terrestrial ecosystems are not. Less than 2% of the incoming solar energy is converted to green stuff. Over 80% of the incoming solar energy is turned into heat or is used to pump water. This is not an example of efficiency. However, from a second law effectiveness point of view, ecosystems have effectiveness ratios that exceed 80%. They are very effective at using their resources.

There are three ways to cope with a changing environment.

1. Take control of the environment.
2. Isolate the system from the environment.

3 Adapt the system to the changed environment by:
   a changing the behavior and role of elements of the system;
   b changing the elements of the system;
   c changing the interconnections between elements.

Natural ecosystems do make use of the first and second of these strategies. (Beavers build dams, tropical rainforests throw up a cloud cover daily, thus limiting solar energy hitting the canopy, streams are isolated from changes in ion concentration in precipitation by the filtering action of adjacent forests.) However, the primary means of coping is the last strategy, adapting. In particular, the loss of elements of an ecosystem (death of individuals and loss of species through displacement) is common (survival of the fittest, Holling four-box model).

On the other hand, humans tend to focus on the first and second strategy as opposed to the third. This is well illustrated by our tendency to bulldoze the natural landscape and replace it with concrete and gardens. The act of building a house or office is about isolating the system (humans) from the environment (Nature). Humans do not tend to focus on adaptability insofar as it means abandoning or radically changing elements of the system. We value human life and try to minimize the "hardships" felt by members of our species. Natural ecosystems have no such concerns.

So, while it would behoove us to take advantage of the collective learning and wisdom that is reflected in the system characteristics of natural ecosystems, mimicry of natural ecosystems must be tempered by an appreciation that humans have a set of priorities that will cause them to find a different balance between the need to make good use of resources while coping with a changing environment.

## *Appropriate biotechnology*

The third design principle is: *whenever feasible, the function of a component of a man-made system should be carried out by a subsystem of the natural biosphere.* This is referred to as using *appropriate biotechnology.* Over the past two decades, in the region of Waterloo, Ontario, a number of experiments using "appropriate biotechnology" have been tried. For example, it is now standard practice to use natural landscapes for storm water management, in place of concrete channels. These natural landscapes include holding ponds and creeks with natural vegetation on the slopes. Our experience is that the capital cost of "natural" stormwater management is about 10% of that of concrete and operating costs are similarly less. Furthermore, these waterways double as aesthetically attractive recreational amenities in the community. Another example is the replacement of turfgrass with natural communities that are self-maintaining. This significantly reduces the cost (both dollars and environmental damage due to chemicals etc.) of maintaining landscapes. Composting has been actively promoted by the regional government, with tens of thousands of composters being distributed, thus diverting solid waste from local landfills. Wetlands (both existing and man-made) have been used for sewage treatment plants and for remediation of mine tailing ponds. Luvall's work on greening US cities has demonstrated how judicious use of trees and other flora can significantly reduce the heat load on a city. The experience with "appropriate biotechnology" has been that it saves much money, both capital and operating costs.

*Renewable resources*

The final design principle is: *non-renewable resources should be used only as capital expenditures to bring renewable resources on-line.* This principle is a corollary to the axiom that we must live within our carrying capacity. Resources are not inherently renewable. It is how we use them that make them renewable. When a resource is used at a rate that is less than the rate at which it can be replenished by natural systems, then the use of the resource is renewable. Be clear that term replenished is not a synonym for produced. Replenished means the that natural system is producing stock of the resource at a rate such that the stock of the resource in the natural system does not decrease. Recycled materials are a renewable resource insofar as the cost of recycling is borne by renewable resource consumption. In the final analysis, unless humans move off the planet, the human population must be such that it can be supported by renewable resources.

## Construction ecology

So far this chapter has sketched out the relationship between industrial ecology and complex systems and ecosystem thinking. The challenge posed by these considerations is how to design an adaptive, resilient, evolving, self-organizing hierarchical human production–consumption system that provides for a sustainable livelihood, whose ecological footprint is minimal, and which interfaces with natural systems in a way that promotes ecological integrity. In this section this challenge is explored in the context of construction ecology. The issue that this poses for construction ecology is not so much how to construct efficiently, but rather how to construct a building system that is resilient and can adapt and evolve while fitting into the natural environment. Currently, construction ecology seems to focus on buildings that deal with a changing environment by being robust enough to be impervious to change. Thus, our only means of adapting or evolving our structures seems to be by tearing them down and starting over. Surely we can come up with a better process for our structures to evolve than blowing them up in a spectacular fashion. Perhaps the notion of mimicry is a place to start thinking about how to do this.

Currently buildings are essentially static structures. If they are to mimic natural ecosystems then they must have the capacity to self-organize, that is to reshape the internal configuration of the building and to change the buildings' connection to the outside world, in response to a changing situation. This would need to be thought through from different types of perspectives at different scales. For example, some of the types of perspectives would be in the context of the different energy flows and material flows through a building and the ability to incorporate different flows as they come on- and off-line.

For example, at the author's university, the need to accommodate different forms of recyclable material has been a major challenge. Over the past fifteen years, there have been significant changes in the market for recyclables, with paper being desirable one year, cardboard the next, and aluminum the year after. This has necessitated frequently putting in place and taking out collection systems for the specific recyclables at some economic and aesthetic cost. We are thinking hard about how new buildings could be built so that they can accommodate the oscillation in the demand for different elements of the solid waste stream. Another type of perspective would be that of the users of the building. Is there a way of reconfiguring the internal layout of the building as user needs change? A number of other perspectives [maintenance, deconstruction, heating,

*Box 3.4* Ecological integrity is the biophysical purpose of industrial ecology

Ecological integrity is about three facets of the self-organization of ecological systems: (a) current well-being, (b) resiliency, and (c) capacity to develop, regenerate and evolve (Kay and Regier 2000). An evaluation of ecological integrity must consider:

1 the current organizational state of the system;
2 the ability of the system to reorganize in the face of environmental change;
3 the system's capacity to continue to self-organize in its normal environment, that is to:
   - continue to develop, i.e. increase its organization relative to an attractor;
   - regenerate, to deal with the birth–growth–death–renewal cycle (i.e. the Holling four-box model), i.e. to deal with the multiple nested dual attractor problem; and to
   - continue to evolve, i.e. switch attractors spontaneously (emergent complexity).

*Box 3.5* Three ways a system can cope with a changing environment

A system can adapt by:

1 taking control of the environment;
2 isolating itself from the environment;
3 changing its internal organization by:
   - changing the behaviour and role of elements of the system;
   - changing the elements of the system;
   - changing the interconnections between elements.

ventilation, and cooling (HVAC), etc.] can be thought through in terms of the integrity (Box 3.4) and adaptability (Box 3.5) of the building.

These issues will also need to be thought through at different scales. There are the obvious scales of the basic units of the building (rooms, offices, etc.) and the building itself, but also the scale of the group of buildings and natural ecosystems which the building is connected to. How does the building fit into the bigger man-made and natural system? What is the hierarchical nesting of holons which it is a part of? What are the feedback loops between the building and the bigger world. Can the connections between the building and the outside world be changed?

For example, at my university we investigated the possibility of recycling all the water on campus. This turned out to be feasible only because all the buildings on campus are connected to an internal campus waterworks. This waterworks is connected to the city works in only two places. So, closing the loop for the campus could be done. If each building had been individually connected to the city waterworks, closing the water loop would have been physically impossible without reconnecting all the buildings on campus. The physical infrastructure at the scale of the campus allowed for an evolutionary strategy that otherwise would not have been practical.

Thus, thinking of a building in terms of a SOHO system model opens up a whole set of design questions related to adaptability and integrity that are not normally considered.

This recasts buildings as evolving dynamic structures which can be reshaped as the situation changes. And, if one accepts Holling's four-box phase model for ecosystems, it begs the question: what is the birth–growth–deconstruction–renewal process for a building, at all scales? This short section only scratches the surface of what I think is a very exciting challenge for the design of buildings. It goes far beyond the issues of efficiency and industrial parks which are usually associated with mimicry of natural ecosystem in industrial ecology.

The next step is to undertake some projects that demonstrate what an adaptable self-organizing building might look like. The only example of this that I am aware of is the FLEX housing project of CMHC (Canada Mortgage and Housing Corporation (Government of Canada) 1997). The key question to be explored is the trade-off between capital costs and efficiency and the overhead of flexibility, redundancy, renewability, and monitoring associated with adaptability. Nature is constantly revisiting this balance, and I suspect that this will also be the case with construction ecology.

In addition to the self-organization issue that ecosystem mimicry brings up, there is also the issue of mimicry of specific strategies of ecosystems. For example, terrestrial ecosystems will capture all the precipitation they need and store it for times of drought. Recently, in Ontario, an apartment block was built using techniques that attempted to minimize the ecological footprint. One strategy incorporated in this building was to mimic Nature by collecting all the precipitation that falls on all surfaces on the site. The collected water is stored in a large reservoir that can meet all non-potable water needs for three months (the longest time without significant precipitation in the area.)

Another example is the work of Lo *et al.* (1997) in greening cities They studied the radiation properties (absorption, reflection, emmisivity, etc.) of forest canopies and applied the lessons learned to desirable characteristics of roofing materials. Forest canopies tend to have high thermal inertia, i.e. it is hard to heat them up or cool them down. This means that the canopy temperature variation from day to night is much less than the variation in the air temperature. By using roofing materials that have similar radiative characteristic as forest canopies, it is quite easy to reduce drastically the HVAC load on a building.

It would be interesting to compile a catalog of such properties of natural ecosystems and their application to building design. Such a document could act as guide for architects and engineers.

The next logical step, beyond mimicking natural systems, is to actually incorporate the natural systems in buildings. This brings us to the subject of appropriate biotechnology for buildings. John Todd's living machines are an example of this (Todd and Todd 1994). He constructs, in buildings, natural ecosystems that transform wastewater into drinking water. At a larger scale this could be used for neighborhood sewage treatment plants. Another example is the use of rooftop gardens to insulate and cool buildings. By picking species with significant transpiration capacity, such gardens can actively cool buildings in the daytime. This can also be accomplished by planting mini-woodlots next to buildings, an idea pioneered by R.S. Dorney of University of Waterloo. Composting is a means of dealing with organic waste. The University of Waterloo is experimenting with in-building composting using worms and other fauna in soil pots distributed throughout a building. (And, no, we have never had a worm escape!) A final example of appropriate biotechnology is the effort to develop assemblages of vegetation that can act as air purifiers in buildings.

A compilation of "appropriate biotechnologies" and a focused research program on the use of these technologies would further their use in construction.

*Box 3.6* A selection of green design standards for buildings (prepared by James Wu)

| | | |
|---|---|---|
| 1 | ASHRAE | Energy consumption benchmark |
| 2 | BREEM | Approximately eighteen criteria (organized as global, local and indoor) |
| 3 | BEPAC | Approximately thirty criteria, with subsets (organized as ozone layer protection, environmental impact of energy use, indoor environmental quality, resource conservation and site and transportation) |
| 4 | C-2000 | 170 criteria targeting commercial construction (energy efficiency, environmental impact, health/comfort/productivity, functional performance, longevity, adaptability, ease of operation and maintenance, economic viability) |
| 5 | Eco-Profile | Criteria are structured in four main areas: energy, indoor environment, pollution, and exterior environment |
| 6 | Embodied energy profile | |
| 7 | Global Environment Impact | Criteria categorized under seven major headings: reduction in greenhouse gas emissions, conservation of tropical rainforest, reduction in gases that reduce acid rain, conservation of water resources, solid waste, reduction in ozone-depleting substances, ecological considerations |
| 8 | Green Builder Program | Approximately sixteen criteria (water, energy, building materials, solid waste) |
| 9 | Green Building Program (City of Austin '96) | Eight-one criteria targeting commercial construction (predesign, programming, schematic design, design development, construction management, commissioning, post occupancy) |
| 10 | LEED (US Green Building Council, Green Building Rating System) | |
| 11 | Life cycle assessment | |

So far in the discussion of construction ecology, the second and third design principles have been discussed. The first and the fourth have been left for the end of the discussion as much has been written about design for efficiency, using renewable resources, and minimizing ecological footprint in construction (Vale and Vale 1991; Todd and Todd 1994; Papanek 1995; Yeang 1995). One of my students, James Wu, has written a 100-page summary of these ideas, and I will not attempt to condense this here. Box 3.6 itemizes eleven different green design standards for buildings that he identified. However, this literature does not deal with the problem of interfacing in an integrated systems way (the SOHO system model), as discussed earlier. There are fundamental theoretical and practical obstacles to be overcome before this can be done in a satisfactory way.

One issue I wish to flag again is that of quality versus quantity, of effectiveness versus efficiency. The difference between these is rarely acknowledged in the literature. As pointed out earlier, there are many situations in which a design is efficient but not effective, and hence opportunities to decrease the ecological footprint are missed.

For example, if one is using natural gas turbines to generate electricity (as is done in the new "green" city hall in Kitchener, Ontario) and using the waste heat from the generators to heat the building or, through an adsorption cycle, to cool the building in summer, the gain from using the waste heat will not show up in an efficiency calculation but will show up in an effectiveness calculation. Similarly, buildings (such as the Ontario Hydro headquarters in Toronto, Ontario) which capture heat generated during the day by workers and machines etc. and store it for use to heat the building at night do not have any advantage from an efficiency point of view, but do from an effectiveness perspective. Effectiveness must become as important a criterion as efficiency as a more effective solution can actually be less efficient while having a smaller ecological footprint (for example, natural gas versus electrical domestic hot water systems).

### *An ecosystem approach*

Central to the design process is the activity of making trade-offs. Choosing between alternatives usually comes down to people's values. This is inescapable, especially when dealing with complex systems (Funtowicz and Ravetz 1993, 1994; Kay and Schneider 1994). So, ethics and values must be incorporated into any discussion of industrial ecology, if for no other reason than that trade-offs between sustainable livelihoods and ecological integrity will have to be made. This leads me to reject the position of Allenby and IEEE.

> This elucidation makes the important point that industrial ecology strives to be objective, not normative. Thus, where cultural, political, or psychological issues arise in an industrial ecology study, they are evaluated as objective dimensions of the problem. ... Whether this is good or bad – whether it "should" be the case – is not properly an issue for industrial ecology.
>
> (Allenby 1999)

It is a waste of time to try and build industrial ecology in the mode of traditional science. The inherent complexity of the subject it deals with means that a "post-normal" science epistemology will be much more fruitful. (I leave it to the reader to explore the literature and debate in this regard.)

I have proposed, with others, an ecosystem approach for planning and decision making for sustainability (Kay *et al.* 1999). In Figure 3.7, this approach is adapted for construction ecology.

Two steps are carried out in parallel at the beginning: identifying the players (stakeholders, actors, users, etc.) and the issues they have about the project; and building a systems description of the situation, preferably cast in the mode of SOHO system model. These two steps generate a set of descriptions (narratives) of how the building project might proceed. These scenarios represent different integrations of science and best practice with people's preferences. They reflect different combinations of trade-offs. A decision-making process must be developed that resolves these trade-offs in a way that is acceptable to all actors. (Which begs the question: who gets to decide?)

Once a resolution is reached, it is necessary to develop an ongoing adaptive management strategy. This involves monitoring and managing the internal organization of the building

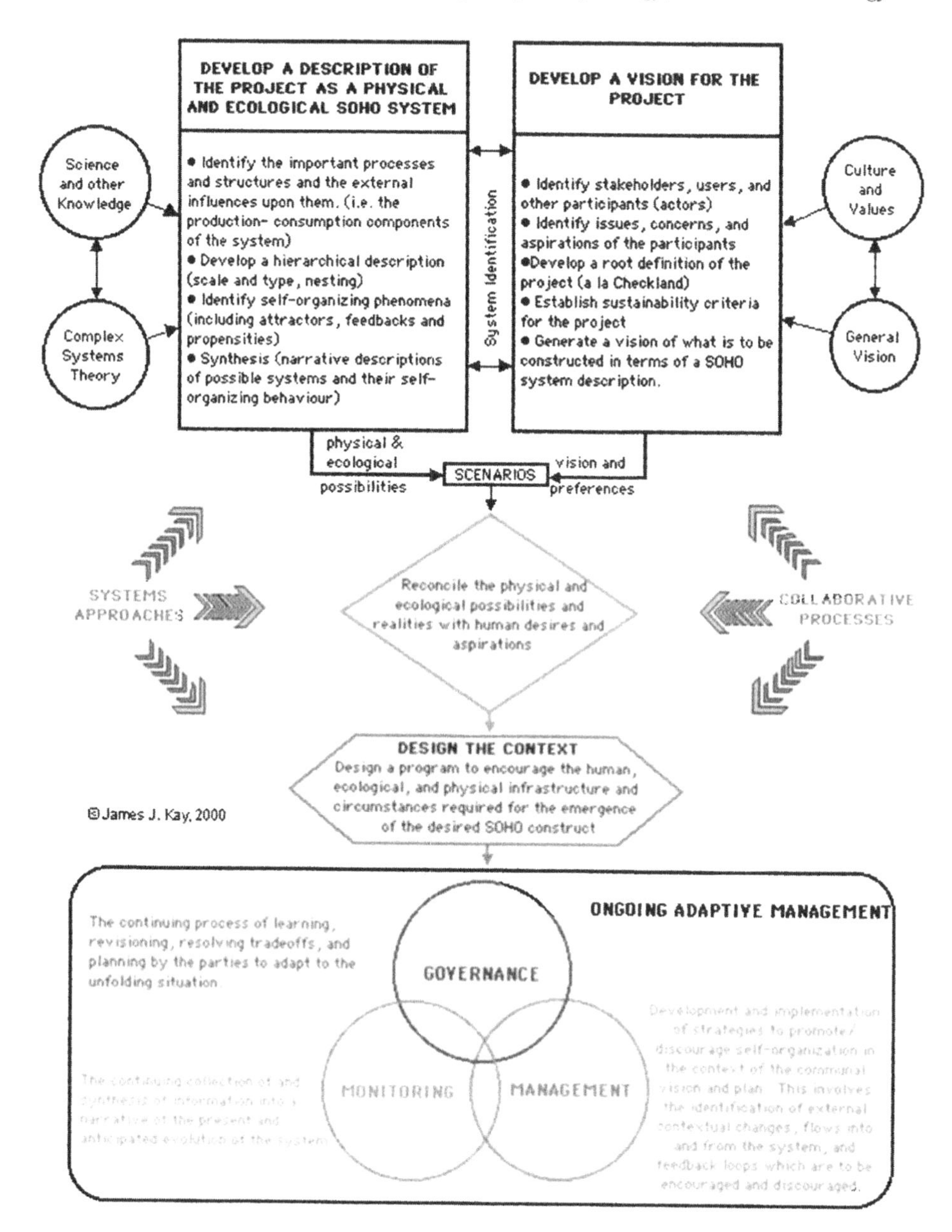


*Figure 3.7* An adaptive ecosystem approach to construction ecology.

and its relationship to the outside so that, if any reorganization is needed, it can be identified. However such reorganization can occur only if an appropriate governance structure is in place to make decisions about reorganization. Without a governance structure, no adaptation is possible.

Fundamental to this approach is a different mindset about design. Design can no longer be seen as finding a solution to a problem, in effect the right answer. Rather, it must be seen as setting in process the evolution of a built environment that evolves to meet the evolving needs of users and which can adapt so as to fit into changing environmental conditions. In an ecosystem approach, design must become about developing dynamic processes rather than static structures. Only in this way can our construction fit into an evolving, dynamic biosphere.

### *Summation*

> Humanity is facing a resource crisis which its present world model, economics, seems to be unable to cope with. It appears that the alternatives are as follows:
>
> 1 Continue our present behavior and hope that we continuously evolve makeshift solutions (i.e. find sufficient new energy and material resources, technical fixes, and places to dump our wastes so that we can continue to grow). All evidence suggests that this would end in disaster.
> 2 Continue our present behavior but the pressures of environmental factors force us to expand outward into space (via development of interstellar flight capabilities). This would be the classical solution, since throughout history whenever humans did not have enough resources they expanded into new territory. This solution would gamble our remaining non-renewable resources on an attempt to develop the technology necessary and then find the new resources.
> 3 Maintain the present system but assign dollar costs to the usage of resources and to the dumping of wastes in the environment. The problem then becomes one of management.
> 4 Humans recognize that they are part of the natural environment and not external to it. They must integrate themselves into this environment. Their behavior must take into account the limitations of the natural environment.
>
> The last alternative seems to offer the best chance for humankind's survival.
>
> (Kay 1977)

In this regard, this chapter argues that industrial ecology is the activity of designing and managing human production–consumption systems, so that they interact with natural systems, to form an integrated (eco)system which has ecological integrity and provides humans with a sustainable livelihood. To accomplish this end requires an ecosystem approach, the application of systems thinking to the analysis and design of biophysical mass and energy transformation systems. In essence, industrial ecology is about designing human ecological–economic systems that fit in with natural ecological systems. Construction ecology is about constructing built environments which have integrity and the ability to adapt.

The practice of industrial ecology and construction ecology must be carried out against the backdrop of the new understanding of complex systems, and in particular ecosystems, which is emerging. The hierarchical nature of these systems requires that they be studied from different types of perspectives and at different scales of examination. There is no one correct perspective. Rather, a diversity of perspectives is required for understanding. Ecosystems are self-organizing. This means that their dynamics are largely a function of positive and negative feedback loops. This precludes linear causal mechanical explanations of ecosystem dynamics. In addition, emergence and surprise are normal phenomena in systems dominated by feedback loops. Inherent uncertainty and limited predictability are inescapable consequences of these system phenomena. Such systems organize about attractors. Even when the environmental situation changes, the system's feedback loops tend to maintain its current state. However, when ecosystem change does occur, it tends to be very rapid and even catastrophic. When precisely the change will occur, and what state the system will change to, are often not predictable. Often, in a given situation, there are several possible ecological states (attractors), that are equivalent. Which state the ecosystem currently occupies is a function of its history. There is not a "correct" preferred state for the ecosystem.

This new understanding requires that the practice of industrial ecology and construction ecology must address the issues of complexity, self-organization, and inherent uncertainty. As suggested in this chapter, this requires a very different framing of design. Design becomes about developing dynamic processes rather than static structures, about setting in motion an evolutionary process. An (eco)system based approach for doing this, which has its underpinnings a self-organizing hierarchical open systems analysis of the situation is sketched out in this chapter. As suggested herein, much work remains to be done to flesh out this approach.

In the end, human socioeconomic systems are utterly dependent on natural systems for their context. As McHarg (1998) put it so eloquently thirty years ago, "human society must fit in with nature." Humans must understand that the integrity of human societal ecosystems are inextricably linked to the integrity of natural ecosystems. Maintaining the integrity of the biosphere is necessary for the continuation of our society. This means that we must design our physical systems so as to maintain the context for the integrity of the self-organizing processes of natural ecosystems that are necessary for the continued existence, on this planet, of self-organizing human ecosystems. This is the task that industrial ecology must accomplish, to design the intertwined ecological–societal system that is emerging on this planet.

## Acknowledgments

I would like to acknowledge the work of two undergraduate students who researched this topic for me, James Wu and Craig Hawthorne. I would like to thank Charles Kibert for drawing me back into this topic after a long absence. Finally, I would like to acknowledge the tremendous foresight exhibited by my graduate supervisor, M. Chandrashekar, in championing research in this area from the mid-1970s on. I would have welcomed his co-authorship of this piece but, alas, he passed away at the untimely age of fifty, in 1997.

## References

Ahern, J.E. 1980. *The Exergy Method of Energy Systems Analysis*, New York: John Wiley.

Akbari, M.H. 1995. Energy-based indicators of ecosystem health. MSc thesis. University of Guelph.

Allen, T.F.H., and Hoekstra, T.W. 1992. *Toward a Unified Ecology*. New York: Columbia University Press.

Allen, T.F.H., and Starr, T.B. 1982. *Hierarchy: Perspectives for Ecological Complexity*. Chicago: University of Chicago Press.

Allen, T.F.H., Bandurski, B.L. and King, A.W. 1993. *The Ecosystem Approach: Theory and Ecosystem Integrity*. Washington, DC: International Joint Commission.

Allenby, B.R. 1999. *Industrial Ecology: Policy Framework and Implementation*. Englewood Cliffs, NJ: Prentice-Hall.

Ayres, R., Ayres, L. and Martin, K. 1998. Exergy, waste accounting, and life cycle analysis. *Energy* 23: 355–63.

Boons, F. and Bass, L. 1997. Types of industrial ecology: the problem of coordination. *Journal of Cleaner Production* 5 (1/2): 79–86.

Boyle, M., Kay, J.J. and Pond, B. 1996. *State of The Landscape Reporting: The Development of Indicators for the Provincial Policy Statement Under the Land Use Planning and Protection Act*. Toronto: Ministry of Natural Resources.

Boyle, M., Kay, J.J. and Pond, B. 2002. Monitoring and assessment as part of an adaptive ecosystem approach to sustainability and health. In *Encyclopaedia of Global Environmental Change*, Vol. 5. Munn, T.E. (ed.). New York: John Wiley (in press).

Brodyansky, V., Sorin, M. and LeGoff, P. 1994. *The Efficiency of Industrial Processes: Exergy Analysis and Optimization*. Amsterdam: Elsevier.

Caley, M.T., and Sawada, D. 1994. *Mindscapes: the Epistemology of Magoroh Maruyama*. Langhorne, PA: Gordon and Breach.

Canada Mortgage and Housing Corporation (Government of Canada). 1997. FLEX housing – homes that adapt. Web page. Available at http://www.cmhc-schl.gc.ca/rd-dr/en/flex/.

Casti, J.L. 1994. *Complexification: Explaining a Paradoxical World Through the Science of Surprise*. New York: HarperCollins.

Chinneck, J. 1983. *Systems Theoretic Overview Models of Industrial Plant Energy Systems Incorporating Exergy*. PhD thesis. University of Waterloo.

Chinneck, J., and Chandrashekar, M. 1984. Models of large-scale industrial energy systems. 1. Simulation. *Energy: The International Journal* 9: 21–34.

Connelly, L. and Koshland, C.P. 1997. Two aspects of consumption: using an exergy-based measure of degradation to advance the theory and implementation of industrial ecology. *Resources, Conservation and Recycling* 19: 199–217.

Dempster, B. 1998. A self-organizing systems perspective on planning for sustainability. Master's thesis. University of Waterloo.

di Castri, F. 1987. The evolution of terrestrial ecosystems. In *Ecological Assessment of Environmental Degradation, Pollution and Recovery*. Ravera, O. (ed.). Amsterdam: Elsevier Science, pp. 1–30.

Edgerton, R.H. 1982. *Available Energy and Environmental Economics*. Lexington, MA: D.C. Heath.

Ehrenfeld, J. 1997. Industrial ecology: a framework for product and process design. *Journal of Cleaner Production* 5 (1/2): 87–85.

Erkman, S. 1997. Industrial ecology: an historical overview. *Journal of Clean Technology* 5: 1–10.

Ford, K.W., Rochlin, G.I. and Socolow, R.H. 1975. *Efficient Use of Energy*. New York: American Institute for Physics.

Fraser, R., and Kay, J. 2002. Exergy, solar radiation and terrestrial ecosystems. In *Thermal Remote Sensing in Land Surface Processes*. Quattrochi, D. and Luvall, J. (eds). Ann Arbor, MI: Ann Arbor Press (in press).

Frosch, R. and Gallopoulos, N. 1989. Strategies for manufacturing. *Scientific American* 261 (3): 144–54.

Funtowicz, S., and Ravetz, J. 1993. Science for the post-normal age. *Futures* 25: 739–55.

Funtowicz, S. and Ravetz, J. 1994. Emergent complex systems. *Futures* 26: 568–82.

Gaggioli, R.A. (ed.) 1980. *Thermodynamics: Second Law Analysis*. Washington, DC: American Chemical Society.

Gaggioli, R.A. (ed.)1983. *Efficiency and Costing: Second Law Analysis of Processes*. Washington, DC: American Chemical Society.

Graedel, T.E., and Allenby, B.R. 1995. *Industrial Ecology*. Englewood Cliffs, NJ: Prentice-Hall.

Hevert, H. and Hevert, S. 1980. Second law analysis: an alternative indicator of system efficiency. *Energy: The International Journal* 5: 865–873.

Holling, C.S. 1986. The resilience of terrestrial ecosystems: local surprise and global change. In *Sustainable Development in the Biosphere*. Clark, W.M. and Munn, R.E. (eds). Cambridge: Cambridge University Press, pp. 292–320.

Holling, C.S. 1992. Cross-scale morphology, geometry, and dynamics of ecosystems. *Ecological Monographs* 62: 447–502.

IEEE Electronics and the Environment Committee 1995. White Paper on sustainable development and industrial ecology. Web page [accessed 2000]. Available at computer.org/tab/ehsc/ehswp.htm.

Jantsch, E. 1980. *The Self-Organizing Universe: Scientific and Human Implications of the Emerging Paradigm of Evolution.* Toronto: Pergamon Press.

Kay, J.J. 1977. *An Investigation into Engineering Design Principles for a Conserver Society*. Master's thesis. University of Waterloo.

Kay, J.J. 1984. *Self-Organization in Living Systems*. PhD thesis. University of Waterloo.

Kay, J.J. 1991. A non-equilibrium thermodynamic framework for discussing ecosystem integrity. *Environmental Management* 15: 483–495.

Kay, J.J. 1997. Some notes on: the ecosystem approach, ecosystems as complex systems. In *Integrated Conceptual Framework for Tropical Agroecosystem Research Based on Complex Systems Theories*. Murray, T. and Gallopin, G. (eds). Cali, Colombia: Centro Internacional de Agricultura Tropical, pp. 69–98 .

Kay, J.J. 2000. Ecosystems as self-organizing holarchic open systems: narratives and the second law of thermodynamics. In *Handbook of Ecology.* Muller, F. and Jørgensen, S.E. (eds). Boca Raton, FL: CRC Press, pp. 135–60.

Kay, J.J. and Regier, H. 1999. An ecosystem approach to Erie's ecology. In *The State of Lake Erie (SOLE) – Past, Present and Future. A Tribute to Drs. Joe Leach & Henry Regier*. Munawar, M. Edsall, T. and Munawar, I.F. (eds). Dordrecht: Backhuys Academic Publishers, pp. 511–33.

Kay, J.J. and Regier, H. 2000. Uncertainty, complexity, and ecological integrity: insights from an ecosystem approach. In *Implementing Ecological Integrity: Restoring Regional and Global Environmental and Human Health.* Crabbé, P., Holland, A., Ryszkowski, L. and Westra, L. (eds). Dordrecht: Kluwer, pp. 121–56.

Kay, J.J. and Schneider, E.D. 1992. Thermodynamics and measures of ecosystem integrity in ecological indicators. In *Proceedings of the International Symposium on Ecological Indicators.* McKenzie, D.H., Hyatt, D.E. and Mc Donald, V.J. (eds). Amsterdam: Elsevier, pp. 159–82.

Kay, J.J. and E. D. Schneider. 1994. Embracing complexity, the challenge of the ecosystem approach. *Alternatives* 20 (3): 32–38.

Kay, J.J., Regier, H., Boyle, M. and Francis, G. 1999. An ecosystem approach for sustainability: addressing the challenge of complexity. *Futures* 31: 721–742.

Koenig, B. and Tummala, R. 1991. Enterprise model for the design and management of maufacturing systems. In *Proceedings Joint US/German Conference on New Directions for Operations Research in Manufacturing*.

Koenig, H.E. and Cantlon, J.E. 1998. Quantitative industrial ecology. *IEEE SMC* 28: 16–28.

Koenig, H.E. and Cantlon, J.E. 1999. Sustainable ecological economies. *Ecological Economics* 31 (1): 107.

Koenig, H. E., and Tummala, R.L. 1972. Principles of ecosystem design and management. *IEEE SMC* 2: 449–459.

Koenig, H.E., Cooper, W. and Falvey, J. 1972. Engineering for ecological, sociological and economic compatibility. *IEEE SMC* 2: 319–331.

Koenig, H.E., Edens, T. and Cooper, W. 1975. Ecology, engineering and economics. *Proceedings of the IEEE* 63: 501–511.

Koestler, A. 1978. *Janus: A Summing Up.* London: Hutchinson.

Koestler, A. and Smythies, J.R. (eds) 1969. *Beyond Reductionism.* London: Hutchinson.

Lo, C.P., Quattrochi, D. and Luvall, J.C. 1997. Applications of high-resolution thermal infrared remote sensing and GIS to assess the urban heat island effect. *International Journal of Remote Sensing* 18 (2): 287–304.

Ludwig, D., Walker, B.B. and Holling, C.S. 1997. Sustainability, stability, and resilience. *Conservation Ecology* 1 (1): Article 7.

Luvall, J.C. and Holbo, H.R. 1989. Measurements of short term thermal responses of coniferous forest canopies using thermal scanner data. *Remote Sensing and the Environment* 27: 1–10.

Luvall, J.C. and Holbo, H.R. 1991. Thermal remote sensing methods in landscape ecology. In *Quantitative Methods in Landscape Ecology*. Turner, M. and Gardner, R.H. Berlin: Springer-Verlag, Chapter 6.

Luvall, J.C., Lieberman, D., Lieberman, M., Hartschorn, G. and Peralta, R. 1990. Estimation of tropical forest canopy temperatures, thermal response numbers, and evapotranspiration using an aircraft-based thermal sensor. *Photogrammetric Engineering and Remote Sensing* 56: 1393–1401.

McHarg, I. 1998. Architecture in an ecological view of the world (1970). In *To Heal the Earth: Selected Writings of Ian L. McHarg*. McHarg, I. and. Steiner, F.R. (eds). Washington, DC: Island Press, pp. 175–185.

Maruyama, M. 1980. Mindscapes and science theories. *Current Anthropology* 21: 589–599.

Moran, M.J. 1982. *Availability Analysis: A Guide to Efficient Energy Use*. Englewood Cliffs, NJ: Prentice-Hall.

Nicolis, G. and Prigogine, I. 1977. *Self-Organization in Non-Equilibrium Systems*. New York: John Wiley.

Nicolis, G. and Prigogine, I. 1989. *Exploring Complexity*. New York: W.H. Freeman.

O'Callaghan, P. W. 1981. *Design and Management for Energy Conservation*. Toronto: Pergamon Press.

O'Rourke, D., Connelly, L. and Koshland, C. 1996. Industrial ecology: a critical review. *International Journal of Environment and Pollution* 6 (2/3): 89–112.

Odum, E.P. 1969. The strategy of ecosystem development. *Science* 164: 262–270.

Papanek, V. 1970. *Design for the Real World*. London: Thames & Hudson.

Papanek, V. 1995. *The Green Imperative: Ecology and Ethics in Design and Architecture*. London: Thames & Hudson.

Peacocke, A.R. 1983. *The Physical Chemistry of Biological Processes*. Oxford: Oxford University Press.

Peusner, L. 1986. *Studies in Network Thermodynamics*. Amsterdam: Elsevier.

Quattrochi, D. and Luvall, J.C. 1999. Thermal infrared remote sensing for analysis of landscape ecological processes: methods and applications. *Landscape Ecology* 14: 577–598.

Regier, H.A. and Kay, J.J. 1996. An heuristic model of transformations of the aquatic ecosystems of the Great Lakes–St. Lawrence River basin. *Journal of Aquatic Ecosystem Health* 5: 3–21.

Saama, P.J., Koenig, B.E. and Koenig, H.E. 1994. Analytical tools for material and energy balance, cash flow, and environmental loads in a dairy cattle enterprise. *Journal of Dairy Science* 77 (4): 94.

Scheffer, M. 1998. *Ecology of Shallow Lakes*. London: Chapman & Hall.

Scheffer, M.S., Hosper, H., Meijer, M.-L., Moss, B. and Jeppesen, E. 1993. Alternative equilibria in shallow lakes. *Trends in Ecology and Evolution* 8 (8): 275–279.

Schneider, E., and Kay, J.J. 1993. Exergy degradation, thermodynamics, and the development of ecosystems. In *Energy, Systems, and Ecology*, Vol. 1, *Proceedings of ENSEC 93*. Tsatsaronis, G., Szargut, J. Kolenda, Z. and Ziebik, A. pp. 33–42.

Schneider, E.D. and Kay, J.J. 1994a. Complexity and thermodynamics: towards a new ecology. *Futures* 24: 626–647.

Schneider, E.D. and Kay, J.J. 1994b. Life as a manifestation of the second law of thermodynamics. *Mathematical and Computer Modelling* 19 (6–8): 25–48.

Schneider, E.S. 1988. Thermodynamics, information, and evolution: new perspectives on physical and biological evolution. In *Entropy, Information, and Evolution: New Perspectives on Physical and Biological Evolution*. Weber, B.H., Depew, D.J. and Smith, J.D. (eds). Cambridge: MIT Press, pp. 108–138.

Szargut, J., Morris, D.R. and. Steward, F.R 1988. *Exergy Analysis of Thermal, Chemical, and Metallurgical Processes*. New York: Hemisphere Publishing.

Tibbs, B.C. 1992. Industrial ecology: an environmental agenda for industry. *Whole Earth Review*, Winter: 4–19.

Todd, J. and Todd, N. 1994. *From Ecocities to Living Machines: Designing for Sustainability*. Berkeley, CA: North Atlantic Books.

Tummala, R. and Koenig, B. 1993. Process network theory and implementation for technology assessment in maufacturing. In *Proceedings of a Joint US/German Conference*. Berlin: Springer-Verlag.

Ulanowicz, R.E. 1997a. *Ecology, the Ascendant Perspective*. New York: Columbia University Press.

Ulanowicz, R.E. 1997b. Limitations on the connectivity of ecosystem flow networks. In *Biological Models: Proceedings of the 1992 Summer School on Environmental Dynamics*. Rinaldo, A. and Marani, A.A. (eds). Venice: Istituto Veneto di Scienze, Lettere ed Arti, pp. 125–143.

Ulanowicz, R.E. and Hannon, B.M. 1987. Life and the production of entropy. *Proceedings of the Royal Society of London B* 232: 181–192.

UNDP 1998. Sustainable livelihoods concept paper. Web page. Available at http://www.undp.org/sl/Documents/Strategy_papers/Concept_paper/Concept_of_SL.htm.

Vale, B. and Vale, R. 1991. *Green Architecture: Energy-Conscious Future*. Boston: Little Brown.

van Berkel, R. and Lafeur, M. 1997. Application of an industrial ecology toolbox for the introduction of industrial ecology in enterprises. II. *Journal of Cleaner Production* 5 (1–2): 27–37.

van Berkel, R., Williams, E. and Lafeur, M. 1997. Development of an industrial ecology toolbox for the introduction of industrial ecology in enterprises. I. *Journal of Cleaner Production* 5 (1–2): 11–25.

von Bertalanffy, L. 1968. *General Systems Theory*. New York: George Braziller.

von Bertalanffy, L. 1975. *Perspectives on General Systems Theory*. New York: George Braziller.

Wall, G. 1986. *Exergy – A Useful Concept*. Gothenburg: Physical Resource Theory Group, Chalmers University of Technology.

Wicken, J.S. 1987. *Evolution, Thermodynamics, and Information: Extending the Darwinian Program*. Oxford: Oxford University Press.

Wong, F.C. 1979. *System-Theoretic Models for the Analysis of Thermodynamic Systems*. PhD thesis. University of Waterloo.

Wong, F.C. and Chandrashekar, M. 1982. Thermodynamic systems analysis. *Energy: The International Journal* 7: 539–566.

Woodley, S., Kay, J. and Francis, G. 1993. *Ecological Integrity and the Management of Ecosystems*. Delray Beach, FL: St. Lucie Press.

Yeang, K. 1995. *Designing With Nature*. New York: McGraw-Hill Inc.

# 4 Applying the principles of ecological emergence to building design and construction

*Timothy F.H. Allen*

As humans are biological entities, it should not be surprising that many human activities, such as building construction, can be understood through analogy to happenings in biological systems. The analogies from emergence of biological organization to human building construction are many and rich, and likely to be helpful at this time. Over recent decades a body of theory has developed in ecology that applies energetics to emergence of organization in ecological systems (Odum 1983; Wicken 1987; Schneider and Kay 1994). This chapter uses recent advances in ecological energetics to inform building design, construction, and use.

I make no apology for the use of analogy. Analogy is a powerful device (Lorenz 1974), particularly in ecology, in which most of the technical terms are easily identifiable as metaphors imported from other disciplines. The notion of ecological "stress" clearly comes from materials science, "invasion" comes from military science, and "competition" is straight from economics, to name just three borrowed terms. In fact, metaphor is so commonly used in biology that terms such as gene pool pass unrecognized as analogies, and have taken on concrete meaning for biological practitioners. Metaphor is organic to biological thinking, such that even mechanists, who might think of analogy as a less than fully scientific device, use it without thinking: human design of a machine is a metaphor that underpins the whole scheme of the biological mechanist. I will use analogies from ecological thermodynamics to transfer insights about ecological emergence to building design and construction.

Modern thermodynamics in biological systems is sufficiently new that I have no single basic reference to cite, as a summary encapsulation of this work [although Schneider and Kay (1994) is fairly complete for my purposes]. Indeed, the following summary will be new even to students of ecological and biological thermodynamics. Before I can apply biological thermodynamic analogies to building design and construction, I must first give a background on the biological and ecological side of the analogy. As an incentive to readers more interested in buildings than ecology and thermodynamics, I promise to return to buildings and apply the ideas on thermodynamics directly. We will see that systems appear to mature through an alternation of pulses of input followed by long periods during which resources only trickle. In buildings, there is the pulse of designing and constructing the structure, and then there is a long phase during which the occupants support the building by paying taxes and making repairs. The life cycle of buildings can be well understood if we first get a clear view of what happens in the life cycle of straightforwardly biological entities, as well as the cycles of human ecology and sociality. The link comes through a thermodynamic approach.

## Thermodynamics in biological and human organization

Thermodynamics is a young science, whose links to ecology were not really substantially developed until late in the twentieth century (Schneider and Kay 1994). Furthermore, thermodynamics is far from a closed book, with laws that everyone can follow. The new ideas contrast systems that are (1) at equilibrium with those that are (2) displaced from equilibrium, or even (3) held away from equilibrium.

The example of a ball in a cup is helpful here to explain the distinctions between equilibrium and the various other possibilities. When a ball comes to rest at the bottom of the cup, it is at equilibrium (case 1). If someone shakes the cup, so that the ball, from time to time, is temporarily moving around in the cup, we need a non-equilibrium model. Disturbance is used to explain the departure from time to time of the ball from the bottom of the cup (case 2). In the third condition (case 3), if someone rotates the cup so that the ball is held on the walls of the cup like motorcycles on a "wall of death," then the model we need is one of far-from-equilibrium. Here the ball is held away from equilibrium by persistent centrifugal force. To keep a far-from-equilibrium system going, there must be a constant input of energy or matter, as when an animal must eat to stay alive. A further complication is that some non-equilibrium or far-from-equilibrium systems behave like a ball jumping out of one cup into another. Once the ball is in the new cup, there is no need to shake the system to keep the ball away from the cup from which it has departed.

While systems that yield to an equilibrium model of thermodynamics are fairly well understood, even physicists studying physical systems see it yet as very early days in the development and application of far-from-equilibrium models (Roydon Fraser, personal communication, 2000). Non-equilibrium and far-from-equilibrium models break with simplistic early notions that systems are generally in balance and at rest. Far-from-equilibrium models offer a much more realistic conception of biological thermodynamics. The thermodynamics of far-from-equilibrium systems in ecology and biology provides a real opportunity to contribute significantly to both ecology and physics.

It may be early days in the thermodynamics of ecological systems, but already significant progress has been made. The earliest work in ecological energetics used the first law of thermodynamics, which states that energy must always be conserved. Ecosystem approaches use the first law when they assert that the difference between inputs and outputs must equal changes in the total of the system. The use of the first law in ecology invokes either an equilibrium model or, at most, an indication of balanced fluxes in homeostasis. The original work on the energy budgets of plants was carried out very early in the twentieth century (Transeau 1926), even before the term ecosystem had been coined (Tansley 1935). The first study on the energetics and mass balance in a natural ecological system was aquatic (Lindeman 1942). A full implementation of ecosystem energetics turning on the first law of thermodynamics had to wait until sufficient computational power became available in the 1970s.

Modern work on ecological energetics is based on the second law of thermodynamics. The familiar statement of this law is that a closed system will run down to equilibrium; it is the law that states that entropy necessarily increases. More helpful than the familiar statement is the interpretation of Schneider and Kay (1994), who point out that running down to equilibrium is not the most interesting aspect of the second law. Schneider and Kay point out that systems running down to disorder is a trivial matter compared with the elaborate things that happen and persist if the system is not allowed to run down. In that sense, their interpretation is the inverse of the normal expression of the second law.

The inverse statement is more encompassing, as it suggests that there is increased resistance as a system is pushed further away from equilibrium by energy or matter inputs. Emergent structure and organization in biology and ecology arise because the system exists away from thermodynamic equilibrium.

The familiar statement of increasing entropy over time is a model for steam engines, which are relatively closed systems. Life, on the other hand, is emphatically an open system that has material passing in and out all the time. The inputs of material and energy have the potential to push the system away from equilibrium, causing structure to emerge as the system balances itself around inputs and outputs. Quite the reverse of emphasizing a system running down, Schneider and Kay's (1994) statement indicates that continuing inputs, such as food, cause living systems to organize. Like living systems, buildings involve accumulating and organizing material and channeling energy. Therefore, the second law of thermodynamics, not the first law, has greater application in both biology and construction ecology. For open systems, such as organisms and buildings, the second law is particularly applicable.

In the above discussion, we have been dancing around the issue of emergence. As matter and energy enter, systems are pushed away from equilibrium. Because of the arrival of new matter or energy, gradients appear. The new material and energy in the system move naturally down gradients toward an equilibrium condition. Flux down these gradients causes the system to become more elaborate. The elaboration is often sudden, and is called emergence.

In emergence, new structures appear, but emergence is more importantly an elaboration of organization. It is organization that leads to emergent structure. The gradients cause directional movement that eventually comes to press against some sort of limit. The contrast of pressure against the limit and the resistance offered by the limit generates the new structure that emerges. Most parts of a system bear symmetric relationships to each other, but the parts of the system involved in the flux have an asymmetric relationship to the parts that limit the flux. Some parts come to control others, and this is the basis of organization. Emergence arises through new relationships of control and constraint. Note that in the reverse of emergence, as in the death of an organism, the body is still made of basically the same stuff, arranged in approximately the same manner. In death, the gradients and fluxes cease to exist. Thus, in death there is little change in structure as the symmetrical relationships persist, but there is a great deal of change in organization. Emergence is the creation of organization.

Allen *et al.* (1999) have worked through an energetic analysis of the emergence of higher levels of organization in ecological and human societal systems. They point out that the key to appraising increasing organization is not to confuse "complicated" systems with "complex" systems. As a system becomes more complicated, there is no increase in flux or organization. Complications are the mere addition of new parts that are in some way equivalent to the old parts. The equivalence between old and new parts in a complicated system exhibits symmetry. The accumulation of symmetric relationships is merely structural. By contrast, as systems become more complex, new gradients appear, along with their attendant asymmetries and propensity to organization.

Allen *et al.* (1999) start from Tainter's (1988) book *The Collapse of Complex Societies*. Tainter identifies societies as problem-solving systems, which elaborate in response to local problems (Figure 4.1). Allen *et al.* (1999) indicate that such elaboration of structure is only a matter of increasing complicatedness. There are more parts and more relationships as the system complicates, but there may not be more organization (Figure 4.2). Emergence

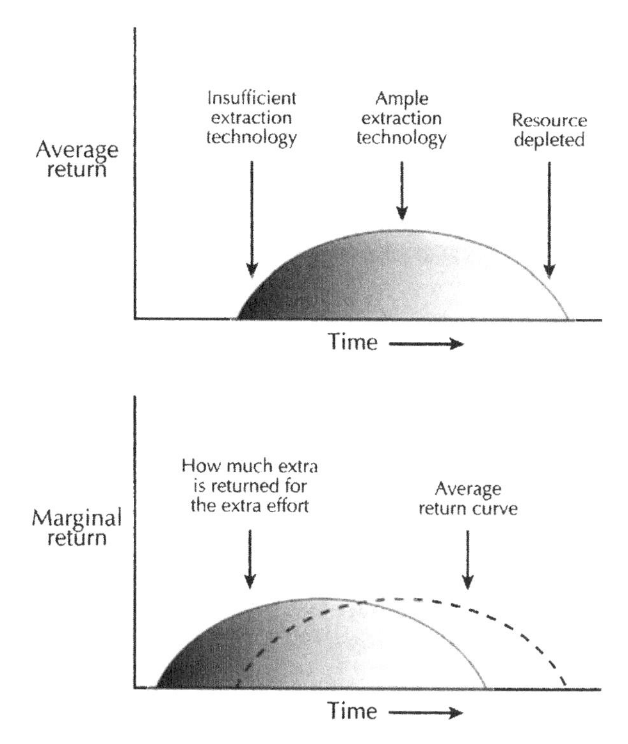


*Figure 4.1* As human economic systems use resources they suffer diminishing returns. The average return comes from improving extraction technology against a diminishing resource. More pressing is the marginal return, the amount of extra resource one obtains for the extra effort in extraction. After Allen *et al.* (1999). This figure is in the public domain.

of higher levels of social organization makes the system more complex, and that is a different matter than day-to-day problem solving at the lower level. Day-to-day problem solving only complicates societies, whereas the emergence of a new society on different terms is a true complexification.

As an example of the difference between complicated and complex structures, consider the change from the Industrial to the Information Age. Progress in the Industrial Age involved bigger, more complicated systems, but the elaborate relationships were all mechanical, and therefore amounted to more of the same. On the other hand, as society changes into the Information Age, there are new relationships that are electronic rather than mechanical. Electronic controls organize industry in a new way that is more complex, not just more complicated. Certainly, buildings were bigger toward the end of the Industrial Age, but they were designed and built as merely bigger and more complicated versions of Victorian bridges. There is an unbroken line between Brunel's bridges, the Eiffel Tower, and the girders in the Empire State Building, for the last is only bigger and more complicated. We are now well into the Information Age, and finally there is the emergence of a new approach to building. Indeed, the very existence of this publication and the meeting on which it is based is a sign of new patterns of organization in design, materials, and construction.

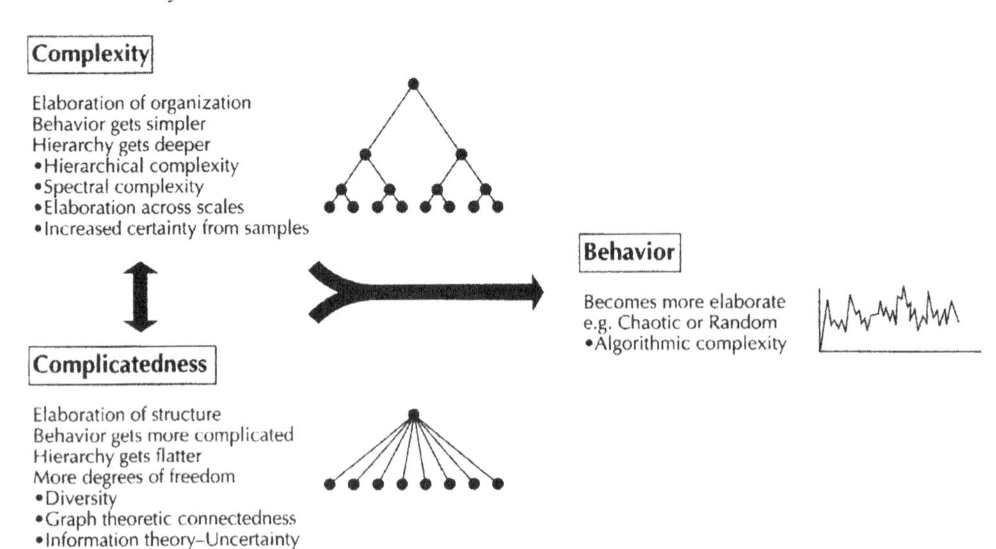


*Figure 4.2* As a system becomes more complicated, more parts are added or more connections develop between parts. When a system becomes more complex, new asymmetric relationships arise, such that some components at a higher level become the context of others at a lower level in the system. After Allen *et al.* (1999). This figure is in the public domain.

Building construction is becoming organized in a new way as it accommodates to larger societal concerns. A truly modern building fits into natural cycles better, as it uses materials that can more easily be recycled. It fits better into its setting and uses less energy than older buildings as it offers human comfort and services inside. Building in this new way is a complexification, not a complication. Counterintuitively, more complex systems are easier to control and exhibit simpler and more predictive behavior. That is the benefit of increased organization. Notice that the forethought that goes into choice of materials in modern buildings will pre-empt sudden unforeseen disposal problems when the building is torn down. When a more complex system is in place, things run more smoothly. The apparent clockwork simplicity of a smoothly running system comes from an expansion of system function, to include what before were societal externalities. By accommodating to the larger context through admitting to new limits and constraints, complications are avoided in the new more organized building cycle.

Having made the distinction between complicatedness and complexity, Allen *et al.* (1999) note that the cost of a system becoming complicated is incremental, and may be encumbered whether or not there is energy to pay for it: deficit spending for doing what must be done. Think of complicatedness increasing in a series of small jumps that are hard to reverse, like a ratchet. By contrast, complexity emerges with one great leap. The cost of elaborating organization through changes in complexity is a one-time encumbrance that must occur at the time of the change in complexity. The process of emergence that is essential to increasing complexity is driven through positive feedbacks that will not perform without the energy input to pay for them. Complication is a ratchet, whereas complexification is a process of feedback.

The benefits of complicatedness amount to the solution to local problems until new problems arise. The average return curve in economics comes from the easy solutions to problems that give the greatest return being implemented first. In the end, only the difficult solutions with minimal returns are available. The cost-to-benefit ratio in solving a class of problem rises over time, offering badly diminishing returns. By contrast, the benefit of increasing complexity is a one-time reset of the cost–benefit ratio of system complicatedness. Remember that an increase in complexity occurs only when there is an energy resource base that can pay for the new organization. The presence of that resource base moves the goalposts and resets the game (Figure 4.3) to a new league level. The entry to the new league is announced by a return of easy local solutions that lead to dramatic initial successes in acquiring resources. However, with time, the new league develops a new class of problems that pose increasing challenges to the emergent organization.

In complication, cost is incremental and keeps rising, whereas benefit falls as the easily captured, great benefits have already been taken. The cost of complexification is a one-time pulse of energy that is used to drive the feedbacks that reorganize the system. The benefit of complexity is a one-time resetting of the cost–benefit ratio of being complicated. The benefit comes from moving the goalposts to a position where there is a return of easy improvements.

A series of examples of societal complications followed by complexifications may serve here to bring the above discussion down to earth. The move from hunting and gathering to farming removes the constraints of nomadic life, mobility being more or less a requirement for hunters and gatherers. Agriculturalists have a different set of problems, and at first they are the easy problems of farming prime land. The first increases in complicatedness represent a process of maturation of the Neolithic Revolution (Figure

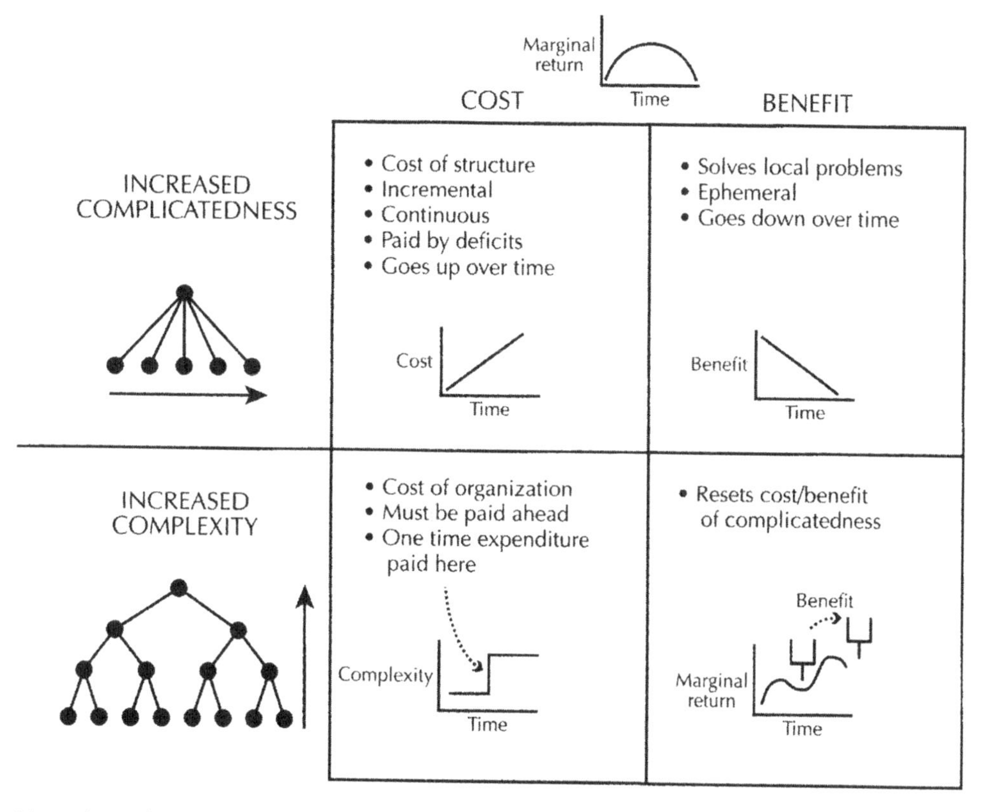


*Figure 4.3* The cost and benefit of complication versus complexification. After Allen *et al.* (2001). This figure is in the public domain.

4.4). But later, an increasing population requires more food production, and so the system must elaborate structure to solve harder problems so as to meet demand for food. The harder problems are solved only with diminishing returns on effort.

A new complexification occurs when the situation becomes desperate through crashing marginal returns on increasing complicatedness. It appears that high-gain, high-quality resources give a boost to complexity as a bridge between long phases of using a poor-quality but universal low-gain resource.

New complexity emerged with irrigation in the Bronze Age: imperialism. Imperialism appeared when agricultural surpluses and metal made standing armies possible for the first time. Imperialism itself has two phases of complication, with a complexification in between. The first is the high gain of mining accumulated capital, the surpluses of a neighbor. The new problems are easy. They change from, "How can we get more production out of our agricultural system?" to "How do I get the loot home?" But soon the pickings become poor, and further conquest with longer, more complicated lines of supply fails to pay for itself. Thus enters a new complexification. The shift is from invasion to administration. The theft becomes the relatively low gain of stealing the sunshine of the conquered people, by taking their land for agriculture or taxing production from the new lands. Not all empires reach the second phase, for example the Mongol Empire in the Middle East degenerated back to local raiding and herding once the plunder was gone. But other empires have moved on to the new low-gain phase. Caesar's conquest of Gaul follows this model, conquest and pillage (high gain) giving way to taxation and five hundred years of administration. After the complexification of imperial tax collection in an agrarian system comes the complexification to an industrial model. Turning to fossil fuel in the Industrial Revolution is clearly another example of a new resource that resets benefits.

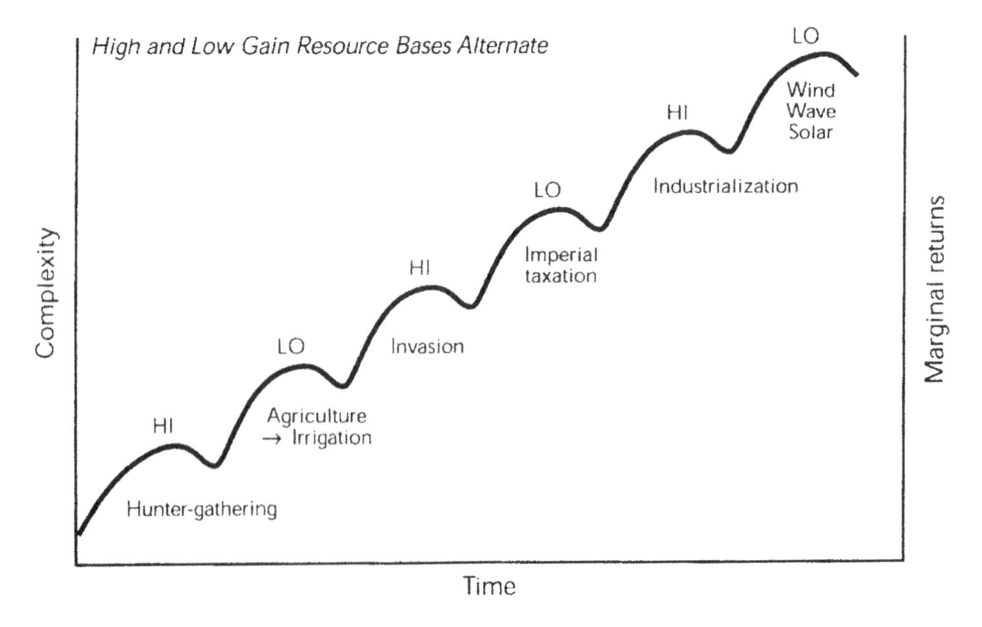


*Figure 4.4* Graph of increasing organization through a series of complexifications. Note the reference to the marginal return curves that work at a lower level of analysis. This figure does not resurrect the grand narrative of history leading to European dominance but rather it identifies a sequence of breakpoints that were far from inevitable. This figure is in the public domain.

In the grand sweep through history described above, there appears to be a general pattern of exploitation of a high-gain material alternating with a phase of use of some low-gain resource. This same pattern appears in some biological systems. There are species of ants in the genus *Atta* that farm fungi on various resources. They gather the resource and eat the fungi that grow on it. Primitive attoid species with this ecology use caterpillar droppings. If one is farming fungi, guano is a high-gain resource base. Species that use caterpillar droppings have small colonies, and are limited by the quantity of fuel available, much as the geographic limits to invading armies are fixed by the supply of gold and other high-quality loot. The most advanced ants use leaves as the substrate for their fungus farms. Leaves are everywhere but are not a concentrated resource for growing fungi. While guano ants exploit a high-grade resource, leaf-cutting ants are clearly exploiting a low-gain fuel. The switch from guano to leaves is the same pattern as the switch from the looting phase of imperial expansion to administration and taxation. The colonies of leaf-cutting ants are massive, and they construct elaborate roadways, even to the extent of over- and underpasses. Similarly, the Roman Empire was massive in its administrative structure. From this example it becomes clear that fundamentally new ways of organizing a system can only be achieved in a big push, be it the use of guano or the invasion of Gaul. In modern times, the big push was industrialization, and it looks as if it will lead to new low-gain resources: wave, wind, and solar power. Much as leaf-cutter ants could not have emerged without guano precursors, the projected hydrogen economy based on renewable resources could not be an option except in the aftermath of a carbon-based industrial system.

The high-gain resource offers raw high-quality energy for so long as it lasts, whereas the low-gain resource offers low-quality energy, but in functionally unlimited supply. The low-gain resource is often exploited in a diffuse manner, whereas the high-gain resource is often exploited in a point-focused manner. Capital is often rapidly accumulated under a high-gain regime, but the resource runs out fairly quickly, precipitating either collapse or a shift to some new low-gain resource (Figure 4.5). In the end, there is an even larger accumulation of capital under the low-gain resource, but it takes a long time.

Between low-gain phases, short pulses of use of high-gain, high-quality resource occur in a short cycle of complexification. The end of a low-gain phase is a system that is too big, with too much demand. The end of a high-gain cycle occurs as the quality resource is depleted. Hunter–gathering (crash and move), invasion, and fossil fuels all appear as high-gain short pulses. Agriculture, taxing agriculturalists, and green energy cycles (our next phase) are long. The sequence leads humans closer to the sun as the ultimate energy source.

## History, accidents, and positive feedbacks

If notions of emergence in complex systems are going to be important, then they must be able to predict outcomes. There is a problem here, in that emergence can be understood in hindsight but is difficult to predict before the event. There are general symptoms of a system with the potential to reorganize through emergence of a new level of organization. However, it is distinctly possible for a system that is a good candidate to show emergence not to undergo reorganization. Beyond that, if a system does reorganize, important features of the emergent system cannot be predicted.

If there is a gradient at which a system is far from equilibrium, then one can expect a new level of organization to emerge. Furthermore, the steeper the gradient and, thus,

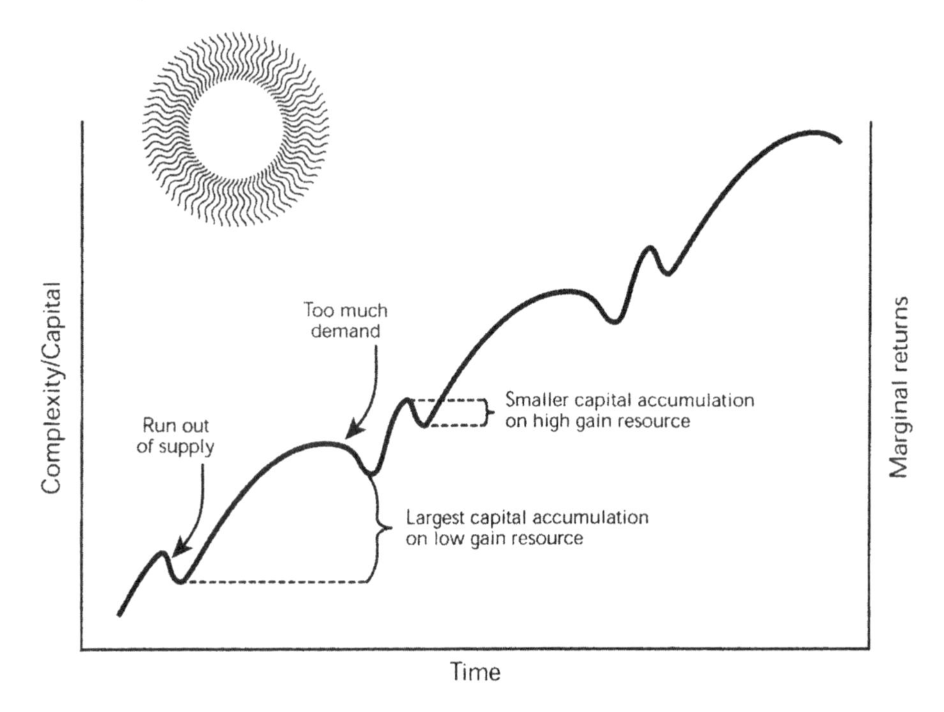


*Figure 4.5* As human systems increase in complexity, they spend a long time using a low-gain, poor-quality resource, which is more or less universal (e.g. sunlight in agriculture). After Allen *et al.* (2001). This figure is in the public domain.

the higher the top end of the gradient, the more likely emergence is to occur. Emergence starts, if it will, with some small irregularity that sets a positive feedback in motion. The positive feedback is driven by, and so consumes, the energy in the gradient. An example could be a supercooled liquid. The gradient in the supercooled liquid is between the normal freezing point of the liquid and the colder temperature at which the supercooled liquid sits. Were the liquid to start freezing, there would be a positive feedback whereby crystals would form upon each other. The process of crystal formation would generate heat, thus warming the liquid and lowering the gradient. The process of freezing would keep on going until the liquid was warmed to its normal freezing temperature and the gradient had been completely dissipated. The colder the supercooled liquid, the more likely it would be that the freezing process would start.

The limiting factor in starting the freezing is some small nucleus on which the first crystals could form. The colder the supercooled liquid, the smaller and less inviting the freezing nucleus needs to be to get freezing to start. As the world of most water is dirty and involves flawed surfaces, water usually starts to freeze at 0°C. If, however, the water is pure, and in a clean container with no flaws on its inner surface, then water can be supercooled to –7°C. At that temperature, the smallest flaw or impurity is enough to start the freezing process. So, the first uncertainty is predicting emergence of crystalline structure is how clean and unflawed is the container. That influences when or even whether the emergence will occur. In the case of water, the limiting factors are well known, but in social systems ready to undergo emergence the uncertainties are much greater.

Note also that no two snowflakes are alike. They are all six-sided and symmetrical, but each is different. The cause of the difference lies in the details of the particle that seeded the freezing process. This is the second source of uncertainty and encumbrance on prediction. If we are completely unable to predict the outcome of a well-understood process such as water freezing, what hope do we have for predicting the outcome of emergence in a social system?

The third reason that prediction is difficult in emerging systems is that the limits against which the positive feedbacks press to give the emergent structure are only potentialities until the emergence happens. In some systems with which we have experience, we are aware of the potential, and can predict accordingly. We are dealing with water and low temperatures and so expect ice and snow. If, however, one had never seen a solid before, let alone a snowflake, predicting snow would be impossible. With the emergence of snow, turbulence, or whirlpools we have no problem, but in social systems undergoing emergence, we have usually never before seen an example of the outcome. I challenge the reader to predict the outcome of the Internet as it connects businesses, homes, and government. We are aware that a process of emergence is under way, but we can only make guesses as to what are the ultimate limits.

Because the expression of positive feedback is steered by accidental details at the outset, history matters in the emergence of new organized structures. The history that matters is that which initiates positive feedback. The rest of the historical details disappear into inconsequence. The positive feedbacks keep driving until they come up against an unforeseen negative feedback that becomes a structure in the new system. The history of the accident that sets the positive feedback going offers much explanation (Allen *et al.* 1999). At a Nazi rally in 1923, a demonstrator was shot dead by police. The demonstrator standing next to the man that was shot could just as easily have been hit instead. That man was Adolf Hitler. The accident matters because of the political positive feedback that took him to the position of the Führer. The unforeseen limit at the end of the whole mess was an Allied victory and the start of the Cold War.

For centuries, science has sought to explain process or structure with mechanisms that are immune to the vagaries of history, and to the noise of chance. However, modern insights into general systems theory suggest that history cannot be ignored. The reason why mechanism is of limited utility in biology is that it does not invoke history. Mechanism has no history in itself, it just is. Mechanism is timeless. The appeal is that mechanism works every time it is applied appropriately. In mechanism, only history that is external to the mechanism applies, that is the actions and decisions of the watchmaker. Thus, mechanism invokes a design principle. Thomas Aquinas observed strict order in what he saw, and sought a design principle; he chose God. If there is no design principle, then mechanism and its predictions are moot. If the explanation for a situation depends only on history, accident, and unforeseen limitations, then mechanism is irrelevant.

Many situations in biology require explanations that are mixtures of design and accident, and what is design as opposed to accident is a matter of level of analysis. The process of natural selection is dynamic, and it depends on accidents. Evolution is not a sentient process that in any way knows where it is going. At one level of analysis, the structures that appear adapted are merely the frozen present in a continuing process. On the other hand, at another level of analysis, evolution produces structures that have been fitted into their environment. For the dynamics that operate in the context of the structural products of evolution, evolution is an externality, much as the watchmaker is external to the dynamic functioning of a watch. The watchmaker is the precursor to the watch and is the origin of its design. In this same way, evolution can be the design principle, in the context of whose products lower-level dynamics and local historical accident happens.

None of this means that evolution works in itself by sentient design, any more than there had to be design in the accident of upbringing that made the watchmaker choose chronometry as a profession.

A critical distinction between design, embodied in mechanisms, and system dynamics is the notion of rate dependence as opposed to rate independence. Dynamics is described as a series of rates of processes that are interrelated. Dynamics depends on rates, and a description of dynamics has to rely on rates for adequate description. Structure is an entirely different matter. While structures can change states at rates of behavior, their existence is not a matter of rate. The question "At what rate is a table a table?" makes no sense. The table just is. The same applies to policies and plans. While a policy might influence rates at which things happen, it does not apply at a rate. Policy is rate independent. Similarly, signs and symbols may describe rates, but the linguistic quality of a symbol is not rate dependent. Rate-dependent factors can be dissected, so that they can be seen in more detail as more local rate-dependent processes. The same cannot be said for structures, policies, or symbols. Dissection below the level of structures, policies, or symbols loses any and all of them, once they are taken apart. For instance, one can read a book quickly or slowly, depending on the rate at which one deals with its ideas and material. However, the meaning of the book is not in itself dependent on rate. The rate at which you read has no effect on the meaning. If one does not know the meaning of a printed word, analysis of the chemistry of the ink will not help. Thus, the notions of design and process are crucially separated from each other by design being rate independent and process and behavior being rate dependent. I will use this distinction to cleave building design from the dynamic aspects of construction, although they come together to create the final structure.

## Applying biological thermodynamics to buildings

After a long discourse on biological and ecological thermodynamics, I can now fulfill my promise to apply these ideas to building construction and the building life cycle. An indication that it is early days in analysis of building thermodynamics is that, like the early days of ecosystem analysis, much of thermodynamics traditionally applied to building turns on the first law of thermodynamics. Using first law reasoning, material and energy are fed into the building, and the aim is to minimize the losses to the environment. In this way, fewer energy resources need to be put into the building to keep the human environment inside the building in a desirable state. Certainly, we should continue to identify safe, durable, and recyclable materials to insulate buildings, but there is much more we can do to take advantage of thermodynamic insights in making buildings effective. The cutting edge of analysis of ecosystems has moved beyond the first law of thermodynamics. Similarly, the process of designing buildings and their use can be facilitated by taking advantage of what we know about the implications of the second law of thermodynamics. One can talk of life as a system with materials entering and leaving, but the discussion that stops there is rudimentary. It is probably more interesting to look at something akin to the physiology of a building. In the present volume, Jürgen Bisch (Chapter 11) refers to "natural metabolism." A view of buildings that uses the second law of thermodynamics is probably superior to one that focuses on the first law.

Slavish adherence to application of the first law is responsible for sick buildings. By conserving energy through sealing buildings tight, insufficient replacement of air leads to accumulation of volatile chemicals from insulation, adhesives, and flooring. Efforts to minimize water consumption through low-flush toilets can deny the effluent stream enough water for it to function properly. Drains can block. Buildings can be viewed as living systems, an extension of the humans that occupy them; they are our thick skins in adverse environments. Process and flux are a natural part of building function, and squeezing processes of input and output to a minimum denies the healthy exchange that is the normal part of the functioning of a biological system.

The second law of thermodynamics refers to exchange and flux, and so offers a more dynamic model for buildings. Organization in biology comes from systems pushed away from equilibrium, but under the second law they organize to offer resistance to further displacement up the gradient. Organization in the functioning of buildings comes not from the static conservation of materials put inside them, but rather from the natural processes. By harnessing the directionality of fluxes, buildings can be better organized. Rather than stifle the flow of energy out of a building, we can benefit from learning to ride the matter energy flux. Elsewhere in this volume, Jürgen Bisch's "ceiling activation" system uses the thermal mass of ceiling and floors to drive work space air temperature either up or down as needed. This is an example of the use of the second, not the first, law of thermodynamics.

In biological systems, the degradation of energy inputs is far more elaborate than in simple physical systems. With energy inputs or concentration of materials, both biological and physical systems move material and energy down the gradient toward regions of lower concentration. The difference is that, unlike simple physical systems, biological systems far from equilibrium do not degrade high-quality energy to low-quality energy in a small number of big steps. Rather they degrade energy in a large number of small steps. If there is free energy left after some work has been done, then biological systems will get more work out of that energy, gradually lowering the quality of the energy so as to extract as much work as possible. In fact, the structure of complex far-from-equilibrium systems often reflects that process of eking out more work for the passage of material or energy down the gradient. Even in physical systems far from equilibrium one can see structures such as whirlpools that use energy to maintain elaborate structure. Far-from-equilibrium systems divide energy into that which degraded so as to organize the degradation of the rest of the energy and that whose degradation is organized. In complex systems, energy is degraded at several different levels.

When there is free energy left as a building functions, then that energy is available for more work. In this spirit there are already buildings that are heated endogenously by the output of heat from computers and typewriters inside the building. However, there are other buildings, such as the United Nations building in New York, that function inefficiently because energy is wasted or misused. The large east and west faces of that building can mean that air conditioning is needed on the east side of the building because of passive heating from the morning sun, while the west side rooms could benefit from heating. A solution with a biological touch would be to use the excess heat on the east side of the building to heat the west side of the building, and use the cool air on the shaded side of the building to cool the warm side. Unfortunately, there is not enough free energy to move, hot to cool, passively, and actively pumping it is not cost effective for the UN building. In biological systems, there are often elaborate countercurrent systems, as when the warm blood leaving the core of a wolf to the feet is run beside cold blood returning from the wolf paw. By the time the blood reaches the foot it is already cold, and so loses less heat to the ice and snow. The blood in the paw has already donated its heat to warm the returning blood. The transfer of heat is passive, and comes about because of carefully organized juxtapositioning of veins against arteries. It is hard to retrofit such elaborations into an existing building.

As the proper function of a building involves flux and energy degradation, putting minimization of energy dissipation at the top of the list of priorities is at odds with full functionality of buildings. A better priority is the efficient dissipation of energy in support of building function. In simple physical systems, energy is degraded in a manner that is fairly incidental. By contrast, the structures associated with the pathways of energy dissipation in living systems are in no way incidental and are the very substance of a biological system. Organizing a building to function in a more biological fashion means that the elaboration inside cannot occur happenstance or as a short-term convenience, and must be carefully designed to take advantage of as many passive processes as possible.

## The design and energetics of the building

Like biological systems, buildings emerge, grow to maturity, and then degenerate. This analogy of buildings as organisms invites dissection of the concepts of inception, growth, maturity, and degeneration.

Design is akin to the genetics of the incipient building. The construction phase is the growing and maturing of the building. The mature building then lives its life. Its services pay for repairs and alterations, such that the building is self-supporting. Eventually the building senesces. The biological basis of buildings involves an association with the people who are part of each phase in the building cycle. It is helpful to view of the building as larger than its simple physical being, so as to include the people involved in the building and to think of the building as a system.

The inception of a building is akin to the biological conception of an organism. In conception, a unit that is principally informational, sperm, is united with another unit that is principally energetic substance, the egg. A sperm has some energetic substance and an egg some genetic substance, so the distinction is a simplification rather than an absolute dichotomy. In building terms, the sperm is the person who conceives of the building, perhaps as a matter of speculation, a dream house for retirement, or some other conception. The egg also contains genetic information, but it is distinct from the sperm in its store of resources to execute the first stages of growth. Its food store is equivalent to the capital that will pay for construction and the people from whom the capital emanates. Bankers do not usually dictate the details of a building, but they do influence those who conceive of buildings. As often as not, the banker's influence is executed through the self-sanctioning of the borrower, so as to be able to present a fundable package. In denying funding for a poorly designed project, bankers will influence the next attempt at design, but only as a very general constraint.

At the point of union of egg and sperm, or of financier and owner, the process of growth takes on a life of its own. The information and the growth potential unite to begin the process of growth. The prevailing paradigm of biology since Darwin has had three players. One is a genetic agent (even though Darwin's model of genetics was pre-Mendelian). The genetics of the organism interacts with the second player, the environment. The interaction of the first two players generates the third player, the phenotype, or a physical form of the organism. While the building concept may be firm, the emergent building will be different from the plan in the final analysis. Externalities will come to influence the growth process, and will leave their marks on the mature structure. Buildings are emergent structures that come into existence through the dissipation of an energy gradient. The owner, with finance in hand, purposefully moves quickly to dissipate the capital. The banker may accumulate money, but the borrower actively dissipates it. As the energy is dissipated, the pathways that form manifest themselves as a far-from-equilibrium structure.

There are several analogies between organisms and building here. In organisms, the structure needs the dissipation to continue, otherwise it dies. Buildings have a greater potentiality than organisms to persist, even if the energy dissipation stops. From time to time, glacial Lake Duluth would lose its ice dam, and burst down the St Croix River channel. So strong were the whirlpools that rocks caught in them would drill holes in the underlying rock. While whirlpools were energy-dissipating structures, and the floods that fed them were gone many millennia ago, the holes drilled in the rock persist today as a record of the whirlpool action. In this way, a building can be considered the mark left behind after the energy-dissipative building process and, like the holes in the rock at St Croix Falls, they persist without energy dissipation to support them. On the other hand, it is possible to consider the building as analogous to an organism that has grown to maturity and now needs much less energy to live through its adult phase. Indeed, a building does need to have its taxes paid, and to have some monies spent on repairs. This is akin to the adult organism , which needs fewer resources to persist than were required to grow to maturity. Teenage boys consume much more food than their overweight fathers. Buildings need fewer resources committed to them once they are built.

In the process of growth in organisms and buildings, the direction and position of the pathways become fixed by the accidental details that pertain at the outset. The process of growth in the emergence of energy-dissipative structures manifests a memory of how history has steered. In evolution, the end product appears to be exquisitely designed, but it is an illusion. Most of what appears to be design is the result of accidents that leave their mark. These characteristics of the evolving system become the context of subsequent, more actively adaptive selection. Most of the specifics of the parts of biological mechanism arose through disconnected, incidental happenings, with no design to them at all. So is it with buildings: the early events in the history of the building and their physical consequences become the template that determines the large form that the building takes.

Many examples show how historical accidents have controlled the expression of design. The house of the president of a leading African university, built in the late 1960s, is open at one side, designed to take advantage of a glorious vista. Unfortunately, the view from this side is of the houses and apartment buildings of the rest of the campus. In contrast, the other side of the house faces a magnificent view of treetops and mountains. At some point, either in design or while laying the foundations, the building orientation was reversed from the original "genetic" conception. Realizing the situation as the house was nearing completion, the site architect was powerless to do anything about it, and so the house "phenotype" looks in the wrong direction. Usually, the final consequences of happenstance in early decisions in design or construction are not as transparent as in this case, but almost all buildings will have a form that is the result of accidental decisions as well as design.

Pattee (1978) identifies two sets of limits on what we see, and they are be used to identify the separate roles that design and process play. First there are rules, and these correspond to design. Rules comes from the observer's decisions as to how to observe and what is significant. Rules are linguistic, local, arbitrary, and rate independent. The second set of limitations as to what the observer will see, Pattee (1978) calls laws. These are aspects of observation that are above and beyond the observer's choices. In a Kantian dualism the laws come from the external world, the observed. A perspective with a softer realism would say laws come from "the other," beyond the observer's volition. Laws are universal (given the universe of discourse), inexorable, and independent of structure and significance. Laws capture dynamics, in that they are rate dependent. Design pertains to rules, whereas the building process pertains to laws.

Both laws and rules offer limits in building construction, but their limitations are very different. Design, being rule-like, can be changed more or less at will, and so is only weakly constrained. Process, being rate dependent, is law-like, and so encounters limits that cannot be changed. The number of bricks laid in an hour by hand, like many building activities, is always fairly close to the limit, as costs and deadlines are always important. By contrast, an architect rarely works for long close to a limit of design. Design, regardless of what it encounters historically, remains the same, and has the same tendencies in bringing about the final condition. In contrast, process is rate dependent and sensitive to history. Process in building construction is usually tightly constrained, and when some historical accident occurs the construction process has to live with it.

While the design is more or less fixed once the process of building starts, there is history in processes of construction. The distinction here is between organization that comes from ideas in design, as opposed to self-organization, which comes from the unfolding of the dynamic emergence of far-from-equilibrium structure. The patterns in the design could have been altered at the last minute, removing earlier rejected patterns. However, it is much more difficult to change the patterns that emerge as the material building is constructed. Once a brick is in place, the builder is more or less stuck with it. Of course, grossly undesirable patterns will be corrected under quality control, but at such expense that it happens rarely and is done with reluctance. Many more changes are made during car design than as a result of manufacturers' recalls. The history in emergence is intrinsic to process in construction. On the other hand, any history associated with design is external to the design, and comes from some outside process.

There is an asymmetry between design and the processes of construction. The limits to construction are the material possibilities in the world. The limits of design are a further set of limitations coming from what the designer allows. Design offers the allowable subset of emergent possibilities. If levitated roofs cannot emerge under construction, there is no point in designing them. This is akin to the relationship between biology and physics. Biology must be bound within the possibilities that physics allows. However, a perfect knowledge of physics does not encompass biology, for physics does not allow a prediction that terrestrial quadrupeds with only one bone in the lower jaw, mammals, all have five digits (Polanyi 1968). Biological systems cannot be properly investigated as merely complicated physics and chemistry. This is because what makes biology biological are the limits that biology puts on physical possibilities. Design cannot violate construction possibilities, but the full set of construction possibilities does not inform design.

There is a lot to be said for being explicit about separating the rate-independent aspect of creating a building from those aspects that are rate dependent. Let the architect design a set of constraints for the processes whose pathways become fixed later. The functions of a building that meet the direct needs of the prospective occupants can be put in first. The design can be made to meet the need for mood setting, human convenience, privacy, or congregation. Once these linguistic assertions are in place, a computer design system is instructed to perform a least path analysis that puts in the ductwork going through the least number of walls. In a factory floor where utility is at a premium, the reverse process can be employed. The most dynamically efficient relative placement of machines can be prescribed first, and then the space for humans and their needs can be designed afterwards. Either way round, the aesthetic design particulars are well placed separate from mechanical functionality by putting one or the other first as a constraint on the other half of the process of total building design.

## Energy in the building cycle

As I discussed the thermodynamics of emergent human systems, I made the distinction between high- and low-gain energy systems. The limit on a high-gain cycle is supply, whereas the limit on a low-gain cycle is demand outstripping supply. Emergence through access to high-gain quality energy has a different character and outcome from that in a low-gain cycle. High-gain growth is sudden, because the high quality of the resource forces the change in the species or human society. The emergence of large-scale human social structure under a low-gain resource is slower because of the poor-quality energy source. Emergence through a high-gain resource produces a system that is relatively short-lived, because the system quickly becomes overly complicated in its attempts to extract the resource from an ever-decreasing source. Industrial use of fossil fuel appears to be a case in point. Although contemporary humans may view carbon-based industry as long term, the Industrial Revolution was sudden, and we are coming to the end of ready carbon-based energy. By contrast, low-gain sources of energy, such as the photosynthetic basis of agriculture, are renewable, and do not run out, although demand may come to outstrip renewable supply. The Industrial Age is passing in a fraction of the time that it took for agriculture to develop fully. This distinction between high-gain speed and low-gain persistence can be helpfully applied to buildings.

The high-gain, non-renewable resource in building construction is the capital that is available to build the building. At some level, the resources have to be available to drive the processes of emergence as construction occurs. In the end, the building process stops when the resource is depleted. Costing a building is important, because there are very undesirable consequences should the money to pay for building run out before a useful structure is completed. Toward the end of cost overruns, less and less satisfactory solutions are found to the problems of finishing the building.

Human societies can be a useful lens through which to view the building cycle. In human societies, each new level of emergence appears to depend on a healthy alternation of shifts between high-gain and low-gain resource cycles. High gain is the important boost that leads to a level of organization that can support exploitation of a new low-gain material. For instance, the coming transition to low-gain universal power is a case in point. Solar, wind, and wave power are all low-gain energy sources. For them to generate electricity that can be used to dissociate hydrogen from water, some sort of high-gain industrial society precursor, such as oil or coal, is crucial. On the other hand, the accumulated capital of a low-gain cycle gives the capacity to start a new round of high-gain exploitation. For instance, looting a neighbor's capital in the high-gain phase of imperialism depends on the sufficient accumulation of agricultural capital to get the high-gain process started. Only with irrigated agriculture did a sufficient surplus of food make wholesale warfare possible. In buildings, it takes the consumption of a mass of capital (high gain) to create a building.

In the case of a building, it is only after construction is complete, and the capital is spent, that people come to live in it. Buildings are the context of their working parts, particularly the people who live or work in them. These people generate materials and energy more effectively if comfortable and fulfilled. In this way, a building, with its parts and occupants, is a living thing. After construction, the building still needs energy to make it function, and the resources to pay for that energy come from the capacity of the building to service its occupants. Thus serviced, the occupants then spend low-gain resources, their pay packages, to support building function.

Tainter (1988) points to the diminishing returns through the life of a social emergent structure, and attributes ultimate decline to diminished returns. At the end of the building's high-gain phase of construction, one could always put more bells and whistles on the project, but with less return on investment. In the end, the building phase stops as capital is all spent or further investment is seen as not worthwhile. In the low-gain phase, the building is supported by income associated with occupation of the building. Repair and small remodeling projects fidget around the building, until it becomes largely a set of patches superimposed on the original structure. The end of this low-gain phase occurs when maintenance is viewed as no longer worth the effort. The structure serves its occupants in a way that is not commensurate with the cost of upkeep. The building is abandoned or demolished. Properly functioning buildings are kept up, and they create an environment for further development in a neighborhood. This can generate a desirable setting that allows a new high-gain phase of building anew. Gentrification can be a new high-gain organization that generates new buildings from old ones. Buildings that are poorly put in their context are not used properly, and so have a short life. In the USA many low-income, high-rise apartments (often dismissively referred to as the "Projects") are being torn down for that reason. Not all high-gain organizations move on to the next long-term, low-gain phase.

## Conclusion

In this chapter I have laid out some of the newest ideas in biological thermodynamics. These apply easily to the design, construction, and life of buildings. Biology has benefited from borrowing ideas from other disciplines for biological metaphors. Construction ecology may not wish to buy into biology wholesale, but there are insights to be gained by those associated with construction casting a critical eye toward modern biological thinking.

C.S. Holling comes from a background in prey–predator systems, a complete zoologist. He has devised a summary (Holling 1986) for intermittent biological happenings (Figure 4.6) that is consonant with the work of Tainter (1988) on the ecological economics of societies. It also fits with H.T. Odum's (1983) powerfully generalized ideas on energy in pulses in ecological systems. Holling (1986) identifies two phases of growth. One is incremental and accumulates a highly organized system that sits on top of a huge capital. The other growth phase is rapid, and depends on spending the accumulated capital of the prior incremental phase. The incremental phase maps onto my low-gain phase, whereas the capital consumption phase maps onto my high-gain phase. Holling (1986) himself maps his two ecological growth phases onto politics, technology, and even psychological phases of logic and feeling. The present chapter provides an entry into that literature for those wishing to map the process of building construction and use onto other familiar processes so as to gain a deeper insight into construction ecology.

The ideas presented here rest on various dualities. It is worth casting building construction and use as sets of periodic behavior. The periods appear dualistic. One period is the high-gain phase, in which organization comes from the high quality of the inputs,

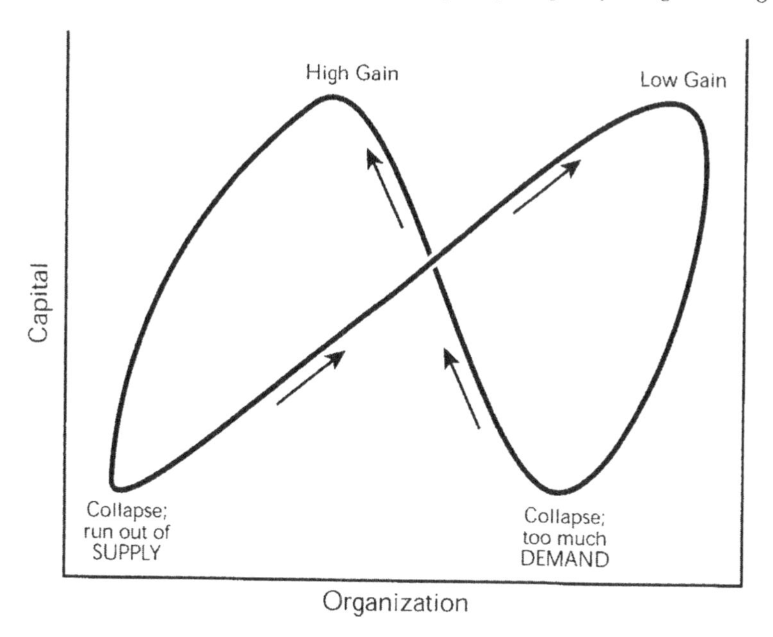


*Figure 4.6* Holling's (1986) lazy eight diagram has been insightful for ecologists, although the nature of the axes on the graph are equivocal. This is solved to an extent by representing the two ascendant paths as high- and low-gain process of emergence. The phases of building construction and use map onto this figure.

namely mobile, liquid capital. The other period is the low-gain phase, in which organization is manifested in the accumulation of capital through continual effort. Another duality I have used in this discussion is between the sentient design process as opposed to the real-time movement of materials. Dualistic approaches emerged first in this century in physics, and have moved to biology through general systems theory and ecological thermodynamics. It is now time to move dualism into human social systems, and considering buildings may be exactly the device that is needed.

Concepts in building design, construction, and use can be informed through analogies to systems cast conventionally as living. On the other hand, buildings represent tangible things that play various roles in social settings. Much as organisms are particularly comfortable and useful in biology because of their tangibility (Allen and Hoekstra 1992), buildings may have a similar role to play in concretizing abstractions about the human social condition. Thus, buildings may be a particularly useful intellectual device for furthering our understanding of energetics and the interface between biology, sociology, and economics.

## References

Allen, T.F.H., Tainter, J.A. and Hoekstra,. T.W. 1999. Supply-side sustainability. *Systems Research and Behavioral Science* 51: 475–85.

Allen, T.H.F., Tainter, J.A., Pines, J.C. and Hoekstra, H.W. (2001) Dragnet ecology, "just the facts Ma'am': the privilege of science in a post-modern world. *Bioscience* 51: 475–85.

Allen, T. F. H. and T. W. Hoekstra. 1992. *Toward a Unified Ecology.* New York: Columbia University Press.

Holling, C.S. 1986. The resilience of terrestrial ecosystems: local surprise and global change. In *Sustainable Development of the Biosphere*. Clark, W.C. and Munn, R.E. (eds). Cambridge: Cambridge University Press, pp. 292–316.

Lindeman, R.L. 1942. The trophic–dynamic aspect of ecology. *Ecology* 23: 399–418.

Lorenz, C.Z. 1974. Analogy as a source of knowledge. *Science* 185: 229–234.

Odum, H.T. 1983. *Systems Ecology.* New York: John Wiley.

Pattee, H.H. 1978. The complementarity principle in biological and social structures. *Journal of Social Biological Structures* 1: 191–200.

Polyani, M. 1968. Life's irreducible structure. *Science* 160: 1308–12.

Schneider, E.D. and Kay, J.J. 1994. Life as a manifestation of the second law of thermodynamics. *Mathematical Computer Modelling* 19: 25–48

Tainter J. 1988. *The Collapse of Complex Societies.* Cambridge: Cambridge University Press.

Tansley A.G. 1935. On the use and abuse of vegetational concepts and terms. *Ecology* 16: 284–307.

Transeau, E.N. 1926. The accumulation of energy by plants. *Ohio Journal of Science* 26: 1–10

Wicken, J.S. 1987. *Evolution, Thermodynamics, and Information: Extending the Darwinian Program.* New York: Oxford University Press.

# 5 Using ecological dynamics to move toward an adaptive architecture

*Garry Peterson*

The Earth's biosphere provides the ecological services that underpin human life. However, as the scope and intensity of human domination over the biosphere have expanded, basic attributes of the biosphere such as the physical movement of materials, numbers and distribution of species, and the arrangement of Earth's landscape are increasingly controlled by human action. In some areas, such as the center of large cities, human transformation is nearly absolute, whereas in other places, such as remote parks, human influence is felt chiefly through the alteration of global cycles. However, no place is free from multiple, interacting human impacts. The reckless manner in which this domination has taken place has led to sustainable development movements that attempt to harmonize human actions and desires with ecological reality. One of these efforts is taking place in the construction industry.

The construction industry uses a significant proportion of the material and energy consumed in the USA (see Chapter 1). Furthermore, buildings have a large impact on ecological processes, both locally and regionally. There appear to be many opportunities to improve the relationship between the built and natural environment while also increasing the usability of buildings. In this chapter, I suggest how construction ecology could be developed based upon concepts from ecology.

In this chapter, I view Nature's functioning from the perspective of systems ecology, which is one of many subdisciplines of ecology. I describe ecosystems using terms such as processes, pattern, organization, and scale.

Ecological processes describe mechanisms that transform ecosystems over time. They include processes such as succession, nutrient cycling, and seed dispersal. Ecological pattern refers to the relative distribution of organisms and physical environment, for example the types of trees in a forest and their spatial arrangement. Ecological organization refers to the connections among ecological pattern and process. A tropical forest represents one type of ecological organization, while Arctic tundra represents another, very different type of ecological organization. Scale is used to describe the domain of space and time over which a process, such as seed dispersal, or organism, such as a mouse, operates. For example, the domain of space and time over which a mouse forages, lives, and moves is smaller and faster than the domain in which people live.

In this chapter, I use these concepts to link ecological theory to construction ecology. The chapter begins by introducing several theories of ecological dynamics. In particular, I discuss ecological resilience, ecological change, ecological scale, and the methods of adaptive ecological management. I continue by applying these theories to the analysis of the ecology of building. I use a theory of ecological dynamics to divide construction into four distinct phases and suggest what construction ecology should focus on within each phase. I argue that adaptive methods used in ecological management could help construction ecology learn from its experiences. I conclude with a consideration of how ecological scaling principles may relate to the built environment.

## Ecological dynamics

Ecological processes occur across a wide range of scales, ranging from the microscopic to the global. The interactions among these processes are both stabilizing negative feedbacks and excitatory positive feedbacks. The tension between these stabilizing and destabilizing forces causes ecosystems to exhibit complex, shifting patterns of ecological organization. These shifts mean that ecological dynamics are difficult to understand and predict. In the following sections I discuss four theoretical constructs that are used to understand ecological structure and dynamics: scale, resilience, the adaptive cycle, and cross-scale dynamics.

### *Scale*

Ecological organization emerges from the interaction of structures and processes operating at different scales. A scale is a range of spatial and temporal frequencies. This range of frequencies is defined by resolution, below which faster and smaller frequencies are noise and above which slower and larger frequencies are background. Processes that operate at the same scale interact strongly with each other, but the organization and context of these interactions are determined by the cross-scale organization of an ecosystem.

Moving across scale reveals that patterns are nested inside other patterns in several different ways. I propose that it is useful to distinguish between spatial organization, temporal organization, and self-organization.

Any area can be considered as a part of a spatial hierarchy. In a spatial hierarchy, larger spatial scales incorporate smaller objects. In the boreal forest, for example, the organization of vegetation changes across scales. At the scale of centimeters, one can resolve pine needles, the crowns of individual trees are visible over meters, forest patches become clear at several meters, stands of even-aged trees appear at tens of meters, and areas of forest become apparent at the scale of kilometers.

Similarly, temporal organization across scales can be analyzed by observing the dynamics of a fixed area over different time periods. Time can be considered as a nested hierarchy. Microseconds occur within seconds, which occur within hours, which occur within years, which occur within millennia, and so on. Ecologists have only recently begun to integrate the paleoecological with historical and experimental time scales to try and understand the relationship between slow processes such as post-glacial vegetation spread and fast processes such as seed dispersal. For example, in a boreal forest the mix of species making up the forest has changed over centuries, outbreaks of the defoliating insect spruce budworm occur once or twice a century, fires occur more frequently, animals forage food moving across the landscape over days or weeks, and the weather changes over periods of hours.

While an observer can examine space during a fixed time period or a fixed space over different time periods, most processes interact across space and time. A landscape records past events, such as floods, fires, and past agricultural practices, in its patterns. These patterns continue to shape ecological processes. Current processes can erode or erase the patterns produced by events in the past, or interactions between pattern and process can result in mutual reinforcement, whereby pattern and process sustain one another. This mutual reinforcement between pattern and process leads to self-organized hierarchies that are more complex than either spatial or temporal hierarchies because they are nested in both space and time.

In a self-organized hierarchy, the spatial–temporal interaction of structures and processes produces an emergent pattern at a larger and slower scale than the scale of the processes and structures themselves. These self-organized patterns and processes interact with larger and slower processes to organize still larger and slower sets of pattern and process. In the boreal forest, for example, biophysical processes that control plant morphology and function dominate small and fast scales. At the larger and slower scale of patch dynamics, interspecific plant competition for nutrients, light, and water interacts with climate to influence local species composition and regeneration. At a still larger scale of stands in a forest, mesoscale processes of fire, storm, insect outbreak, and large mammal herbivory determine structure and successional dynamics from tens of meters to kilometers and years to decades. At the largest landscape scales, geomorphological and biogeographical processes alter structure and dynamics over hundreds of kilometers and millennia. These processes organize the landscape, but then are constrained by the pattern that they have organized. These interactions appear to have organized a hierarchical structure in space and time.

Herbivore foraging, mesoscale disturbances, and atmospheric variation influence vegetation structures. Contagious mesoscale disturbance processes provide a linkage between macroscale atmospheric processes and microscale landscape processes. Scales at which different species, such as deer mouse, beaver and moose, choose food items, occupy a home range, and disperse to locate suitable home ranges vary with their body size (Figure 5.1).

Distinguishing between spatial, temporal, and self-organized hierarchies clarifies the varieties of cross-scale interaction. While the strongest interactions lead to self-organization, linkages across time and space determine the context in which the tightly coupled self-organized dynamics occurs.

### *Resilience*

Ecological resilience is a measure of the amount of change or disruption that causes an ecosystem to switch from being maintained by one set of mutually reinforcing processes and structures to an alternative set (Holling 1973). For example, a clear, nutrient-limited lake may shift to being a murky, nutrient-rich lake as a result of disturbances as diverse as a flood, a large pulse of nutrients, an increase in disturbance in the lake's sediment, or a decline in the number of predators (Carpenter *et al.* 1999). In this case, the amount of change a lake can withstand before changing determines its resilience.

Ecosystems are resilient when ecological interactions reinforce one another and dampen disruptions. Such situations may arise as a result of compensation when a species with an ecological function similar to another species increases in abundance as the other declines, or as one species reduces the impact of a disruption on other species. Resilience emerges from both cross-scale and within-scale interactions (Peterson *et al.* 1998).

Cross-scale resilience is produced by the replication of ecological function at different scales. Ecological disruption usually occurs across a limited range of scales, allowing ecological functions that operate at other, undisturbed scales to persist. Ecological functions that are replicated across a range of scales can withstand a variety of disturbances. For

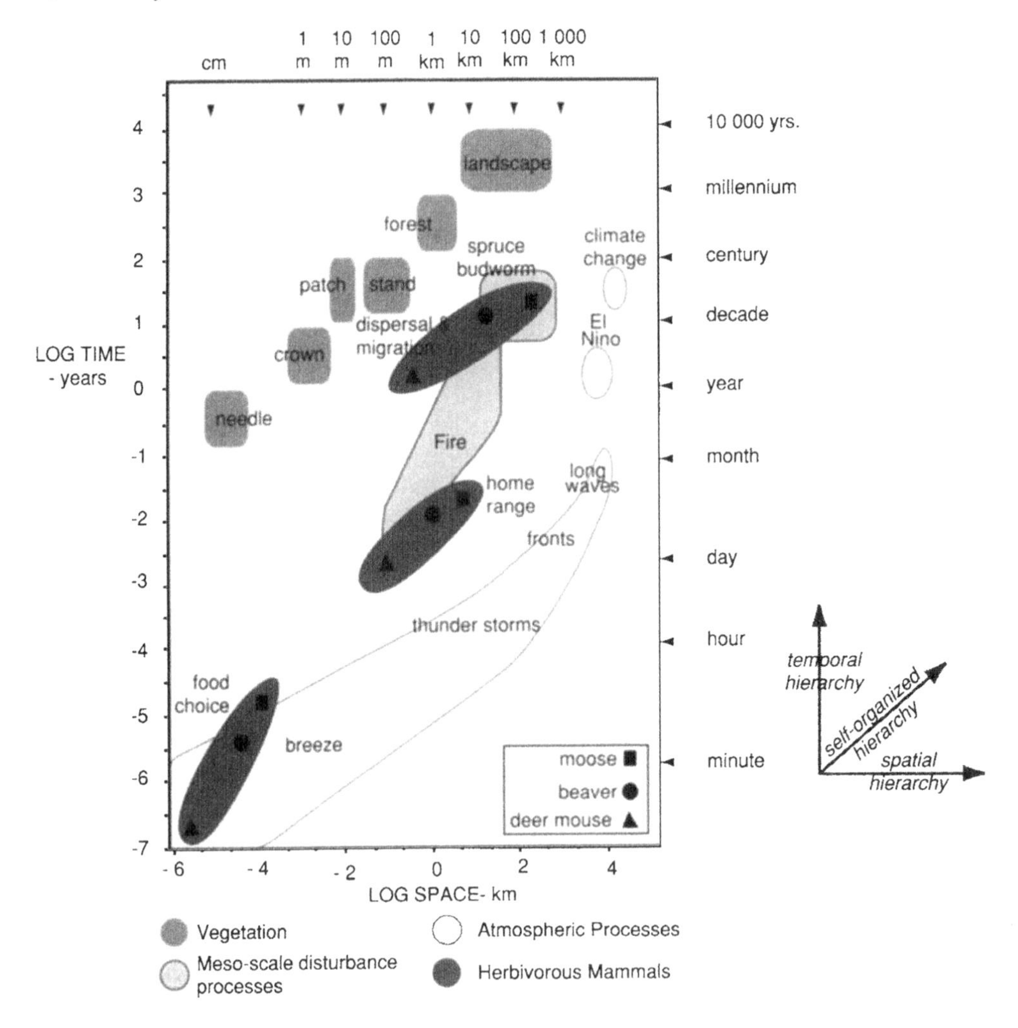


*Figure 5.1* Time and space scales of the boreal forest. Adapted from Peterson *et al.* (1998).

example, if a number of differently sized species disperse the same seeds, the movements of the species will disperse seeds at a variety of scales, allowing a plant population to persist despite disturbances. Small-scale dispersal allows patches to spread, while large-scale dispersal bypasses poor local habitat to create new plant populations (Peterson *et al.* 1998).

Within-scale resilience complements cross-scale resilience. Within-scale resilience is produced by compensating overlap of ecological function between similar processes that occur at the same scales. Variation in environmental sensitivities among species that perform similar functions allows functional compensation to occur. In Australia, for example, when the dominant species of grasses on ungrazed rangeland were suppressed by grazing, they were replaced by functionally similar species that were less abundant prior to grazing (Walker *et al.* 1999).

Ecological change or management that reduces the number of ecological functions and reduces ecological functions at many scales reduces ecological resilience. The concept of resilience allows ecologists or managers to focus upon the likelihood of transitions among different sets of organizing processes and structures.

### *Adaptive cycle*

Holling (1986) has developed a general model of ecological change that proposes that the internal dynamics of ecosystems cycle through four phases: rapid growth, conservation, collapse, and reorganization. Traditionally, ecologists have focussed upon ecological growth and conservation. Explicitly including the processes of destruction and reorganization provides a more complete view of ecosystem dynamics that links together ecosystem organization, resilience, and dynamics.

The model proposes that, as weakly connected processes interact, some processes reinforce one another, rapidly building structure or organization. This organization channels and constrains interactions within the system. However, the system becomes dependent upon structure and constraint for its persistence, leaving it vulnerable to either internal fluctuations or external disruption. Eventually, the system collapses, allowing the remaining disorganized structures and processes to reorganize (Figure 5.2).

The arrows in Figure 5.2 indicate the speed of the cycle. The short, closely spaced arrows indicate slow change, whereas the long arrows indicate rapid change. The cycle reflects ecological change in the amount of accumulated capital (nutrients, resources) stored by the dominant structuring process in each phase and the degree of connectedness within the ecosystem. The exit from the cycle at the left of the figure indicates the time at which ecological reorganization into a more or less productive and organized ecosystem is most likely to occur.

As the phases of the adaptive cycle proceed, an ecosystem's resilience expands and contracts. The r, or exploitation, stage begins a process of growth, resource accumulation, and storage. Initially, ecological resilience is high: ecosystem components are weakly connected to one another and the ecosystem's state is only weakly regulated. During this period, species or processes can grow rapidly as they utilize disorganized resources. The ones that thrive develop interrelationships that reduce the impacts of external variation

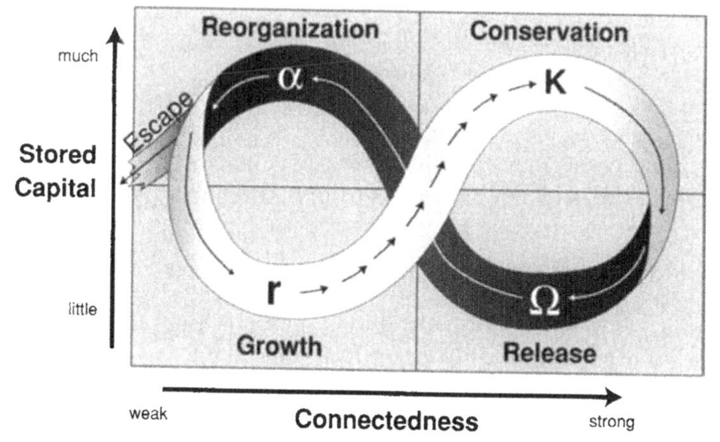


*Figure 5.2* The dynamics of an ecosystem as it is dominated by each of the four ecological processes: rapid growth (r), conservation (K), release (omega), and reorganization (alpha). Adapted from Gunderson *et al.* (1995).

and reinforce their own expansion. This process corresponds to vegetative control of microclimate or the increased closure of nutrient cycles. These processes of organization increase an ecosystem's efficiency at the cost of its flexibility, decreasing its resilience.

As an ecosystem becomes more organized, the competitive advantage shifts from actors that are able to grow rapidly despite environmental variation (i.e. r-selected species in ecology), to those that can effectively manage and benefit from interactions with other actors (i.e. K-selected species). The diversity of actors in the ecosystem begins to decline as intense relationships, both competitive and facilitative, among actors squeezes out actors that do interact well with others within the increasingly interactive ecosystem. As the ecosystem evolves toward the conservation or K-phase, connectivity among the flourishing survivors intensifies, and it is increasingly difficult for new actors to invade. In ecology, an example of such a system is a mature forest. The future dynamics of ecosystems in this state appear to be gradual, constrained, and predictable.

The progression from the exploitation to the conservation phase increases an ecosystem's available capital and connectivity but decreases its ecological resilience. The ecosystem's increasing dependence upon the persistence of its existing structure leaves it increasingly vulnerable to any process or instability that begins to release its organized capital. In ecosystems, this occurs when an ecological disturbance processes, such as fire, floods, ungulate grazing, or disease outbreaks, destroys existing ecological structures. This release or collapse phase is termed the Ω phase, because it indicates the end of an existing organization. This disturbance state is transitory. Disturbance destroys ecological structure and disperses ecological capital, but this process eliminates the conditions that allow disturbance to persist. For example, a forest fire extinguishes itself when there is no more fuel left to burn.

Ecological disturbance is followed by ecological reorganization, the alpha phase. Disturbance typically eliminates some ecological actors, while others persist with few available resources. During this phase, an ecosystem can easily lose resources, and new actors can enter it. The lack of control allows novel organizations to form. Such an ecosystem has little resilience and can easily be reorganized by small inputs. This is the time when exotic species of plants and animals can invade and dominate an ecosystem. It is the time when accidental events can freeze the direction of the future. Out of these interactions a new ecosystem organization emerges; this organization may be a rebirth of a past ecosystem organization or something entirely new. This process of emergence rapidly creates a loose, resilient organization.

While progress from r to K represents a prolonged period during which short-term predictability increases, the shift from Ω to α represents a sudden explosive increase in uncertainty. The consequence of this alternation between long periods of somewhat predictable behavior and short periods of unpredictable, weakly constrained behavior results in ecosystems that alternate between accumulating organized resources and experimenting with alternative organizations. The process generates novelty and tests diversity of species in an ecosystem or functions in an organization. The weakly constrained fashion in which ecosystems reorganize during the alpha phase make it next to impossible to predict what organization will emerge from this phase.

The adaptive cycle probably arises from perpetual change in an ecosystem's internal components and environment. Such changes make it impossible for a given organization to remain static and stable. The adaptive cycle illustrated in Figure 5.2 has two very different stages. One, from r to K, is the slow, incremental phase of growth and accumulation. The other, from Ω to α, is the abrupt, rapid phase of reorganization leading to renewal. The first is predictable with a high degree of certainty. It maximizes production and accumulation. The consequences of the second stage are unpredictable and highly uncertain. It increases disorder and novelty. This alternation between stages may represent a necessary tension between invention and efficiency. It appears that neither of these processes is stable alone. Efficiency undercuts its own ability to persist by reducing the ability of a system to respond to change, and successful innovations grow while unsuccessful innovations vanish.

The adaptive cycle can temporarily break down as a result of either the loss of ecological capital or the creation of a very robust ecological organization. In the first case, the removal or destruction of ecological resources eliminates the possibility of an ecosystem reorganizing. This could occur when an event, such as an intense forest fire, destroys a site's soil, removing the ecological substrate for the old ecosystem. While life still persists at such a site at the fine scale, local ecological dynamics have been derailed. In the second case, an ecosystem that manages to organize into a very robust ecological organization may be able to survive disturbance and environmental variation and prevent species turnover. However, such a system is probably possible only in situations in which the ecosystem experiences a limited amount of environmental variation, and is isolated from species or other ecological impacts from neighboring ecosystems, and when interactions between biotic and abiotic processes strongly reinforce each other. However, because no ecosystem is completely isolated from other ecosystems, the systems that are trapped in either poverty or stability will eventually be altered by the movement of organisms or materials from surrounding systems, or regional or global changes produced by dynamics in those systems.

## *Coping with change*

The dynamic nature of ecosystems has important consequences for the species that inhabit them. A plant species that lives within a fire-dominated forest ecosystem (Figure 5.3) can adopt a number of strategies that favor its survival.

Fire kills trees, opening the forest canopy and making light and nutrients available to other plants. Survivors and new plants begin to grow and, after much growth, competition, and environmental modification, a mature forest is established.

Specific strategies correspond to different phases in the adaptive cycle (Figure 5.4). A plant species may adopt a K-phase strategy of resisting the damaging effects of fire if it evolves fire resistance attributes, such as thick bark. Alternatively, it could adopt an r-phase strategy by being able to grow quickly following a fire, for example by storing energy in its root system to invest in growth following the loss of above-ground biomass. Another strategy is to focus upon the alpha or reorganization phase, to ensure that the plant species is present immediately following fire. An example of this strategy is serotinous pine cones, which open to release their seeds only following the heat of a fire. A complex strategy is the creative destruction, Ω-phase strategy of modifying the qualitative character of fire to the benefit of a species. For example, longleaf pine sheds flammable needles that burn readily, encouraging frequent low-intensity fires. Longleaf pine can survive these fires, but oaks, an ecological competitor, cannot.

Each of these strategies focusses a plant's energies on adapting to an ecological phase of the fire-dominated ecosystem. Often, plants in such a system will adopt a suite of such strategies, but trade-offs among strategies, such as between thick bark and rapid growth, inhibit a plant from dominating during all phases of an adaptive cycle. As plants or other

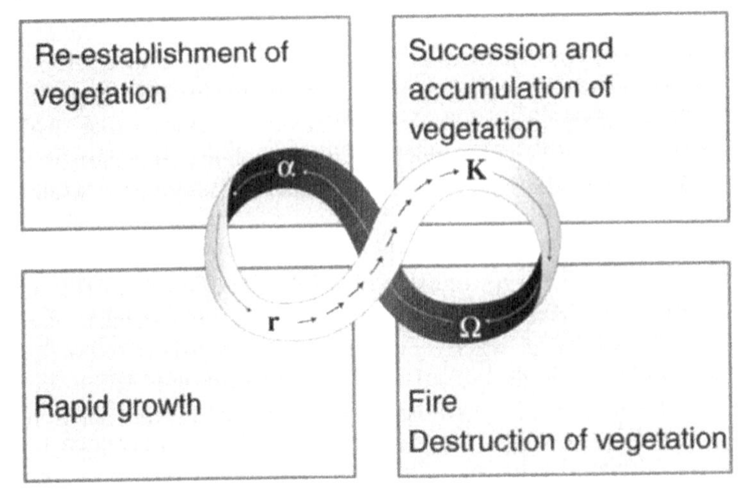


*Figure 5.3* The adaptive cycle in a forest that experiences fire.

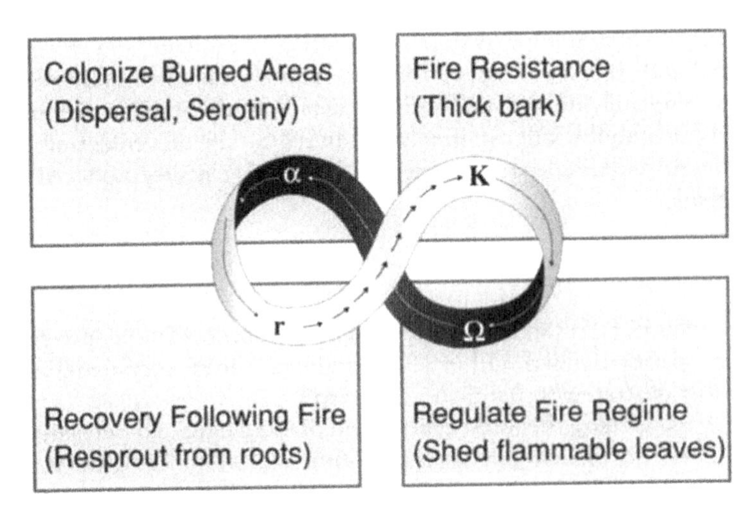


*Figure 5.4* The fire adaptations of plants respond to different phases in the adaptive cycle.

animals adapt to the ecological dynamics they inhabit, they also modify those dynamics by inhibiting or accelerating different phases in the adaptive cycle.

The response of plants to fire is a specific example of general strategies for dealing with change within a dynamic system. There are four possible active responses to change: learning, insurance, resistance, and management. There is also the passive strategy of doing nothing in response to change. A learning strategy attempts to understand system dynamics so that knowledge can be used to reconfigure a future system. This corresponds to the plant strategy of colonizing newly burned areas. Insurance is a strategy of investing in alternative strategies so that, when a system reorganizes, quick regrowth is possible. This strategy corresponds to the plant strategy of being able to grow quickly following a disturbance. Resistance attempts to control systems dynamics to prevent disturbance from happening. A plant version of the resistance strategy is to invest plant resources in thick bark.

Management is a strategy of trying to control the timing and nature of change or disturbance rather than trying to prevent it. For example, the practice of prescribed fire attempts to control ecological dynamics by burning a landscape at times that people choose rather than letting wildfires occur. Plants use this strategy when they modify a fire regime by shedding fire-encouraging material. The comparison between the specific responses of plants to fire and general strategies of coping with change is shown in Table 5.1.

These general strategies are different approaches that a component of a system can take to deal with systemic change at a larger scale. Ecosystems containing species that embody a wide diversity of alternative strategies will be more resilient than those that contain a less diverse group. A broader mix of strategies reduces the susceptibility to a broader range of environmental variation or disturbance.

### *Cross-scale dynamics*

Any ecosystem exists within an environmental context and is itself composed of subsystems. The dynamics of a system are shaped by its interactions with other systems at other scales. Using the adaptive cycle to describe each of these systems allows the sensitivity of a system to change to be assessed by considering which phase in the adaptive cycle the system currently occupies. Similarly, the degree of impact that the transformation of a system has upon its environment or its subsystems depends upon the state of those systems (Figure 5.5).

As a system changes over time, its openness to change from above or below also changes.

1. A system in the "r" or growth phase is resilient. Its development is constrained and guided by the larger system within which it is embedded and lower-level variation of the system's subsystems has little effect.
2. A system in the "K" or conservation phase is "brittle," possessing little resilience. In this state, a system is extremely vulnerable to any change in its subsystems, as a small disturbance (a spark) can rapidly propagate through the entire system. Similarly, the system is vulnerable to sudden change of the system in which it is embedded. A reorganization of this higher level system can trigger the collapse of the system.
3. A system in the "Ω" or release phase is in the midst of a disturbance – its old structure is being destroyed and reorganized. This destruction can trigger reorganization of higher level systems, and of lower-level systems, if they are vulnerable to disturbance.

*Table 5.1* Strategies that plants use to respond to different phases of a fire regime and generalized strategies for dealing with change

| *Phase in adaptive cycle* | *Reorganization ($\alpha$)* | *Growth (r)* | *Conservation (K)* | *Creative destruction ($\Omega$)* |
|---|---|---|---|---|
| Strategy for coping with fire | Colonize burned areas | Regenerate quickly | Resist fire | Regulate fire regime |
| General strategy for coping with environmental change | Learning | Insurance | Resistance | Management |

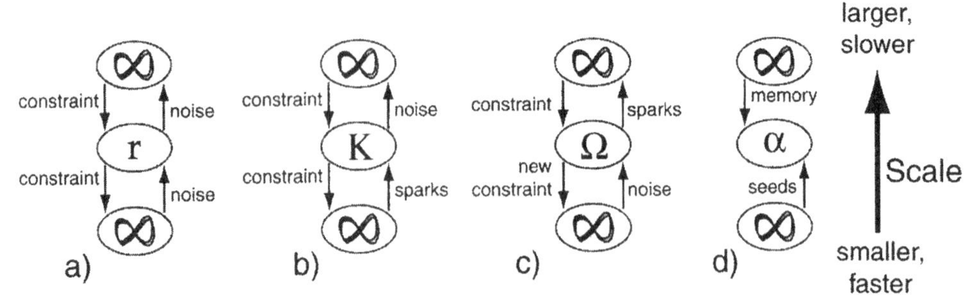


*Figure 5.5* An idealized map of the relationships between a system and the hierarchical levels above and below it – that is the system in which it is embedded and its subsystems. Each level in a hierarchy changes over time. These changes can be represented by Holling's four-phase cycle (Holling 1986).

4 A reorganizing system in the "alpha" phase only weakly constraints its subsystems, making it prone to organize around any "seeds" of order that emerge from change in the lower levels. The reorganization is also defined by the system in which it is embedded. The higher level system can provide a memory of the reorganizing system's past organization as the higher level system has itself emerged from interactions with this past organization. This structure can provide a template around which a reorganizing system can be structured.

I propose that there are four ways that change propagates through dynamic hierarchies. First, change at a higher level alters a lower level as a result of the constraints that it places upon it. For example, an increase in the atmospheric concentration of carbon dioxide increases the ability of most plants to conduct photosynthesis. Second, reorganization at a higher level can trigger reorganization at a lower level. A plague of defoliating insects that kills large areas of forest can cause significant local changes. By opening the forest canopy, the forest floor would experience increased temperature variation, decreased moisture, and increased light levels. These changes could cause shade-tolerant vegetation to be replaced by vegetation that can grow quickly in high light levels and tolerate temperature variations. Third, a small-scale disturbance can trigger a larger-scale collapse if the larger system is in a brittle stage in its adaptive cycle. For example, a lightning strike can ignite a tree. If this tree is surrounded by dry vegetation a fire may spread out from the tree to burn a large area. Fourth, following the collapse of a system, small-scale and surrounding large-scale systems provide the components and constraints out of which a system reorganizes. An example of this type of cross-scale change is provided by the recovery of south Florida's mangrove forests following Hurricane Andrew. The hurricane killed many large trees, but young flexible trees survived. These young trees occurred in gaps in the forest produced by frequent, small-scale fires produced by lightning strikes. Recovery of the forest from large-scale disturbance was facilitated by the diversity produced by past small-scale disturbances (Smith *et al.* 1994).

### *Summary of ecological dynamics*

In the above section I have focused on the dynamic nature of ecosystems, and how these dynamics both control and are controlled by cross-scale connections. These interactions are at times stabilizing, at other times disruptive. The tension between these forces causes ecosystems to exhibit complex, shifting patterns of ecological organization. This dynamic complexity makes designing or even managing ecosystems quite difficult.

## Managing ecosystems

Ecosystems, like any complex system, are difficult to manage. Many attempts to re-engineer ecosystems have been unsuccessful. In this section I discuss the difficulties with engineering ecosystems, and then describe one approach to ecological management that attempts to cope with the complexities of ecological engineering.

### *Engineering ecosystems*

Engineering ecological systems is dangerous. People currently lack the skills, understanding, and data to manage ecosystems competently. Ecosystems are evolutionary and self-modifying, rather than stable and static. Furthermore, ecological processes are not constrained to the same scales as human action – air and water move across the landscape and animals migrate thousands of kilometers. This openness makes it difficult for management to predict how changes in the external environment will affect the managed system.

Traditional engineering uses physical rules to manipulate the world and, consequently, it has benefited from advances in physics. Physicists have had great success in developing scaling rules that explain the behavior of physical systems over a wide range of scales. These systems tend to be homogeneous, ahistorical, and unchanging; however, ecosystems are diverse, evolving, historical entities. These differences cause ecological systems to vary across scales. Different processes dominate at different spatial and temporal scales, making it difficult to predict how processes will interact across scales, or how ecological understanding can be transferred from one scale to another (Levin 1992).

These scaling differences mean that ecological engineering has more diffuse and less tractable negative externalities (e.g. irrigation leading to the spread of river blindness) than does traditional engineering (e.g. a bridge collapsing as a result of an unexpected load). In traditional engineering systems, failure is often local, abrupt, and catastrophic, whereas "failure" in ecological systems often occurs gradually over larger areas. The chronic and diffuse nature of ecological degradation makes dealing intelligently and equitably with the unpredictable, the unknown, and surprising aspects of ecosystems difficult, but this does not remove the need to make decisions and manage these systems.

Engineers traditionally have attempted to avoid disaster by "fail-safe" design (e.g. using beams that are several times stronger than calculations suggest they need to be), but we know that there is no "fail-safe" strategy in managing any partially known complex non-linear system. This difficulty is further compounded by the tendency of ecological dynamics to become increasingly complex in human-dominated ecosystems. Ecological management emphasizes the importance of understanding and adapting to the diverse and shifting connections between people and Nature. These approaches may be useful in construction ecology, because construction dramatically transforms ecosystems by embedding human-created environments, such as roads and buildings, within ecosystems.

### *Human-dominated ecosystems*

It is important to include human behavior in ecological analyses because humans are

Earth's dominant species. It is also important to realize that there are fundamental differences between human-constructed and natural systems. Humans, individually or in groups, can anticipate and prepare for the future to a much greater degree than ecological systems (Brock and Hommes 1997). People use mental models of varying complexity and completeness to construct views of the future. People have developed elaborate ways of exchanging, influencing, and updating these models. This creates complicated dynamics based upon access to information, ability to organize, and power. In contrast, the organization of ecological systems is a product of the mutual reinforcement of many interacting structures and processes that have emerged over long periods of time. Similarly, the behavior of plants and animals is the product of successful evolutionary experimentation that has occurred in the past. Consequently, the arrangement and behavior of natural systems is based upon what has happened in the past, rather than looking forward in anticipation toward the future. The difference between forward-looking human systems and backwards-looking natural systems is fundamental. It means that understanding the role of people in ecological systems requires not only understanding how people have acted in the past, but also how they think about the future.

Economists generally expect that "rational expectations" of future behavior will stabilize system behavior. However, the difficulty and high cost of understanding novel situations, such as those created by technological change, shifts in human values, and ecological change greatly reduce the ability of "rational expectations" to stabilize system dynamics, because high uncertainty makes many alternative views equally "sensible" (Brock and Hommes 1997).

The complexity of human-dominated ecosystems has led to a number of different schools of ecosystem management. One approach advocates various "safe-fail" approaches to management based upon the precautionary principle of engaging in ecological modification only when you are certain that it will not cause serious harm. Another approach is adaptive management, which argues that ecological management should center on increasing understanding of how particular human-dominated ecosystems actually work.

### *Adaptive management*

Adaptive management is an approach to ecosystem management which argues that ecosystem functioning can never be totally understood. Ecosystems are continual changing as a result of internal and external forces. Internally, ecosystems change because of the growth and death of individual organisms, as well as fluctuations in population size, local extinction, and the evolution of species traits. Ecosystems are also changed by external events such as the immigration of species, alterations in disturbance frequency, and shifts in the diversity and amount of nutrients entering the ecosystems. To cope with these changes, management must continually adapt. Management becomes adaptive when it persistently identifies uncertainties in human–ecological understanding, and then uses management intervention as a tool to test strategically the alternative hypotheses implicit within these uncertainties.

Adaptive management is quite different from a typical management approach, which uses the best available knowledge to generate a risk-averse, "best guess" management strategy, which is then changed as new information modifies the "best guess" (Holling 1978; Walters 1986; Lee 1993; Gunderson *et al.* 1995).

Adaptive management identifies uncertainties, and then establishes methodologies to test hypotheses concerning those uncertainties. It uses management as a tool for learning about the system as well as means of altering the system. It is concerned with the need to learn and the cost of ignorance, whereas traditional management is focused on the need to preserve and the cost of knowledge. There are several processes, both scientific and social, that are vital components of adaptive management:

1. linkage of management to appropriate temporal and spatial scales;
2. retention of a focus on statistical power and controls;
3. use of computer models to build synthesis and an embodied ecological consensus;
4. use of ecological consensus embodied in models to evaluate strategic alternatives;
5. communication of alternatives to political arenas for negotiation.

The achievement of these objectives requires an open management process that seeks to include past, present, and future stakeholders. At a minimum, adaptive management needs to maintain political openness, but usually it needs to create it. Consequently, adaptive management must be a social as well as a scientific process. It must focus on the development of new institutions and institutional strategies just as much as it must focus upon scientific hypotheses and experimental frameworks. Adaptive management attempts to use a scientific approach, accompanied by collegial hypotheses testing to build understanding. This process also aims to enhance institutional flexibility and encourage the formation of the new institutions that are required to use this understanding on a day-to-day basis.

Management is a process grounded in the local. It depends upon local constraints, the present state of local institutions, and the personalities of key people. Any attempt to manage adaptively should transfer knowledge and understanding to local individuals, but it must do more than that. It must also develop institutional flexibility by encouraging the formation of networks of individuals that bridge institutional boundaries. These groups of individuals can act as agents of reform within their institutions, and the nucleus around which new institutions can form.

## Moving toward construction ecology

The ability of people to manage ecosystems appears to be fundamentally limited by uncertainty and cross-scale change. Ecological management that has adopted approaches that explicitly attempt to understand and manage both natural and human uncertainty has had some success. I argue that similar approaches are necessary to understand construction ecology. In the following sections I apply ecological theory and adaptive management to construction ecology.

### *Suggestions for construction*

I have discussed the importance of resilience, ecological dynamics, and scale in ecosystems, and the problems that arise in managing these complex systems. I attempt to apply these ideas to construction. I describe design approaches that could be used to create construction ecologies that embrace dynamism and change. I then discuss the multiple time scales at which buildings change, and how these cross-scale dynamics can be incorporated in construction ecology.

### *Dynamics of construction*

The ecology of construction can be considered from the perspective of the adaptive cycle. I propose that the four phases of the adaptive cycle in construction ecology are: design, construction, maintenance, and deconstruction of a building (Figure 5.6). The design phase organizes information about the purpose of the building, the site on which it will be built, and the appropriate building techniques and materials into a plan for a building. The construction phase uses material and energy to construct a building. Once a building is constructed it enters an operation and maintenance phase. Operating a building requires energy, while maintenance and periodic renovation require materials. The building remains in the operation and maintenance phase until it is demolished. At this time energy is required to disassemble the material, and this material is either released from the building, if the building is demolished, or it is reused, as part of a new design, if the building is renovated.

In the construction phase, r, materials and energy flow into the system. In the operation and maintenance phase, K, operation requires energy and maintenance requires energy and materials. In the deconstruction phase, Ω, energy and materials can leave the system – if they are not reused. In the design phase, α, information flows into the system, determining the new configuration of a new burst of construction.

As in the example of a fire-dominated forest, there are adaptive strategies that a building can adopt to enhance its abilities during each of these phases. However, an overemphasis on any one of these phases may result in a building that functions well during some portions of its life cycle but fails during others. The phases of construction and the design approaches I suggest for each of them are summarized in Table 5.2.

### *Design phase*

The process of design is quite similar to adaptive ecosystem management. Both require learning from past experiences and anticipating the future while knowing full well that the world cannot be completely understood. I argue, from experience with adaptive management, that design requires assessment, design documentation, design for use, and design for learning.

Design is an information-intensive process. A wide variety of information must be collected and integrated through an assessment process. This process needs to include analyzing the site, the needs of the building's users, construction methods, and possible

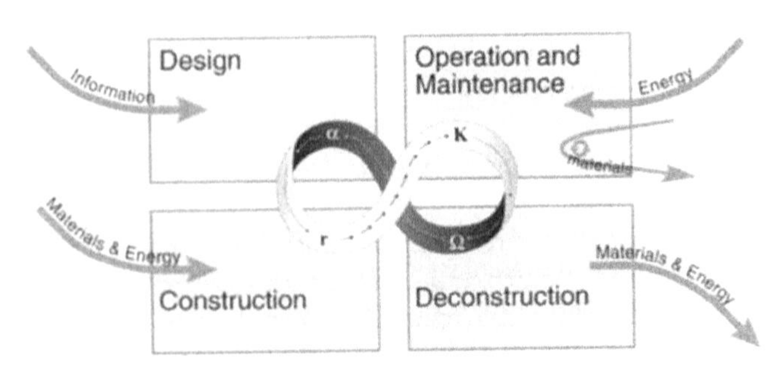


*Figure 5.6* In the adaptive cycle of construction ecology the connections between construction ecology and its external environment vary across phases.

*Table 5.2* The phases of the adaptive cycle in construction ecology are design, construction, operation, and deconstruction. These phases have different types of resource flows, and should be guided by different design principles

| | *Phase in adaptive cycle*<br>*Reorganization* ($\alpha$) | *Growth* (*r*) | *Conservation* (*K*) | *Creative destruction* ($\Omega$) |
|---|---|---|---|---|
| Construction phase | Design | Construction | Operation and maintenance | Deconstruction or demolition |
| Resource flows | Information flows in | Materials and energy flow in | Energy flows in<br>Low flow of material in and out | Material released |
| Design principles | Assessment of situation<br>Design for learning | Design to use resources effectively<br>Design for expansion and adaptation | Design for effective energy use<br>Design for evolution<br>Design for decay | Design for disassembly<br>Design for reuse<br>Design for decomposition |
| Potential strategies | Assessing environment<br>Assessing potential uses components<br>Design documentation<br>Experimental architecture | Simple beginning<br>Allow opportunities for adaptation<br>Add excess capacity<br>Embody information in building | Energy effectiveness<br>Monitoring<br>Design for inhabitant modifiability<br>Accessible and replaceable subsystems<br>Age gracefully | Modular design<br>Reusable<br>Recyclable components<br>Biodegradable components |
| Cycle of learning | (Re)formulate hypotheses | Set up experiments | Conduct monitoring | Test hypotheses |

alternative ways in which the users and the site could change. These issues are discussed in greater detail in other sections of this book.

Design is an energy-intensive process, and sustainable construction should attempt to reduce the amount of design that is required in the future. A reduction in design effort requires learning what materials and methods work well, and which do not. If there is no learning from design successes and, most importantly, from failures, there can be no hope of developing more sustainable construction practices.

Learning requires looking toward the past and the future. Looking into the past is necessary to develop theories about how a building should be built. Looking into the future is necessary to design ways in which the theories embodied within a building can be tested. Documenting the design of a building and the ideas behind it is perhaps the minimal requirement for learning to occur, but learning can be greatly enhanced by active experimentation. Active experimentation is necessary to advance understanding in many areas, especially complex cross-scale processes such as the interaction between the built and the natural environment.

Experimentation can be divided into active and passive experiments. As no two situations are exactly alike, every building is in some sense an experiment, whether or not that is the intention. Even if it is not the intention, every building can be treated as a passive experiment. Data can be collected on how well building techniques perform, for example energy use following construction. Indeed, it appears that getting architects involved in the evaluation of buildings after they have been occupied offers great opportunities for learning (Brand 1994).

Active experimentation consists in formulating a building that is designed to test a hypothesis. To engage in such an extensive, long-term effort is expensive and difficult to organize. A building used for active experimentation may be awful to live in, or it may fail to achieve its goals, but it will at least test the hypothesis. If the risk of having a building fail as a building is unacceptable then passive experimentation can be adopted. Any experimental approach requires an extra investment in experimental design, data collection devices, data collection, and data analysis. This is expensive, however it is necessary if we are to understand how construction ecology functions. The number of active experiments that are incorporated in a building should be at least partly determined by the degree to which a building is novel.

Experimentation requires different efforts during each phase of the construction ecology. Hypothesis formulation occurs during the design phase. The construction phase is when experiments designed to test these hypotheses are set up. The operation and maintenance phase is when data are collected via monitoring devices. The destruction period should be when the data are analyzed to test the hypotheses, leading to the generation or reformulation of hypotheses. Such a complex process may be difficult to enact, but the resulting integrated information would likely greatly improve designers' understanding of how buildings are used, modified, and impact the environment.

Experimentation needs to combine a diversity of approaches within an experimental setting. Diversity reduces the cost of any individual failure, and an "experimental" setting allows people to analyze the consequences of specific design choice. Experimental management requires replicated treatments that can be compared against controls to test competing hypotheses. In the short term, building that follows "best practices" is cheaper than experimental building techniques, but the absence of experimentation removes opportunities for learning and adaptation.

### *Construction phase*

The construction phase of a building is the one that brings in most of the materials and a lot of energy into a site. It is the stage in which the ultimate form and possibilities for future growth and adaptation are locked in. I argue that construction methods that allow future growth and adapatation are critically important. Other chapters in this volume discuss the many issues surrounding the choice of construction-appropriate materials.

Designs that are simple allow additions and modifications; they provide space, into which the building occupants can organically add complexity as they need it. Buildings that are complex, for example constructed out of many obscure materials shaped into strange forms, greatly restrict the possibility of future modifications and additions.

Leaving portions of a building unfinished, or not completely finished, allows the occupants to learn and develop effective ways of using space. It is unrealistic to expect an architect or designer to anticipate all the possible future uses of a structure. The needs and requirements of a building's users change over time as a result of social and technological change, as well as shifts in who owns and occupies the building. Leaving space and excess capacity, such as extra space underneath floor paneling, and leaving time for local adaptations to occur, such as not finalizing some portions of the design until building occupants have been able to inhabit and use another portion, provide opportunities for adaptation to occur.

Finally, the process of construction will almost inevitably depart from the original plans. It is important to record and make accessible the actual materials and methods used in constructing a building. The availability of this information decreases the difficulty of future modifications and expansions, while also providing information that can be utilized by builders of other structures. Ideally, this information should be embodied within the building, both physically connected to components of a building and virtually available to the owners and operators of the building.

### *Operation and maintenance phase*

The operation of a building requires energy, while its maintenance requires both materials and energy. The skill with which a building is operated and its structure and subsystems maintained and upgraded determines its use of energy and generation of material waste. Additionally, the degree to which a building can be cheaply operated, maintained, and upgraded is likely to determine its lifespan. Based upon these characteristics I propose that design for this phase should focus on energy effectiveness, evolution, and decay.

The people who use a building and how they use it will change over time. The physical, ecological, and social environment in which a building exists will also change. In ecosystems, the processes of adaptation and self-organization take time. The process appears to be similar in buildings. Brand (1994) argues that time gives people the opportunity to tinker with and adapt a building to both the general human needs of its inhabitants and the specific needs of its current users, and that this tinkering gives rise to loved and admired buildings. It is impossible to anticipate all forms of change that may occur. For example, few people in the 1980s would have predicted the need to wire buildings for the Internet. However, some types of change are completely predictable, such as the need to rearrange office space. Design should lead to buildings that are able to change.

Systems theory proposes that one of the key aspects of flexibility is maintaining a loose coupling among a system's components (Allen and Starr 1984). This principle can be applied to buildings by designing structures so that parts or components of a structure can be modified and replaced without destroying the integrity of the structure as a whole. For example, service networks should be accessible so that wiring or ducts can be maintained and upgraded without destroying the building. Loose coupling between service networks and structure existed in many old industrial buildings, such as warehouses and garages. The separation they maintained between simple service networks and their structure allowed many of them to be successfully converted to other uses, such as stores or apartments (Brand 1994).

The operation of a building can consume huge amounts of energy unnecessarily. This aspect of a building's operation has long been a concern of those interested in reducing energy use. An important aspect of design should be to focus upon the effective, rather than the efficient, use of energy, as discussed by Kay (see Chapter 3). This distinction emphasizes the amount of energy required to perform a particular service, rather than the efficiency with which energy is converted from one form to another. It focuses upon the role of energy within the building as a whole rather than within a particular task. This approach argues that effective design is based not on economizing, or designing a system with the absolute minimum requirements, but rather on effectively integrating the systems of a building. Considering a building's environmental context allows environmental sources of energy to be utilized and sources of environmental stress to be avoided. Such intelligent design accounts for the changes in the angle of the sun to shade the building during in summer, while allowing heating during winter.

A more neglected aspect of effective energy use is monitoring. Energy use depends upon the actions of building users and operators as well as the designers of the building. Therefore, energy conservation could be facilitated by integrating well-designed monitoring systems into the building. Establishing feedback loops from the environment to the building's controllers and users would make the built environment more resemble an ecosystem.

People can see, hear, smell, and touch physical things, but most environmental changes are invisible. The Viridian design movement (Sterling 1999) argues that good environmentally friendly design should make these invisible flows visible. For example, rates of energy use could be indicated by a sculpture that becomes more beautiful at low rates of energy use and uglier as more energy is consumed. The Sustainability Institute in Vermont recently sponsored a design contest for such a "Viridian electricity meter" (Sterling 2000). A more direct, but less precise, sensor of environmental condition would be ecosystems that utilize waste flows from a building. For example a building's wastewater could flow into a nearby wetland, both for treatment and as a means of measuring the building's impact on the environment.

Maintenance is a process that requires energy and materials. Designing structures that gracefully decay reduces the environmental impacts and cost of maintenance. For example, some architects use copper roofs because the roofs look beautiful as they age. This requires consideration of the process of degradation and the consequences of removing degraded building components, such as its roof, ducts, and lights, and service networks. If something decays gracefully, the costs of maintaining and replacing it are reduced. Complicated systems are more difficult to maintain as everything in them is connected to everything else. However, a well-designed complex system that reduces the number of cross-connections among subsystems and separates systems design into relatively autonomous systems components may degrade more slowly than a complicated system. For example, if building temperature is regulated by a number of passive and active mechanisms which operate at different scales, the failure of one of these mechanisms will not render the building uninhabitable. However, if a building is dependent upon a single air-conditioning plant, the failure of that plant makes the building uninhabitable.

### *Deconstruction phase*

The deconstruction or destruction of buildings is typically neglected in their design. However, the lifespan and fate of used building materials is a key component of their cumulative environmental impact. Three design objectives should provide the focus of this phase: design for disassembly, design for reuse, and design for decomposition.

*Design for disassembly* means that buildings are designed in such a way that they can be easily taken apart and reused. Reducing the costs of disassembly requires reducing the sum of the energetic and material costs of disassembly. Easy disassembly is particularly valuable if it allows portions or segments of buildings to be removed without requiring wholesale destruction.

The cost of disposing of used building materials is the sum of the potential value of the materials and the cost required to dispose of them. This cost can be reduced either by reusing the building's components, decreasing the value of the materials, or by reducing the disposal costs.

Currently, very little of buildings is reused. *Design for reuse* is an extension of design for disassembly. Reuse can range from the recycling of metals into new products, to the reuse of entire components of a building – ranging from bricks, to doors, to entire structural components. In ecosystems, the creative part of destruction occurs when destruction allows the components of a system to be reassembled into a new, potentially improved organization. Reuse could allow this type of experimentation in buildings.

One way to reduce the costs of destruction is to reduce the volume and toxicity of waste from buildings. Currently, the economic costs associated with disposing of toxic building materials are not large. However, as societies increasingly recognize the ecological costs associated with these practices, it is likely that the financial costs of these unsound disposal practices will increase.

Ideally, a building should break down into components that are either biodegradable or recyclable. This may be an unrealizable goal in the short term, but nonetheless it is a worthwhile goal. The consideration of the ultimate destination of all building components would go a long way toward decreasing the burden of the built environment on ecological processes. The reuse of components could be facilitated by minimizing the use of complex synthetic materials that are uniquely joined to specific functions or surfaces and therefore cannot be reused. Paint is an example of such a complex material. Its use interferes with the reuse of other materials, providing another argument for loosely, rather than tightly, connected building functions.

## Scale

The dynamics of buildings can also be apply to the subcomponents of a building. Just like ecosystems, buildings operate at multiple scales. Stewart Brand, in his wonderful book *How Buildings Learn*, divides building structure into a nested hierarchy consisting of site, structure, skin, services, span plan, and stuff (Brand 1994). The site is the location at which a building is built. The structure is the frame of a building. The skin is a building's covering, which is the material that separates the inside of the building from the outside. Services consist of the networks, such as water, heating, and phone lines, that distribute services throughout a building. The space plan is how the interior of a building is arranged – the walls, dividers, and other components. Stuff is a category that includes the location of furniture, lamps, desks, chairs, and the other mobile objects. Each of these building subsystems appears to operate at its own characteristic time scales (Figure 5.7).

This scaling of building functions suggests that sustainability is an issue at all scales. Furthermore, it suggests that the connections that link these different scales are key points for determining sustainability across scales. For example, if a building's space plan is tightly linked to a building's structure, then changing the space plan may also require altering the structure. In the following sections I briefly discuss several of these scaling issues.

### *Sustainability is an issue at all scales*

The fact that different aspects of a building change at different rates suggests that construction ecology's focus on the building is important, but a broader transformation also needs to occur. The ideas of construction ecology can and need to be applied across scales to all building subsystems, as well as to the larger systems within which buildings are embedded, such as road networks, power distribution systems, and watersheds.

The cross-scale nature of sustainability should be viewed as an opportunity. First, construction ecology can benefit and learn from the techniques used for design, construction, operation, and deconstruction in other related fields, such as industrial design or process engineering. Second, techniques and methods developed by construction ecology practitioners may have broader applicability than strictly within the field of construction.

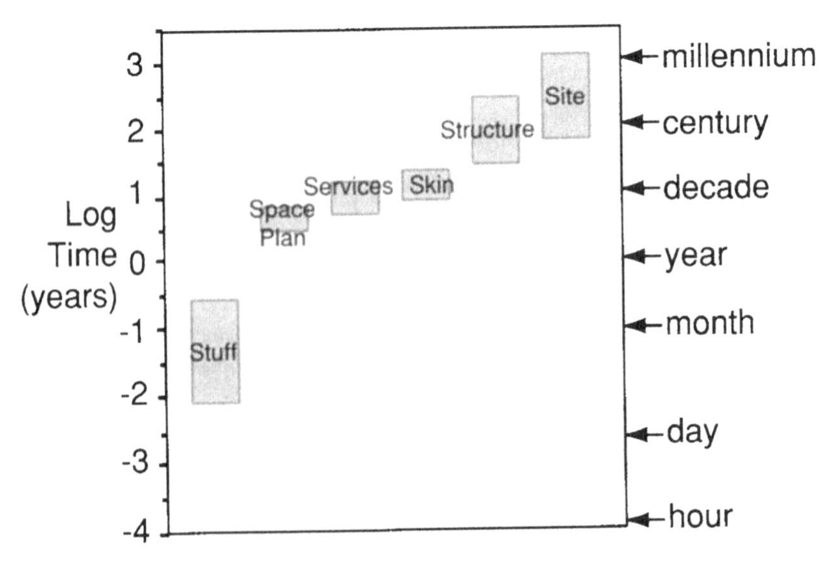


*Figure 5.7* Temporal scales of a building. Data taken from Brand (1994).

### *Scale matching*

Effective use of energy and material does not necessarily require long-lasting buildings. Because sustainability is determined by the impact of a building over time, a building that has a low environmental cost over a short period of time can be equal to a building that has a high environmental cost over a long period of time. For examples, igloos never last longer than one season, and sometimes last for only a few days, yet they are rapidly constructed from local materials and are completely degradable. In a similar fashion, tents or other portable buildings can be moved to where they are needed, set up, used, taken down, and then reused somewhere else. Cheap, short-lived buildings have several advantages: they do not permanently transform the site, and their shorter construction cycle allows innovation in design, construction, and materials to be incorporated more quickly. Short-lived buildings are likely to be appropriate for situations in which human needs are rapidly changing. This could include temporary accommodation, for example during festivals or to house refugees, and experimental buildings.

## Managing disturbance

Buildings that are built in areas subject to periodic disasters, such as floods or hurricanes, can respond it different ways, just as a plant responds to fire. Buildings can be designed to resist the disaster, manage the disaster (if this is possible), or rapidly recover following a disaster. The strategies of rapid recovery, via insurance and resistance, via reinforced walls and dikes, are common.

Along with insurance to help recovery following a disaster, limiting the cost of a building in high-risk areas can both reduce the need to pay insurance premiums and increase the ease with which it can be rebuilt.

Less common is managing disturbance, which is possible with fire but more difficult with floods, and next to impossible with hurricanes or earthquakes. Managing fire could involve periodically burning the area around a house to create a fire barrier. Managing floods could involve designing a house that can tolerate some flooding and preserving wetlands to make flooding more frequent, but less severe.

An alternative approach is to design buildings that can dynamically respond to their environment. A typical example is the use of flexible materials, such as wood, that are resilient to stress, rather than those that fail catastrophically, such as brick. Many animals respond to disturbance by either moving away or hiding, for example in an underground burrow. It may be possible to adopt similar strategies for buildings, or at least some buildings. For example, if a house can be rapidly deconstructed, it could be deconstructed prior to a disturbance – a controlled, rather than destructive form of deconstruction, and then rebuilt afterwards. Similarly, a building that can be deconstructed or moved could be either disassembled and/or moved out of harm's way.

Coping with environmental natural disturbance is an example of the externally driven change that buildings experience. A far more common form of change is social change that changes the use and context of a building. For example, economic change has altered what is manufactured in the USA. The fate of old factory buildings depends upon how well they can be adapted to other uses and how much society values their design. In the following section I describe how construction ecology can begin to design for these types of change.

### *Cross-scale resilience*

My work on cross-scale resilience suggests that ecosystems are resilient as a result of compensation among ecological functions within scales and their replication across scales. This type of functional arrangement makes ecosystem functions robust to a wide range of environmental variation. I suggest that an approach to the functions of construction ecology that attempts to provide sources of cross-scale and within-scale resilience will produce more resilient buildings.

Cross-scale resilience is produced by the replication of function at different scales. In a building, for example, temperature regulation could be performed by a combination of whole-building thermal regulation and local temperature regulation. In this example, local thermostats would be more effective in responding to local variation in thermal input, for example when the afternoon sun shines on only one corner of the building. Whole-building thermostats would be the main source of response to daily temperature variation. The combination of two approaches could act as fine-tuning mechanisms for a building whose temperature variation is principally controlled with thermal mass, which changes only at a large, slow scale.

Within-scale resilience derives from functions that perform similar, but not identical, functions. For example, cooling could be provided by an adjustable shade canopy and the ability to open and close windows. Both windows and the shade canopy can be used to adjust temperature, but they have different strengths and weaknesses that in combination compensate for one another.

### *Cross-scale connections*

The fact that different component subsystems within a system change at different rates means that the connections between these subsystems should be designed very carefully. Otherwise, much faster subsystems may drive the life cycle of entire buildings. For example, a building can remain quite adaptable to other uses when bookcases are built into the walls. However, conversion to another use might be severely hampered when water conduits are embedded in concrete. The building's lifespan is reduced to that of its own service network when that network is too strongly linked to the building's structural elements.

### *Surprise*

The impact of large and slow social changes on building functionality can be manifested in surprising and unpredictable changes on building design. The possibility of surprise makes overoptimizing a structure for a specific use a risky design strategy. It becomes difficult to answer questions such as: How much energy should a building consume? What materials should be used? A sensible approach to this type of uncertainty is to try and develop designs that work well, or at least reasonably well, across a number of different possible future scenarios.

Inventing a desired future, rather than attempting to predict a likely future, is an alternative approach to design. Such an approach would try to construct structures that articulate and help generate a positive future. Such a building can change people's ideas of what a building can or should be, and by doing so acts to reduce the uncertainty of the future toward a vision that the builder desires. In a sense, all buildings do this, by showing what is possible. But a truly visionary building will have much more attention paid to it – both its perceived successes and failures. Such a building, by influencing the mental models people have of the future, can help create the future.

## Conclusion and summary

Ecological interactions across a wide range of scales cause ecosystems to exhibit complex, shifting patterns of ecological organization. Integrating constructed and natural ecosystems requires developing a construction ecology that harmoniously connects to ecological processes across a broad range of scales. I have proposed a number of principles based upon the cross-scale dynamics of ecosystems that may help develop such construction ecology.

I use the adaptive cycle model of ecological dynamics to divide construction ecology into four distinct phases: design, construction, operation, and deconstruction. The strategies that integrate the built and natural vary across these phases. Design should focus on assessing the context and planning for learning. Construction should be planned to use materials effectively and include opportunities for future growth, whereas operation should be organized for effective use of energy and for adaptation. Building strategies that incorporate decay and disassembly strategies are appropriate for the deconstruction phase.

Second, I argue that because construction ecology is attempting to engineer complex systems, there is a lot of uncertainty and a lot to learn. In such a situation construction ecology can profit from the experience gained from adaptive management's attempts to manage complex ecological systems. Adaptive management suggests that designers should explicitly include opportunities for learning in their designs as well as in their design processes. Learning requires hypotheses, experiments, monitoring, and evaluation. These can range from simple evaluations of a building's success to controlled replicated experiments.

Finally, the systems in which buildings are embedded, as well as the subsystems they contain, are dynamic but change at different scales. These dynamics must also be harmoniously integrated with ecological systems. This process can be helped by insuring that different scale systems are not too tightly integrated and have the ability to adapt independently of one another.

## References


Allen, T.F.H. and Starr, T.B. 1982. *Hierarchy: Perspectives for Ecological Complexity*. Chicago: The University of Chicago Press,.

Brand, S. 1994. *How Buildings Learn: What Happens After They Are Built*. New York: Penguin.

Brock, W. and Hommes, C. 1997. A rational route to randomness. *Econometrica* 65: 1059–1095.

Carpenter, S., Brock, W. and Hanson, P. 1999. Ecological and social dynamics in simple models of ecosystem management. *Conservation Ecology* 3: 4. L.H.

Gunderson, L.H., Holling, C.S. and Light, S.S. (eds) 1995. *Barriers and Bridges to the Renewal of Ecosystems and Institutions*. New York: Columbia University Press.

Holling, C.S. 1973. Resilience and stability of ecological systems. *Annual Review of Ecology and Systems* 4: 1–23.

Holling, C.S. (ed.) 1978. *Adaptive Environmental Assessment and Management*. Chichester: John Wiley.

Holling, C.S. 1986. The resilience of terrestrial ecosystems: local surprise and global change. In *Sustainable Development of the Biosphere*. Clark, W.C and Munn, R.E. (eds). Cambridge: Cambridge University Press, pp. 292–317.

Lee, K.N. 1993. *Compass and Gyroscope*. Washington, DC: Island Press.

Levin, S.A. 1992. The problem of pattern and scale in ecology. *Ecology* 73: 1943–1967.

Peterson, G.D., Allen, C.R. and Holling, C.S. 1998. Ecological resilience, biodiversity and scale. *Ecosystems* 1: 6–18.

Smith, T., III, Robblee, M.B., Wanless, H.R. and Doyle, T.W. 1994. Mangroves, hurricanes and lightning strikes. *BioScience* 44: 256–262.

Sterling, B. 1999. Viridian Note 00003 (summary) – Viridian Design Principles. Available on-line at: http://www.viridiandesign.org

Sterling, B. 2000. Viridian Note 00199 – Meter Contest Winner. Available on-line at: http://www.viridiandesign.org

Walker, B., Kinzig, A. and Langridge, J. 1999. Plant attribute diversity and ecosystem function: the nature and significance of dominant and minor species. *Ecosystems* 2: 95–113.

Walters, C.J. 1986. *Adaptive Management of Renewable Resources*. New York: McGraw Hill.

## Part 2

# The industrial ecologists

*Charles J. Kibert*

Industrial ecology is a young science that emerged in the late 1980s. It caught the public's attention and imagination as a consequence of the realization by industrial managers at the Kalundborg Industrial Park in Denmark that their materials and energy waste streams were in fact not necessarily waste but potential resource inputs for the processes of their neighbors. The resulting exchange of former waste streams by the industries in Kalundborg prompted worldwide investigations into the "ecology" of industry and the effects of industry on ecology. Some promoters of industrial ecology would define it as "the science of sustainable development". Ernie Lowe of Indigo Development in California views industrial ecology as an approach to managing human activity on a sustainable basis by seeking the essential integration of human systems into natural systems; minimizing energy and materials usage; and minimizing the ecological impact of human activity to levels that natural systems can sustainably absorb. He states that the objectives of industrial ecology are to preserve the ecological viability of natural systems; ensure acceptable quality of life for people; and maintain the economic viability of systems for industry, trade, and commerce. It is clear from this approach to industrial ecology that it can be applied to the broadest range of human activities, among them the creation and operation of the built environment. The question to be answered in the next few chapters is: What are the major lessons that can be learned from the experience of industrial ecology over the past decade that are applicable to the built environment? As is the case with examining ecology as to how it can and should inform construction, the difficulty at this point in time is to extract non-trivial, immediately useable principles and approaches for adoption by designers, builders, operators, and disposers of the built environment.

In this part, four industrial ecologists provide their thoughts on deriving construction ecology from what has been learned by industrial ecology. Robert Ayres (Chapter 6) initiates the discussion by focusing on emissions from the built environment and establishing strategies for their eventual elimination. He notes that the operation of buildings over their comparatively long lives is the main source of their emissions, producing not only environmental hazards but also threats to human health in the form of fungicides, insecticides, rodenticides, and combustion by-products such as particulates and carbon monoxide. Additionally, there are substantial emissions and hazards created in the manufacturing of construction materials, from dust in mining and quarrying operations to a wide range of gases emanating from the power plants providing energy to extraction and processing operations. For example, he notes that about 5.5 tons of fuel are needed to manufacture 1 ton of cement, which is used in the production of the dominant construction material, concrete. The production of each ton of cement requires, in addition, about 1.8 tons of material inputs and releases about 0.5 tons of carbon dioxide.

Similarly, production of each ton of steel involves the removal of 1.1 tons of overburden, the production of 1.5 tons of ore concentration waste, and the release of 1.1 tons of carbon dioxide. Ayres points out that the recycling of metals produces not only significant energy savings but significant movement of overburden, waste production, and carbon dioxide production. A single ton of recycled iron saves 12.5 tons of overburden, 2.8 tons of iron ore, 0.8 tons of coal, and prevents 1 ton of carbon dioxide and 1 ton of sulfur dioxide from being released.

Ayres also suggests that the consumption of fuel by the built environment is the second major cause of environmental damage, second only to the impacts of wastes produced in extracting and manufacturing materials for construction. Buildings now consume 30% of primary energy in the USA and 20% of all electricity consumption. He notes that several technological trends will increase the dependence of buildings on electricity even while they reduce the per capita consumption of energy. For example, photovoltaics are now becoming cost-effective for residential installation, and he expects that 20–25% of total electricity demand will be met by photovoltaics by 2050. Another area of technical innovation has been in the increasing efficiency of appliances. Swedish refrigerators, for example, have improved in efficiency per unit volume by a factor of 25 over the past forty years. The introduction of compact fluorescent lighting has produced a fourfold improvement in lighting efficiency in the past two decades. New microwave cooking technologies and improvements to heat pumps are providing similar increases in energy efficiency.

In short, Ayres is optimistic that the emissions from the built environment can be dramatically reduced. The recycling of metals and other materials will substantially reduce mass materials movement, and the energy and emissions from processing the used materials into new products will be substantially lower than if virgin resources were employed. With respect to energy, the emergence of photovoltaics and dramatic increases in the energy efficiency of appliances and heating/cooling systems will continue to decrease emissions caused by the built environment.

Iddo Wernick (Chapter 7) begins his contribution by posing the question: What strategies does industrial ecology offer for smart design and systems thinking about the human built environment? He begins by suggesting that industrial ecology should embrace the strategy of minimizing the use of materials resources and disturbance to natural systems. This point includes integrating structures into the natural environment and creating built environment systems that work with Nature, not against it. With respect to mass flows of materials into the built environment, the introduction of materials into buildings and infrastructure has increased by a factor of about 20 in the past century, or by a factor of 6 when population increase is taken into account. Energy inputs to construction materials are significant, especially for metals, cement, glass, and plastics. However, some materials have dramatically more energy investment than the average. For example, aluminum production requires eight times more energy input per ton than steel and fifty times more energy per ton than plastics if all use virgin resources.

Wernick does suggest several strategies for the built environment that would reduce resource consumption and impacts on natural systems. Dematerialization, or a reduction in materials use per building, could be achieved through efficient design of structures, which to some extent has already been achieved through better modeling techniques. Systematic recovery of materials is another strategy. Current recovery rates, even for metals such as copper (15%) and aluminum (22%), are still low and could be dramatically improved. Significant recovery of materials is highly dependent on the initial value of materials used. Relatively low-value materials, in the sense of recycling value, will result in cascading, whereby materials are used in lower value applications in successive lifetimes. In the USA, concrete aggregates, for example, when extracted by crushing from used concrete, are generally reusable only as road sub-base and not in new structural concrete. The same is true for a wide range of products, such as carpet and drywall. The worst case is composite materials that have no potential for recycling or reuse and must be disposed of in landfills at the end of their useful life.

Land is both a critical resource and the plane upon which ecological systems exist. Wernick suggests that the use of land should be evaluated through the use of area-based measures such as the sustainable process index or the ecological footprint. He also notes that, in addition to quantity of land, the quality of land should also be taken into account, and he recommends using the net primary production (NPP) of given land areas as part of a decision systems that would determine where best to site the built environment. Together with examining the animal habitat and hydrology of the site, NPP could be factored into decisions on priorities for development or protection. The assumption here is that productive land, in the biological sense, should be protected. Using land for several purposes, for example by making more use of high-rise structures, could combine living, working, and educational space in a single structure and could decrease the physical footprint of the built environment. Underground buildings could also provide habitat while minimizing the alteration of surface ecological systems.

Wernick also addresses the application of ecology to the built environment and draws upon the work of James Kay (see Chapter 3 of this volume) by suggesting that the design of the built environment should consider (1) the interface with Nature; (2) the model provided by Nature; and (3) the direct use of natural systems.

In Chapter 8, Stefan Bringezu states that, in attempting to bridge the gap between sustainability theory and professional practice, the central question for the built environment is how to increase resource efficiency. It is clear that at this point in human history we are faced with the paradox of how to reduce environmental burdens while increasing material welfare and services to people. The approach he suggests is twofold: considering the materials intensity per service unit (MIPS) and implementing integrated resource management (IRM). He notes that, from an economic perspective, several management rules for sustainability have been created:

1 The use of renewable resources should not exceed their regeneration rate.
2 Non-renewable resources should be used only if physical or functional substitutes are provided, i.e. man-made capital can be substituted for natural capital.
3 The waste assimilative capacity of Nature should not be exceeded.

Although it is clear that sustainability requires a four- to tenfold increase in resource productivity, the demand for materials for the built environment continues to increase, and the land area occupied by buildings and infrastructure is now greater than the area affected by extraction of mineral resources such as mines and quarries.

Bringezu proposes the strategies of "rematerialization" and "dematerialization" to reach the goal of increased resource productivity. Rematerialization is the closing of the materials cycles by reuse, remanufacturing, and recycling and includes the cycling of biomass by agriculture and forestry. The consequences of rematerialization are to reduce the need for primary resource inputs (together with the accompanying "ecological rucksack" or upstream hidden flows of overburden and extraction waste) and to diminish final disposal quantities without decreasing the overall flow of materials needed by humankind. Dematerialization aims to reduce absolute materials flows by having industry become more efficient in its use of energy and materials and includes more efficient use of products and energy by the consumers.

Bringezu introduces us to the concept of material intensity analysis to relate total material requirements to the service provided. He refers to the results of this analysis as materials intensity per service unit (MIPS). In this analysis, the primary input is aggregated into five categories: abiotic raw materials, biotic raw materials, soil, water, and air. Applying this approach to Germany, he and his colleagues determined that in Germany the highest materials demand came from housing, followed by food and leisure. In applying this thinking to construction, he presents the golden rules of ecodesign:

1 Potential impacts to the environment should be considered on a life cycle-wide basis.
2 Intensity of use of processes, products, and services should be maximized.
3 Intensity of resource use (material, energy, and land) should be minimized.
4 Hazardous substances should be eliminated.
5 Resource inputs should be shifted toward renewables.

With respect to construction, he suggests that the following parameters impact its materials and energy intensity: type of construction, materials chosen, durability, repairability, and dismantlability. Achieving a fourfold increase in construction resource productivity could be achieved by reducing the mass of new construction by one-third; choosing materials that reduce the material requirement by one-third; increasing durability by one-third; and improving repairability and dismantlability to reduce life cycle materials use by one-third.

Materials management focuses on opportunities for rematerialization and detoxification, the latter addressing the presence of toxic substances that interfere with recycling. Like Wernick, Bringezu notes the problem of cascading chains of value for recycled construction materials, indicating that it is necessary to main the value as materials are reused and recycled. Presently, most demolition waste is recycled into road sub-base, for which it is in fact far preferable to use mineral products; glass should be recycled into more glass products, and so on to maintain the chain of value. Planning for infrastructure, particularly in the provision of renewable energy sources, needs to be greatly improved. Bringezu mentions a German study that showed that the material requirements for construction and maintenance of wind generators were lower than for grid electricity. Unfortunately, wind generators require the use of non-renewable materials. With respect to photovoltaic installations, he notes that, although they use a renewable energy source, their materials intensity is very high and their emission intensity for carbon dioxide exceeds that of nuclear power plants by a factor of 10. So, even though the price of photovoltaics is rapidly falling, a dramatic reduction in materials intensity if needed to make photovoltaics sustainable in all senses.

Water utilization is also reviewed by Bringezu, and he states that water use in German households has been reduced to 100 liters per day by improved plumbing fixture technologies. Rainwater harvesting may have some additional benefits, but the materials intensity of these systems makes them most suitable for regions with a deficit of potable water. He also reviews possibilities for decentralized wastewater treatment, which has significant materials advantages over conventional, centralized municipal wastewater treatment systems.

The concept of integrated resource management (IRM) is introduced by Bringezu to round out his discussion of materials intensity as it applies to the built environment. IRM "... comprise(s) a minimization of resource requirements and the control of critical emissions together with an optimal use of financial and social resources." For the built environment he presents a method to benchmark buildings that was developed by his colleagues at the Wuppertal Institute. This method uses input-oriented indicators such as materials productivity and energy requirements and output-oriented parameters such as pollution pressures and toxic effects.

Chapter 9, the final chapter in this section, was written by Fritz Balkau. He focuses on implementing and operationalizing industrial ecology through management and policy instruments. In reviewing the concept of industrial ecology, he suggests that it might be defined as the study of materials and energy flows, population dynamics, and the operational rules and interrelationships of the entire production system. The challenges in implementing this strategy are insuring that the industrial ecology concept is complete so that it addresses all policy areas and that an effective combination of management instruments are available for applying the concept in real situations. The main elements of industrial ecology that have been suggested are industrial metabolism, industrial ecosystems or associations, materials cycles in Nature and industry, and the evolution of industrial technologies. These, in turn, have resulted in a number of concepts for operationalizing sustainability: the precautionary principles, the prevention principle (cleaner production and ecoefficiency) life cycle management, the zero emissions concept, dematerialization (the factor 10 concept), and integrated environmental management systems. He suggests that we have not yet seen a mature industrial ecosystem in which management systems have evolved sufficiently to produce a true artificial ecology. However, a number of management elements have appeared which give us hints as to how these management systems may eventually appear. Among the existing dynamic management elements are corporate decisions on sustainability; the adoption of environmental management systems (EMS); the practice of supply chain management; central infrastructure management; cooperative environmental programs; and government industrial development policy. The challenge is to combine these management instruments in an intelligent and systematic fashion.

The dynamics of large industrial estates are being examined by the United Nations Environmental Program (UNEP) for the purpose of developing technical guidelines for ecoindustrial parks. In addition to reducing the environmental burden of individual industries, synergies in operating and support services can benefit the participants. It is also important to realize that environmental and concerns are evolving, as are social concerns, and industrial ecology must similarly evolve to address the changing sustainability landscape.

Balkau suggests that the construction sector also needs to stay abreast of emerging environmental problems and adapt the design, operation, and disposal of the built environment to address new issues. The construction industry also needs to be more aware of the secondary impacts of its activities, i.e. the damage done during the extraction of the resources needed for creating the products that constitute buildings and infrastructure. Quality of life as affected by construction also need to be included in the array of issues for industry awareness and possible action. For example, congested transportation systems, increased noise, and increased municipal solid waste are also outcomes of construction activity. He concludes by suggesting a management framework for construction ecology. A wide variety of instruments from environmental standards to building codes and financial criteria can be applied to construction ecology and assist its implementation. However, the primary prerequisite for creating a framework of management instruments is the definition of environmental goals. To accomplish this, the construction industry itself must come up with a common view of its environmental agenda to include parameters such as energy efficiency.

The four chapters in this section provide a wide range of answers to the question posed at the start of this section: What are the major lessons that can be learned from the experience of industrial ecology over the past decade that are applicable to the built environment? There are perhaps three salient lessons that are provided by industrial ecologists for the construction industry: resource productivity, minimal emissions, and management systems for implementation.

Resource productivity, particularly energy, water, and land, could be significantly improved by a factor of 4–10 as suggested by Stefan Bringezu through better planning and new technologies. With respect to materials, the situation is not as clear. In some cases, higher mass aids passive energy design and the typical materials used for infrastructure development (sand, gravel, Portland Cement concrete, and asphalt concrete) do not lend themselves to maintaining value in recycling. Nonetheless, his suggestion that resource efficiency can be achieved through rematerialization and dematerialization is very applicable to construction in general. In addition to the rules suggested by the authors with respect to ecodesign, an additional rule should be that only materials with demonstrated recycling potential back to their original use should be permitted in building products. This would eliminate low-value materials such as drywall (unless the industry is willing to develop systems that bring used product back for recycling) and composite materials, such as wood laminated with plastic, that cannot be disassembled or recycled. One further note is that there are actually two recycling routes for building materials: (1) through the biosphere for biologically derived products such as wood, cork, or hemp; and (2) through the technosphere for synthetic products such as concrete, steel, and plastics. The latter route, recycling as we know it, is well developed and requires better management and regulation to make it more effective. The development of the former route, biosphere recycling, deserves significant attention because it provides the potential for turning many building products into nutrients for ecosystems via composting and other routes. However, at present, there are no large-scale systems that can take dimensional lumber, oriented strand board (OSB), and a wide variety of other biological products and return them as nutrients to the biosphere. This is clearly an area for major exploration both in terms of technology and policy.

The second lesson provided by the industrial ecologists is to minimize emissions. The built environment produces, or causes to be produced, a wide range of emissions over its life cycle, including solid, liquid, and gaseous substances. There is waste in the materials extraction process, manufacturing, installation, and maintenance. Emissions from energy systems that are needed to support the functioning of the built environment ultimately dominate the overall quantity of emissions, simply because of the long lifetime of buildings and their large continuous demand for energy. The solutions for reducing emissions are straightforward but difficult to achieve and rely on better manufacturing and construction processes as well as rematerialization: reuse, remanufacturing, and recycling. Maximizing the rematerialization of building materials greatly lessens the impact of materials extraction and its accompanying waste stream by reducing the quantity of materials extracted from mines, quarries, and forests. Greatly improving the passive design of buildings and shifting to hyperefficient appliances and heating/cooling systems would greatly reduce their demand for energy. Shifting to effective and efficient photovoltaic systems would address the supply-side emissions problems of energy systems. Unfortunately, conflicting advice is provided by the authors with respect to photovoltaic systems. Robert Ayres suggests that use of photovoltaics would help quickly remove our dependence on fossil fuels, whereas Stefan Bringezu contends that the ecological rucksack of materials resulting from the production of photovoltaics make their present-day application dubious. It is clear that part of the solution must rest on the shift to renewable energy systems. Also, emerging photovoltaic technologies, such as direct conversion, may substantially reduce the ecological footprint of this renewable energy strategy.

The final lesson provided by the industrial ecologists is that implementation of both industrial ecology and construction ecology must be carried out using the appropriate policy instruments by a variety of entities to include government, corporations, and developers. An environmental agenda that the construction industry can agree to is particularly important as it would set the parameters for behavior of the many actors in the construction process. Coordination in the application of policy instruments such as building codes and standards for building products would help orchestrate a steady march toward a system of creating the built environment that pays careful attention to resource and environmental issues. Coherent action is important to be able to produce change and the establishment of an agenda to integrate policy and technical issues is needed to create this coherency.

# 6 Minimizing waste emissions from the built environment

## Toward the zero emissions house

*Robert U. Ayres*

The chapter reviews the material inputs to the construction industry (i.e. those materials that are ultimately embodied in structures) and the wastes associated with materials processing. It also reviews the emissions associated with energy use for purposes of heating, air conditioning, lighting, cooking, and other energy services normally provided within structures. Of the two categories, the latter is more significant in terms of environmental damage. As regards abatement, the most promising strategy for reducing the damages associated with mining and processing the structural materials themselves is increased recycling, especially of metals. Two primary strategies suggest themselves with regard to reducing the environmental impact of energy services. The first is increased end-use efficiency, e.g. by improved design, better thermal insulation, and more efficient equipment, such as refrigerators and compact fluorescent lights. The second primary strategy is to shift as quickly as possible from dependence on fossil fuels for heating and cooking to electricity, especially by utilizing photovoltaic (PV) rooftop units together with heat pumps and microwave cookers. Government intervention may accelerate this shift in various ways, but policy issues are not discussed in this chapter.

### Background

The construction sector, together with the production of associated durable goods and consumables needed for purposes of maintenance and operation, constitutes by far the largest end use for materials that are embodied in products – especially structures. In 1993, 2.13 billion tonnes of materials were embodied in new structures in the USA alone, compared with around 60 million tonnes of machinery and equipment (producer durables) and 20 million tonnes of consumer durables (some of which, such as carpets and kitchen appliances, are essentially components of residential housing).

The materials in structures and durable goods are, for the most part, relatively inert and unreactive. By far the largest share consists of sand, gravel, and crushed stone, which are natural mineral products processed only by grinding, sorting, and washing. The next largest share consists of mineral substances further processed only by heating for purposes of melting, sintering, calcination (driving off carbon dioxide), or simply dehydration. These products include glass, bricks and ceramics, Portland cement, and plaster of Paris. The construction sector and its satellites also consume fairly large quantities of organics such as lumber, paper, plastics, rubber, and textiles. Finally, it is a major user of metals, especially steel and aluminum. Although metals constitute a very small part of the total, they do account for quite a large share of the indirect pollution associated with structures, as noted below.

Only the organics are inflammable and perishable. This means that they can become pollutants if they burn or escape in other ways. Chlorinated plastics such as PVC, widely used in structures and for insulation of electrical wiring, can generate dioxins in fires. Plywood, paper, plastics, and textiles also embody significant amounts of other chemicals, including adhesives, wood preservatives, fire retardants, coloring agents, plasticizers, fungicides, and so on. The chlorofluorocarbons formerly used in foam insulation and compressors, and the polychlorinated biphenyls (PCBs) that were formerly used in electrical transformers and capacitors are further examples.

Surface protection of structures involves additional chemicals that can escape into the environment during application or use. Undeniably, there are significant environmental hazards resulting from some of the materials used in structures themselves. (Asbestos and formaldehyde are two examples.) The residual toxicity resulting from past use of lead-based paints as well as mercury-based fungicides and anti-mildew agents, especially indoors, and the more recent pollution problems arising from solvents, adhesives, wood preservatives, and plasticizers illustrate a few of the problems.

However, beyond doubt, the most significant environmental hazards arise from waste materials associated with metals and chemicals production or consumed in system operation, maintenance, and repair. (Indoor air pollution is increasingly recognized as a more serious health hazard than outdoor air pollution, by a considerable margin.) Fuel consumption for space heating is only the most obvious example. Inefficient combustion, especially in kitchens and wood-burning fireplaces, generates a witches' brew of pollutants, including carbon monoxide, unburned hydrocarbons, and microparticulates (smoke). Solvents used in paints constitute another well-known hazard. Agents used for termite control or as fungicides, insecticides, or rodenticides leave residual toxicity. (Lead-based paints are still a hazard in some older tenement buildings.) Another less well-known example is the chromium-based algicides that are used in large commercial air-conditioning systems.

In terms of absolute priorities, the research objective should be lower operations and maintenance costs, longer life of metallic subsystems, and renewability, in that order. However, it is important to recognize that, although the total quantities of building materials used in our society per capita have not increased remarkably (Figure 6.1), there is a significant trend toward the use of more sophisticated materials involving more complex manufacturing processes. In particular, the use of aluminum and plastics in place of wood and other simpler materials has grown rapidly (Figures 6.2 and 6.3). This means that indirect pollution associated with manufacturing these materials is now much more important than it was a few decades ago. Therefore, I consider next the indirect pollution resulting from mining and manufacturing processes for construction materials. Next, I discuss household energy services, excluding transportation services (even though the two are not completely independent). In the remainder of this chapter I consider potential emissions reduction strategies and speculate a bit about future trends.

### *Life cycle emissions from manufacturing construction material*

As noted above, mineral products are consumed in very large quantities by the construction industry, especially sand and gravel, stone, clay, and derivative products such as cement, brick, glass, and plaster. As such, most of these are comparatively harmless. Apart from dust and noise, largely associated with trucks, this is also true of their extraction processes. Quarrying wastes are, in most cases except for clay, modest in comparison with total

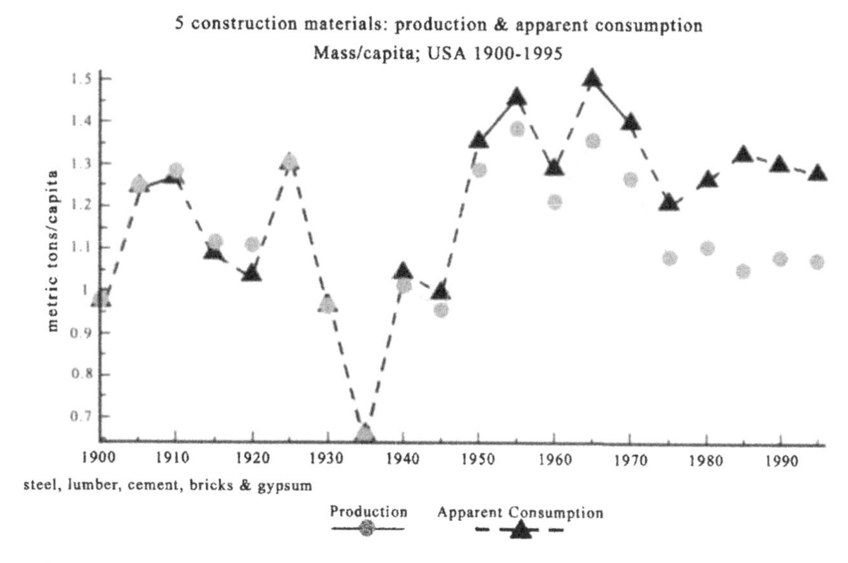


*Figure 6.1* Five construction materials: production and apparent consumption mass per capita, USA, 1900–95.

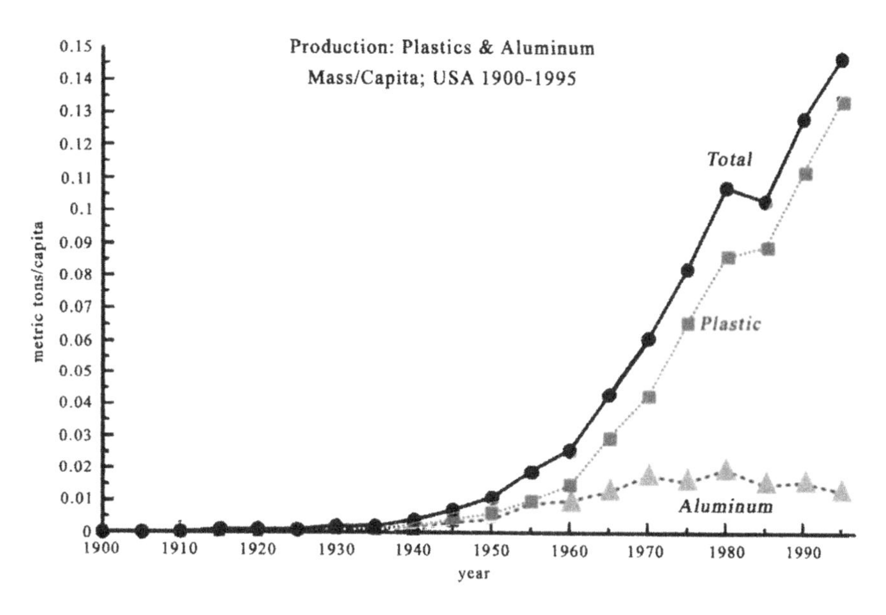


*Figure 6.2* Production of plastics and aluminum mass per capita, USA, 1900–95.

quantities produced. Emissions associated with downstream processing are overwhelmingly of two kinds, namely (1) dust (particulates) from crushing and grinding and (2) combustion wastes due to fossil fuel usage for thermal processing (calcination). Most dust from open-air mining and quarrying operations can be reduced to tolerable levels by simple methods, such as water sprays. (Silicaceous dust, however, is the exception; it is a serious health hazard — albeit mainly to workers. Suppression requires special measures.) Dust from

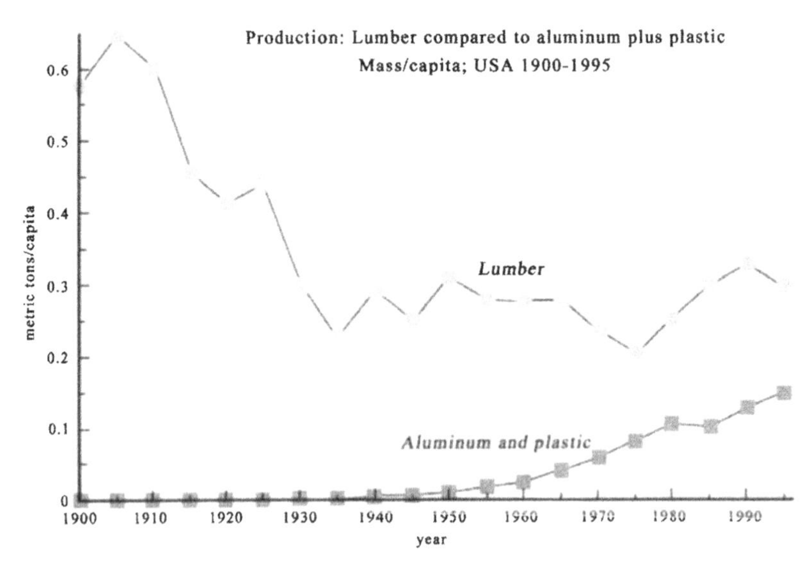


*Figure 6.3* Production of lumber compared with aluminum plus plastic mass per capita, USA, 1900–95.

most grinding operations, cement plants, and other manufacturing operations can be captured quite efficiently by electrostatic precipitators.

Fuel consumption is the source of the most important emissions, mainly of carbon dioxide, carbon monoxide, and nitric oxides. Portland cement plants are by far the major consumers of fossil fuels (among construction material-producing industries), with glass and brick kilns a somewhat distant second. The cement industry of the USA consumed a little less than 12 million metric tons (MMT) of fuel (mainly coal) in 1993, to produce 66 MMT of Portland cement from 117 MMT of mineral inputs, of which about 33 MMT was emitted as carbon dioxide from calcination of input carbonates (United States Bureau of Mines, 1993, Portland cement). Allowing for ash, nitrogen, water, and other non-combustibles, coal is approximately 75% carbon by weight; therefore, fuel consumed in cement plants contained about 9 MMT of carbon and generated about 33 MMT of carbon dioxide in that year, adding another 33 MMT of carbon dioxide, or 66 MMT altogether, or 1 ton of carbon dioxide per ton of cement. Essentially, all of this can be attributed indirectly to the construction industry.

It is worth mentioning, by the way, that the sulfur and ash emissions normally associated with coal burning are not problems in the case of cement plants. In fact, in 1993 US cement plants safely burned over 70 kMT of scrap tires, 90 kMT of other waste solid fuels, and 670 kMT of waste engine and other lubricating oils.

Bricks and tiles produced in 1993 for the construction industry consumed about 13 MMT of common clay, to produce a somewhat smaller mass of finished (dry) bricks and tiles (United States Bureau of Mines 1993, Clay, Table 6.8). Data on fuel consumption by the brick and tile manufacturers are not readily available, but it is not unreasonable to assume that the fuel required per ton of carbon-free input material (adjusting for the carbon dioxide content of limestone used by the cement industry) is similar to that for cement manufacturing, as similar temperatures are involved. This being the case, the brick and tile industry must have consumed about 1 ton of fuel for 7 tons of clay, or a little less than 2 MMT altogether. If the original clay had a water content of 30%, or 4

MMT, it would follow that 9 MMT of finished bricks and tiles were produced, entirely for the construction sector. Assuming that the fuel was mostly natural gas (75% carbon), the carbon content would have been 1.4 MMT, generating 5.1 MMT of carbon dioxide in the processing, or about 0.55 tons of carbon dioxide per ton of bricks.

Approximately 15 MMT of glass is produced each year in the USA, of which roughly two-thirds is used for containers and part of the remainder is used for vehicles (40 kg per car), TV screens, computer monitors, and so on. The quantity used by the construction industry for windows and doors is probably between 3 and 4 MMT. Detailed data on fuel consumption by the glass industry are not available. However, higher temperatures are needed to melt glass than for baking bricks or producing Portland cement. On the other hand, the calcination contribution in glassmaking (from soda ash, or sodium carbonate) is proportionally smaller. The glass used in construction accounts for roughly its own weight of carbon dioxide emissions, or 3–4 MMT.

Roughly 18 MMT of calcined gypsum products (mainly plaster wallboard) were produced and consumed for construction purposes in the USA during 1993 (United States Bureau of Mines 1993, Gypsum, Table 4). This includes the weight of other minor materials incorporated in the products, such as paper and metal. Calcination involves low-temperature heating (up to 350°F) to drive off part of the water of hydration of the gypsum. This water is later added back to the finished material. There are no data on specific fuel consumption, but it must be considerably less than for brick kilns. Lacking other data, we estimate that heat energy requirements for plaster products (by weight) would be one-quarter to one-third of the requirements for bricks. This implies carbon dioxide emissions of around 3 MMT, give or take 0.5 MMT.

Detailed calculations of life cycle emissions generated by metallurgical and chemical processes involved in the manufacture of construction materials are much more complex and cannot be reproduced here in detail. An overview of the US steel sector for 1993 is shown in Figure 6.4 (Ayres and Ayres 1998, Chapter 5). Wastes generated within the sector per ton of steel (including ore mining and concentrating) amounted to about 1.1 tons of overburden, 1.5 tons of concentration waste, and around 1.1 tons of carbon dioxide (allowing for some small contributions such as carbon dioxide emissions during prior calcination of lime, from limestone). Minor wastes include steel slag (about 0.05 tons), ferrous sulfate, or chloride "pickling" wastes, and so on. (Note that iron slag from blast furnaces is no longer considered a waste, nor is blast furnace gas, which is recovered and burned as a low-grade fuel.)

Shipments direct to the construction sector accounted for 15% of the total in 1993, but this is an underestimate, as it takes account only of structural steel such as girders and reinforcing bars, as well as cast iron pipe, purchased by very large contractors. The largest single "consumer" of steel products, accounting for 26.6% of total output, are "distributors and service centers," which resell steel products to other sectors, including construction (United States Bureau of Mines 1993, Iron and steel, Table 3). It is reasonable to assume that, overall, construction uses also account for at least 20% of the sales of distributors and service centers, and probably more. This raises the total to at least 21%. But this figure still does not include small items of hardware such as fasteners, hinges, and locks or steel embodied in machinery installed in structures, such as elevators, kitchen equipment, and heating/ventilation equipment. When these are taken into account, it seems likely that between 25% and 30% of all iron and steel products end up in structures. On this basis, we estimate that the construction sector consumed roughly 26 MMT of iron and steel, with an uncertainty (plus or minus) of 4 MMT.

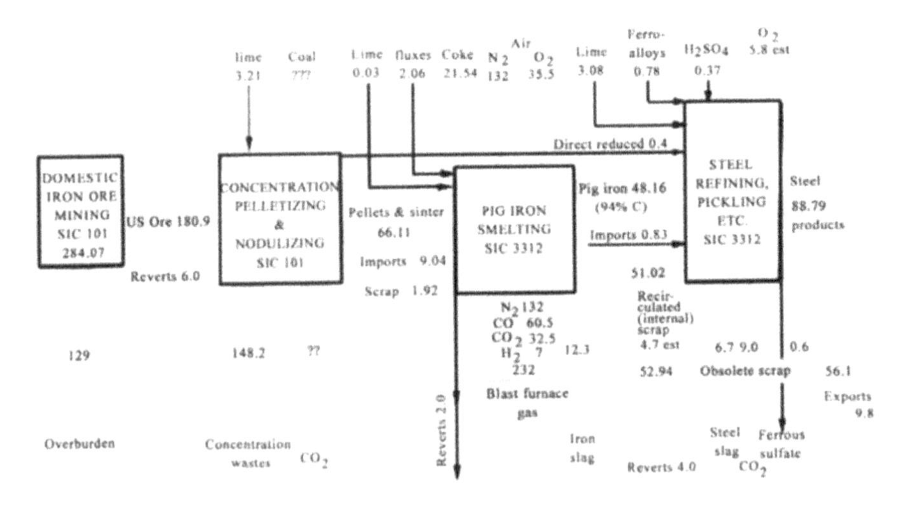


*Figure 6.4* An overview of the US steel sector for 1993.

Translating into wastes and emissions, it follows that iron and steel used for construction purposes carry with them an environmental burden amounting to approximately 30 MMT of overburden from mines, 40 MMT of ore concentration wastes, 30 MMT of carbon dioxide, 1.3 MMT of steel slag, plus waste pickle liquors, coal-washing wastes, coke oven quenching wastes, and other minor contributions.

Aluminum, copper, zinc, and lead are other metals used by the construction sector. Aluminum is used for window and door frames, roofing and "curtain walls" for some large office blocks. These uses accounted for 15% of total US shipments in 1993 (United States Bureau of Mines 1993, Aluminum, Table 6.7]. Copper is used in structures for wiring and water pipes, and as a constituent of brass for hardware. The construction industry accounted for 42% of copper and brass end uses, while electric and electronic products accounted for another 24%, of which a significant portion is also embodied permanently in structures as wiring (United States Bureau of Mines 1993, Copper, p. 331). Zinc is used mainly for protective coating (galvanizing) of water pipes, gutters, or sheet steel as a roofing or siding material. It is also a constituent of brass. Coatings accounted for 54% of demand for zinc metal in 1993, while brass alloys took 14% (United States Bureau of Mines 1993, Zinc, p. 1281). The major end uses included construction, transportation equipment and machinery. A detailed breakdown is not available, but as much as half of all galvanized metal and brass may have been embodied in structures. Lead was formerly used extensively for water pipes and paint and is still used in small quantities for soundproofing, roofing (in a few applications), and in solder. Quantities are relatively minor, but toxicity is still a significant environmental hazard.

Detailed accounts of mine wastes, concentration wastes, smelter wastes and so on due to extraction and processing of these metals cannot be undertaken here. A composite overview is shown in Figure 6.5. More details can be found in Ayres and Ayres (1998, Chapter 5). However, one point worth emphasizing here is the importance of metal recycling as a way of reducing the environmental impact of metal use.

Every ton of metal that is reused, remanufactured, or recycled – or whose use is avoided by more efficient design – replaces a ton that would otherwise have to be mined and smelted, with all of the intermediate energy and material requirements associated with those activities. This is already very significant for iron and aluminum. Each ton of iron recycled saves 12.5 tons of overburden (coal and iron mining), 2.8 tons of iron ore, 0.8 tons of coal (exclusive of its use as fuel), and a variety of other inputs. It also eliminates at least a ton of carbon dioxide pollution and significant additional pollution of air and water from coking, pickling, and other associated activities (see Figure 6.5). In the case of non-ferrous metals, the indirect savings are much larger, of course, although much depends on the original ore quality.

To the extent that the smelting is carried out in less regulated countries outside North America and Europe, recycling or remanufacturing a ton of copper or zinc also saves 1.0 tons of sulfur dioxide that would otherwise be emitted into the air. In the USA, Canada, and Europe sulfur from non-ferrous metal smelters is recovered as sulfuric acid, which is subsequently used for the increasingly important heap-leaching process in North America, but this is not being done extensively elsewhere. A third factor is also increasingly important from an industrial ecology viewpoint. It is the fact that the gross supply of many minor metals, especially some of the most toxic ones such as arsenic and cadmium, is determined not by direct demand *per se* but by the demand for more important metals such as copper and zinc, with which they are normally associated as minor by-products of ore processing and smelting. Therefore, the more copper is recycled, the less needs to be mined, thus reducing the aggregate arsenic supply. Similarly, the more zinc that is recycled, the less cadmium will be produced to pollute the soil or find harmless uses for.[1]

Table 6.1 shows calculated savings (indirect raw materials and other inputs not used) when a metric ton of metal is recycled (Ayres 1997). It is, therefore, the calculated *difference* between inputs per ton of semifinished metal produced from raw materials *vis-à-vis* the inputs per ton of secondary scrap recycled. Because of the complexity of the system, I have assumed simplified process–product chains as follows:

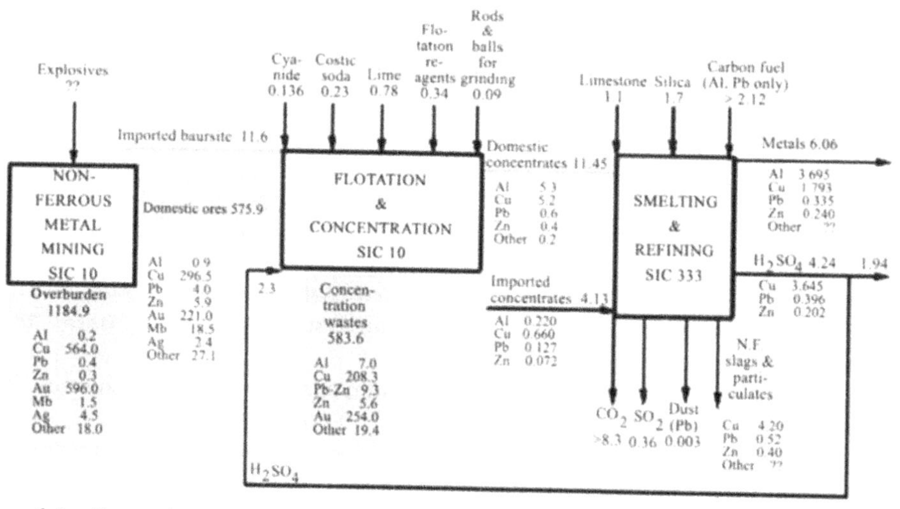


*Figure 6.5* Composite overview of the US non-ferrous metals industry: mine wastes, concentration wastes, smelter wastes, etc. due to extraction and processing.

*Table 6.1* Recycling savings multipliers (tons/ton product)

| | *Iron/steel* | *Aluminum* | *Copper* | *Lead* | *Zinc* |
|---|---|---|---|---|---|
| Ore percent | 52.8% Fe | 17.5% Al | 0.6% Cu | 9.3% Pb | 6.2% Zn |
| Major source | McGannon (1971) | | Gaines (1980) | PEDCo-Environmental (1980a) | PEDCo-Environmental (1980b) |
| Energy used (GJ/t) | 22.4 | 256 | 120 | 30 | 37 |
| Water flow in/out (t/t) | 79.3 | 10.5 | 605.6 | 122.5 | 36.0 |
| *Material inputs* | | | | | |
| Air | 1.9 | 0.3 | 1.6 | 4.4 | 5.8 |
| Solids | 17.3 | 11.0 | 612.1 | 126.2 | 55.8 |
| Total material inputs | 19.2 | 11.2 | 613.7 | 130.5 | 61.6 |
| *Material outputs* | | | | | |
| Product | 0 | 0 | 0 | 0 | 0 |
| By-products | 0.2 | 0.1 | 1.0 | 6.7 | 4.9 |
| Depleted air | 1.5 | 0.2 | 1.3 | 1.2 | 2.4 |
| $CO_2$ | 0.5 | 0.8 | 0.02 | 0.03 | 0.03 |
| $SO_x$ | 0.01 | 0.06 | 1.47 | 0.005 | 0.01 |
| Other gaseous material | 1.18 | 0.002 | 0.15 | 0.28 | 0.03 |
| Potential recycle | 0.6 | | 3.2 | 0.1 | 0.5 |
| Overburden | 12.5 | 0.6 | 395.4 | 72.5 | 37.3 |
| Gangue | 1.1 | 6.1 | 211.0 | 44.6 | 16.3 |
| Other solid material | 1.5 | 1.4 | 0.1 | 2.5 | 0.1 |
| Sludges, liquids | 0.1 | 1.9 | 0.1 | 2.6 | 0.1 |
| Total material outputs | 19.2 | 11.2 | 613.7 | 130.5 | 61.6 |

Notes

Major byproducts include sulfur dioxide used for sulfuric acid production and saleable offgas. Depleted air = air from which all oxygen has been taken for combination with other materials. Potential recycle includes slag, scrap, etc. potentially usable in the process chain, but not used. Overburden = that portion of solid material extracted during the mining process that is not part of the ore. Gangue = that portion of the ore extracted during beneficiation that is not part of the concentrate. Sludges and liquids do not include dilution water. Source for aggregate energy values: Forrest and Szekeley (1991). The materials/energy costs of producing this energy are not considered in this table. Source for water flow values: Lübkert *et al.* (1991). Source for overburden & gangue percentages: Adriaanse *et al.* (1997). Other sources used: Davis ( 1971, 1972a,b), Battelle–Columbus Laboratories (1975), Thomas (1977), Lowenbach and Schlesinger (1979), Bolch (1980), McElroy and Shobe (1980), PEDCo-Environmental (1980c), Forrest and Szekely (1991), Lübkert *et al.* (1991), and Masini and Ayres (1996).

- Raw solids (*iron mining*) → 53% iron ore (*beneficiation*) → concentrate (*sintering/pelletizing*) → sinter/pellets (*blast furnace*) → pig iron (*electric arc/basic oxygen furnace*) → steel.
- Raw solids (*bauxite mining*) → 17.5% aluminum ore (*Bayer process*) → alumina (*Hall Heroult process*) → aluminum.
- Raw solids (*copper mining*) → 0.6% copper ore (*beneficiation*) → concentrate (*smelting*) → blister copper (*refining*) → copper ingots.

- Raw solids (*lead mining*) → 9.3% lead ore (*beneficiation*) → concentrate (*sintering*) → (sinter) (*blast furnace*) → (lead bullion (*drossing*) → drossed lead (*refining*) → lead.
- Raw solids (*zinc mining*) → 6.2% zinc ore (*beneficiation*) → concentrate (*sintering*) → sinter (*smelting*) →slab zinc.
- Recycling scrap, all metals (*melting*) → metal product.

All numerical values in Table 6.1, except as noted specifically below, are calculated by matching the elemental composition of outputs at each stage of processing, element by element, to the required elemental/chemical composition of inputs to the next stage. These elemental compositions are normally given as *ranges* in published process descriptions (for sources, see Table 6.1). The final results are therefore consistent with published process descriptions while satisfying the mass balance requirement. For convenience, I have neglected the *indirect* contributions to processing of all other input materials used (explosives, acids, flocculants, fluxes, anodes, etc.). Only the actual mass of these secondary input materials themselves was counted.

One final class of materials worth mentioning is plastics. Reliable recent data on uses in the building sector are not readily available. However, in 1987 the construction sector accounted for 4.23 MMT or 19.8% of all plastics (resins) consumed in the USA, by weight. Similarly, 18% of all plastics production in Europe in 1991 was used in construction (Ayres and Ayres 1996, Chapter 12, Figure 12.2). Polystyrene foam is used mainly for thermal insulation; polyethylene sheet is used for many miscellaneous purposes, including waterproofing. However, the most important plastic used in structures is PVC, which is used for window and door frames, flooring and siding materials, electrical insulation, and water pipes. These applications accounted for 2.47 MMT or 71.7% of all PVC consumption in the USA in 1987.

The major raw materials for PVC production are ethylene and elemental chlorine, which is a co-product (with caustic soda) of the electrolysis of sodium chloride (as brine). PVC accounts for roughly 25% of all chlorine produced. The intermediate products are vinyl chloride monomer and ethylene dichloride, both of which are either toxic or carcinogenic. PVC itself is not harmful, but it is not easy to recycle (mainly because of the variety of different additives used), and disposal is something of a problem. In particular, incineration of PVC is a likely source of dioxins, which are classed as persistent organic pollutants (POPs) because of their affinity for animal fats and resulting tendency to be concentrated in the fatty tissues of animals as they move through the food chain. Several of the dioxins are known to be very potent carcinogens for laboratory rats and dioxins are regarded as probable human carcinogens.

In summary, most structural materials, by weight, are harmless or nearly harmless to the environment throughout their life cycles. The principal exceptions, which account for very small total quantities, are metals and the plastic PVC. In the case of the metals the harm is mostly associated with mining and primary smelting operations, although secondary recovery of metals also tends to be a rather dirty process. In the case of PVC, both the manufacturing process and disposal by incineration generate some harmful emissions. However, given that these uses are all relatively long-lived, it can be argued that the environmental damage per unit of service to consumers is comparatively small in all of these cases.

## Household energy services

The other major source of environmental damage associated with structures is fuel consumed in supplying energy services. Globally, households consumed 27% of total commercial energy in 1995 (Raskin *et al.* 1998, p. A-13). For the USA in 1979, one of the last years for which detailed census figures on energy were published, residences (R) accounted for 19% of the total consumption in that year and commercial establishments (C) accounted for 7.8%; detailed breakdowns by category of use are shown below (Ayres 1989, Appendix A). (Other broad categories were agriculture, industry, transport, and commercial). Residences accounted for roughly half of the energy consumed by buildings, the other half being divided equally between commercial establishments and other buildings, including offices and factories. Energy consumption by these categories of users has increased modestly since 1979 (whereas industrial energy consumption has declined), and most of that increase has been in the 1990s.

As a matter of some interest, detailed statistics on energy consumption, both direct and indirect, have been carried out for The Netherlands for 1990 (Vringer and Blok 1995). In the Dutch study, energy use was allocated to households in terms of direct energy carriers (fuels) and indirect consumption via other purchased products and services. According to the Dutch data, 25% of total energy consumption was for home heating by oil or gas, 12.7% was for residential electricity consumption, and the same amount (12.7%) was consumed indirectly as products and services used for the residence itself or in "household effects." It should be noted that Dutch households require more heat and much less air conditioning than US households.

Energy services consumed by residences (R) and commercial establishments (C) in 1979, as percentages of the total, can be classified as follows (Ayres 1989, Appendix A):

1 space heating and ventilation (R-6.22, C-2.51);
2 space cooling/air conditioning (R-0.55 C-1.84);
3 water heating (R-1.79, C-0.285);
4 cooking (including stoves, toasters, microwave ovens) (R-0.78, C-0.88);
5 refrigeration and freezing (R-1.39, C-0.31);
6 laundry and dishwashers (R-0.34, C-0.95 including water and sewer services);
7 lighting (R-0.66, C-1.88, including street lighting);
8 radio, television (R-0.515);
39 other electronic detection and signal processing devices, including fans, vacuum cleaners, mixers, electric tools, electric toothbrushes, hair dryers, shavers, sound recorders and amplifiers, video recorders, IR (fire) detectors, battery rechargers, personal computers, etc. (R-0.29).

Since 1979, energy consumed for space heating has increased as a result of building activity, although delivery efficiency has also improved modestly. The same is true for air conditioning, refrigeration, lighting, and other standard appliances, for which both the increased demand and the efficiency improvements have been more dramatic. Energy consumed for cooking has probably declined slightly, at least in residences. However, energy consumed by miscellaneous devices, especially personal computers, has undoubtedly grown substantially. However, these devices are not essentially part of the built environment, as such, and can be neglected hereafter.

### *Technological trends and future possibilities*

It is evident that, in many of the energy service categories (notably categories 5–9) electricity has almost no competition. Electricity is also the only source of energy for purposes of ventilation, air conditioning and cooling, and microwave cooking. Electricity competes with fossil fuels (gas or fuel oil) to supply hot water and space heating, by means of heat pumps (see below). Thus, electricity can – and eventually will – provide essentially all energy services in the future for both residential and commercial buildings. For all types of electrical services, there is a major future potential for using roof-based photovoltaic (PV) cells as a partial or total source of electricity supply. (Rural households may also utilize water power or wind.) The cost of PV systems is declining rapidly (Figure 6.6). The mean price of a PV module has declined nearly tenfold since 1976, partly because of technological advances and partly as a result of scale economies and manufacturing efficiencies. Further cost reductions foreseeable, depending on market penetration and total installed capacity, range from five- to fiftyfold. However, PV rooftop installations currently cost around $5000/kW, or around $17,500 for an average American house requiring 3.5 kW of capacity. There is a high probability that this upfront investment cost will drop to $5,000 or less within twenty years. If rooftop systems can be integrated with the national power network, so that diurnal and seasonal fluctuations in supply and demand can be smoothed out by utilizing other sources, such as water power and wind power (and also nuclear or fossil fuel power), rooftop PV systems will be very competitive. It is estimated that 20–25% of total electricity demand (and a considerably higher percentage of household demand) can be met in this way by 2050.

It is true that household demand for energy services may increase over time as new applications (such as category 9) are found. The introduction of electric washing machines, dishwashers, and driers is a case in point (Figure 6.7). Obviously the substitution of electricity for fossil fuels for cooking, water heating and space heating has also increased

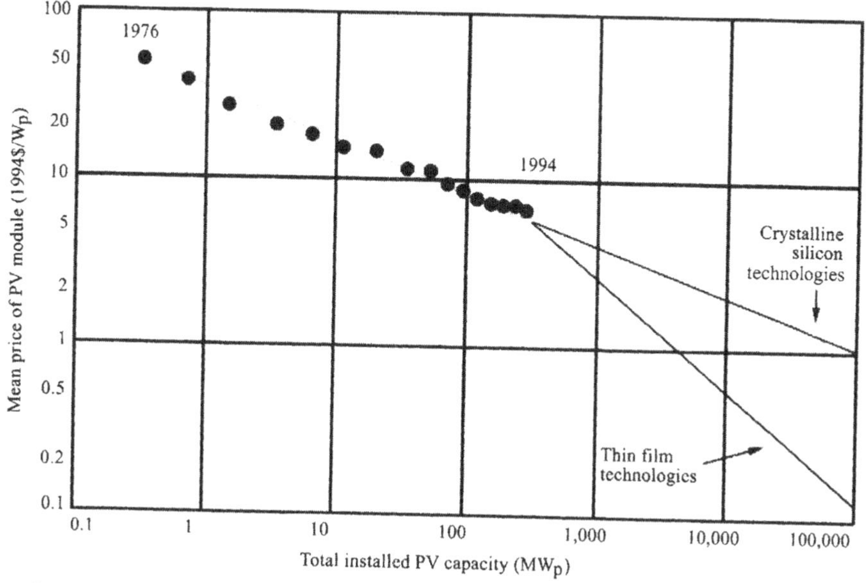


*Figure 6.6* The cost of PV systems.

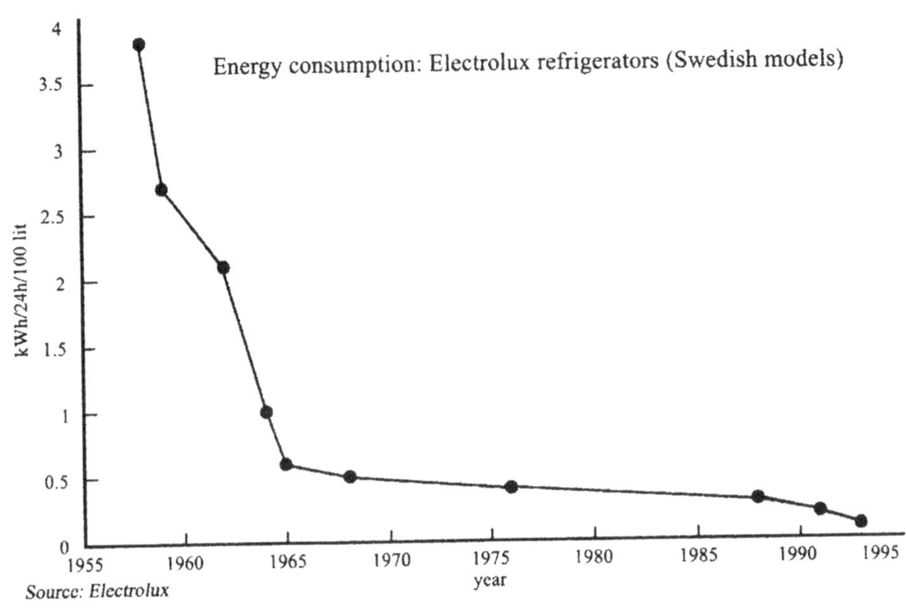


*Figure* 6.7 Household demand for energy services.

electricity demand — at the expense of fossil fuel use — in households and commercial establishments. This trend will continue.

At the same time, increased technical efficiency is also cutting power requirements in most applications. Refrigerators are a case in point. Power consumption by Swedish models, measured in kWh over 24 hours, per 100 liters of volume, have fallen from nearly 4 units in 1958 to 1 unit in 1962 and about 0.15 units in 1993 (Figure 6.8). The declining trend continues. Another example of this phenomenon is lighting, which accounts for a significant share of electric power consumption. But, because of the growing use of improved, compact fluorescent (CF) lights in place of conventional incandescent lights, the quantity of electric power consumed for illumination purposes may decline (Lovins and Sardinsky 1988; Gadgil *et al.* 1991). Other examples of increased efficiency include space heating using more efficient heat pumps[2] and cooking by means of newly developed and improved microwave hot air cookers that can fry, roast, or bake and thus compete with conventional gas or electric ranges.

The three main areas of potential substitution of electricity for other sources of energy are considered in more detail below. It is worth mentioning at the outset that all of these applications, except air conditioning, would result in significant environmental benefits in comparison with the major alternative. The main alternative to air conditioning, however, is lack of air conditioning. The latter requires less energy but may have adverse health impacts on some people living in areas where summer temperatures are very high. Hot spells appear to be increasing in frequency, possibly as a consequence of climate warming. Conditions of extreme heat and humidity can also result in irritability and increased likelihood of accidents, not to mention social discord.

## *Space heating/cooling (climatization)*

Space heating, ventilation, air conditioning, and water heating are jointly the largest

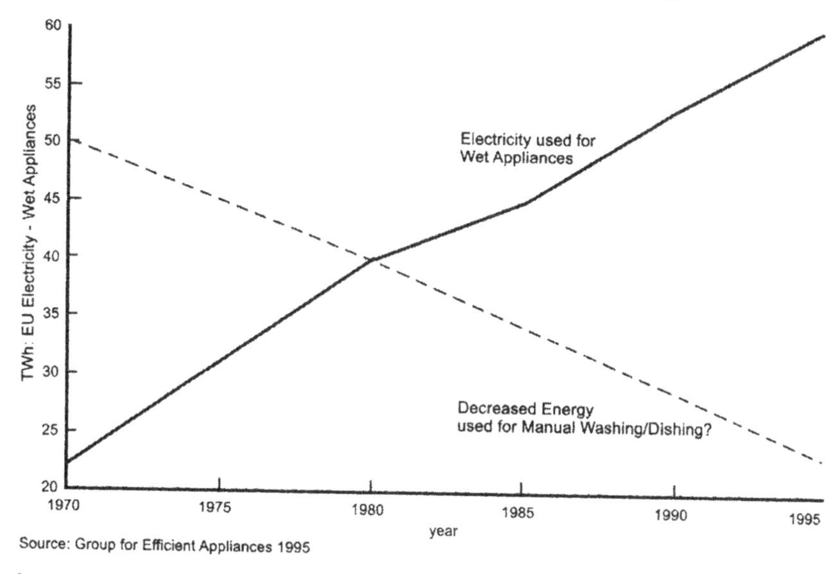


*Figure 6.8* Power consumption.

energy consumers in households. It must be said immediately that the entire process, as currently practiced, is extraordinarily inefficient. Both heating and air-conditioning requirements could be cut enormously – by 90% or more – by utilizing better insulation in walls and, especially, through the use of high-technology double or triple windows. This technology is well known and entails a straightforward trade-off between construction materials for energy. As it happens, the additional costs of optimal insulation would be rapidly recovered by reduced operating costs. Unfortunately, the majority of consumers are, or appear to be, more concerned with minimizing upfront investment costs than downstream operating costs.

An extreme form of the trade-off between capital and operating costs is underground construction, utilizing the insulation capacity of the Earth itself. This concept is unlikely to achieve wide acceptance, however, for two reasons apart from cost. One is that underground spaces are incompatible with natural ventilation (and, in many locations, there would also be a build-up of radon gas from natural decay processes in granite). More important, perhaps, is the difficulty of combining underground construction with natural daylight. Humans are not troglodytes.

Heating systems in general are also typically very inefficient. New houses and apartment buildings today generally incorporate gas-fired central heating plants without central air conditioning. However, many older houses still lack central heating. In fact, in rural areas wood-burning stoves and fireplaces are common, with electrical resistance heaters also widely used as a supplementary heat source because of their low capital cost. However, it need hardly be pointed out that wood burning results in serious winter air pollution problems in many valleys, while electrical resistance heating is inefficient and therefore unnecessarily costly.

Direct solar heating is a straightforward technology that is known to be cost-effective by itself in few locations. It requires special designs, with south-facing windows and wide overhanging eaves, as well as massive internal heat storage facilities, either dry or wet. The major disadvantage is that direct solar heat can rarely be a perfect substitute for other heat sources during extended periods of cloudy, wet, or cold weather. For such periods, either fireplaces, wood-burning stoves, gas heaters, or electric resistance heaters are needed. However, solar heating (including water heating) can be a significant enhanced by the use of heat pumps and vice versa. It is well known that heat pumps can be several times more efficient than either resistance heaters or gas heaters, especially if some of the resistance heat generated by the pump motor is also recovered. A current rule of thumb is that one unit of electricity can produce three units of heat by means of a heat pump, compared with one unit by means of a resistance heater. Heat pumps can be driven by electric motors or (for larger units) by diesel engines.[3] The units can last as long as 60,000 hours, or twenty-five years, without maintenance.

The low-temperature heat source can be either outside air or water from a well, a lake, or a river, or – if combined with solar heating – from a specially designed internal heat storage unit. Water is generally preferable, as it requires a smaller heat exchanger. Of course, the water is cooled by the heat pump so it is important to operate above the freezing point, 0°C. This is normally no problem for groundwater, which typically has a temperature of about 13°C. Heat pumps are available for either closed- or open-loop operation with water. The former recycles the water (e.g. from a well). The latter takes water from an external source, such as a public water supply, and discharges it back into the sewers. This may be undesirable in areas where the water supply is constrained.

Air is a satisfactory heat source at typical daytime winter temperatures in much of North America (5–10°C), but it becomes inefficient (essentially equivalent to a resistance heater) as outside air temperatures fall below freezing. Of course, this means that an air-to-air heat pump, which is the easiest type to retrofit in an older house, is least efficient in the coldest weather, when it is most needed. Also, at very low temperatures, it may be necessary to use resistance heat to prevent ice build-up on the outside heat exchanger. Thus, heat pumps are far more attractive if built in, together with solar heating of water, with a suitable internal storage facility, at the time of building construction.

Air conditioning can be provided at very low marginal cost by a heat pump operated in reverse. This capability adds very little to the cost of the equipment. If groundwater is used for heat exchange, the summer use for airconditioning would warm the groundwater a few degrees, partially compensating for winter cooling (or a two-well system can be used with "summer" and "winter" wells).

The economics of heat pumps – with or without air conditioning – *vis-à-vis* other heating systems obviously depend upon local conditions, including climate. The current cost of installing an open-loop water-based system in the USA (Indiana) in an average house is $8,000. A closed-loop system would cost about $2,200 more to install. These prices can be compared with $7,000 for a high-end propane furnace.[4] Evidently, even the closed-loop heat pump is not dramatically more expensive than a conventional furnace burning fossil fuels. Operating cost savings would make the heat pump even less costly than the conventional furnace on a life cycle basis. Moreover, the capital cost of a heat pump can be offset, to a significant degree, by improved insulation.

As noted, there is significant potential for combining heat pumps with solar water heating systems, or even PV solar power with heat recovery. In summer, a water-cooled heat pump air conditioner could feed warm water to the water heater rather than dumping it, while a solar space-heating system could raise the temperature of the external heat exchanger – and hence the efficiency of the heat pump – in winter. These possibilities have received little study, thus far, but more detailed analysis seems worthwhile.

A few recent news items are of interest.[5] Recently, the Norwegian parliament instructed the government to set up a plan to introduce more heat pumps in the country, for purposes of reducing energy consumption. The city council of Christchurch, New Zealand, offers an incentive package with grants of up to $500 to replace open fires by other systems, including heat pumps. In Gothenburg, Sweden, four large heat pumps are used to extract and concentrate waste heat from refineries and a municipal waste incinerator for district heating. In Germany, demand for heat pumps in 1997 showed an increase of 40% over the previous year. In Switzerland, one out of every three new buildings now includes a heat pump.

The enhanced value of services to consumers is clear. First, heat pumps are much more energy efficient than any alternative source of space heating. This would translate into real monetary savings for most households. Second, the option of air conditioning at very little extra capital cost is also attractive, to some households at least. Third, and very important, the increased use of heat pumps in place of wood-burning stoves or fossil fuel-burning furnaces would significantly reduce both local air pollution (especially smoke, but also other pollutants such as carbon monoxide and nitrogen oxides) and the emission of greenhouse gases contributing to global warming, notably carbon dioxide.

### *Microwave cooking*

Until now microwave cooking has been largely limited to "fast foods" and frozen foods. Thus, the microwave oven is an add-on to the average kitchen, rather than a major alternative to conventional methods of cooking. The reason is that, although it is much faster and more energy-efficient than conventional cooking (because no energy is wasted heating water, crockery and pans, or the oven itself), it heats the "object" uniformly. It is, however, unsuited to the majority of types of cooking. Up until now it has not been possible to use microwaves to simulate non-uniform "outside-in" methods of cooking, such as toasting, pan-frying, roasting, broiling, or baking. In particular, it has not been possible to make foods crisp.

Combination units have been available for some time, combining microwave cookers with conventional resistive heating elements (electric grill) and convective hot-air units for baking. However, this combination is mainly a space saver.

The fundamental disadvantage of all conventional heating elements, whether electric or gas-fired, is that they heat everything within range, which takes time and costs energy. They also generate significant indoor air pollution. The first problem has been overcome by a more sophisticated design pioneered by Whirlpool Corp.[6] The new prize-winning system utilizes a combination of forced air convection with a quartz plate like a frying pan located on the roof of the oven. This grill element absorbs microwave energy and reaches a temperature of 200°C in about two minutes. Like most new units, it can be programmed for different types of food and degrees of crispness. For instance, when cooking meat the crust can be created (by searing) at the beginning to prevent loss of juices. On the other hand, when cooking bread and pastry the crust can be created at the end to prevent water vapor in the air from being reabsorbed to make the product soggy.

In short, a sophisticated programmable microwave system with the whole range of capabilities is now available for both homes and restaurants. In a few years, as the news spreads and costs come down, the new technology will replace the existing gas stove and oven, *as well as* the existing electric stove, grill, and microwave oven, not to mention the large number of stainless-steel, aluminum, or cast iron pans found in the average kitchen. (All of these cooking utensils will eventually be replaced by ceramics, glass, or so-called engineering plastics.)

This development constitutes a significant cost saving for domestic consumers. It will save significantly on the use of fossil fuels, but it will also increase electricity consumption somewhat. Second, it will save time. And, third, it can sharply reduce householders' exposure to certain types of indoor air pollution from cooking, especially particulates (smoke) from frying over open burners.

### *Rooftop PV systems*

The declining costs of PV systems have already been noted. The chief drawback of rooftop PV systems as a substitute for central station power is energy storage. This technology is currently being pushed by would-be developers of electric vehicles (EVs). However, even the most efficient storage batteries under development or under consideration (lithium chloride) would only store about 400 watt-hours (Wh) per kg at 100 watts per kg (W/kg) specific power output. The highly touted nickel–metal hydride (NiMH) battery would store about 55 Wh at 100 W/kg of power output. This is already a great improvement over lead acid batteries, of course. But the EV developers essentially all agree that the future belongs to either sodium sulfur or lithium ion batteries, either of which would increase performance by another 50% or so. However, this is probably the upper limit, and such batteries will not be fully developed for at least several more years, even if the EV market grows rapidly.[7] A possibility that looks attractive technologically, but has received virtually no attention to date for the built environment, would be local energy storage in large structures, or even suburban neighborhoods, by means of high-speed flywheels using modern high-tech materials. The use of flywheels has been promoted as a means of energy storage in heavy vehicles, such as buses, but essentially discarded for safety reasons. However, the safety problems would be much less severe if counter-rotating horizontal flywheels were confined in sealed, evacuated underground chambers. The capability of rapid stored energy dissipation in case of emergency, by flooding with water and release of energy as steam (through a stack), could easily be made part of such a system.

## Summary

The materials used in structures are (local problems notwithstanding) relatively inert and not especially hazardous to the environment in extraction, use, or disposal. Some non-trivial pollution problems are associated with the extraction and processing of metals and chemicals. However, many of these problems are already being addressed.

Moreover, sharply increased recycling of most metals seems both feasible and environmentally beneficial. The major unresolved problems are associated with dissipative materials used for purposes of maintenance and some construction materials, especially plastics, such as PVC, that are difficult to recycle.

By far the major opportunities for reducing environmental damages associated with the construction sector and the built environment are to be found in reducing fossil fuel energy consumption, both directly (for space heating, water heating, and cooking), and indirectly, for electricity generation. (The latter possibility is not discussed in the chapter.)

Fortunately, the technological potential for reducing the use of fossil fuels in the built environment is extremely promising, especially given the existing trend on the demand-side toward electrification of all non-mobile energy services. Specific examples include the increasing use of heat pumps and microwave cooking. Another specific opportunity can be found on the supply side, notably the widespread application of rooftop PV systems.

All of these technological changes could be accelerated by a variety of government policy measures. However, this set of issues is not discussed in this chapter.

## Notes

1 Luckily, more and more arsenic is being used to make gallium arsenide for the electronics industry, thus reducing the supply available for pesticides and wood preservatives. Again, it is fortunate that nickel–cadmium batteries are now taking up most of the available supply. These batteries can be recovered and recycled, although this is not yet happening on a significant scale.

2 In the early post-war period electric utilities in the USA actively encouraged consumer "electrification" by cooperating with appliance manufacturers to advertise and demonstrate "electric kitchens" and other appliances.

3 One of the major research projects on heat pumps is (or was) conducted at the Centre des Etudes Nucleaires in Grenoble.

4 Price data from *The News-Sentinel*, Fort Wayne, Indiana, 16 May 1998.

5 All news items taken from *Reuters* Business Briefings, Spring, 1998.

6 Cuthbertson, I. "Ian Cuthbertson Weighs Up Microwave Ovens" *Nationwide News Proprietary Ltd.*, Australia, 9 May 1998.

7 The most important advantages of EVs, apart from energy efficiency, are silence and zero emissions. (In fact, silence could be a safety issue.) It was the emissions that prompted the California State Legislature to decree that 2% of motor vehicles sold in the state in 1998 should be zero-emissions vehicles, which meant, in practice, electric vehicles (EVs). Moreover, similar laws have been adopted by New York State and Massachusetts. The California mandate was recently reduced to 1% and delayed to 2003 and Massachusetts agreed, but as of late 1997 New York had not (yet) accepted any change or delay. A slightly different and less "hard line" approach has been taken in Europe, but with potentially similar impact. The Zero Emission Urban Society (ZEUS) project has initiated extended tests of EVs in several European cities. The Alternative Traffic in Towns (ALTER) project is an initiative launched by the transport and environment ministers conference of the EU. Six cities (Athens, Barcelona, Florence, Lisbon, Oxford, and Stockholm) have agreed to set aside special areas from which all but ZEVs will be excluded. These may be historical city centers or sensitive areas. These cities will start to renew their bus and service fleets with ZEVs. At a conference in Florence in October 1998, all 1,400 cities in the EU were invited to consider joining the project.

## References

Adriaanse, A., Bringezu, S., Hammond, A., Moriguchi, Y., Rodenburg, E., Rogich D. and Schütz, H. 1997. *Resource Flows: The Material Basis of Industrial Economies*. Washington, DC: World Resources Institute, with Germany: Wuppertal Institute; The Netherlands: National Ministry of Housing; and Japan: National Institute for Environmental Studies.

Ayres, R.U. November, 1989. *Energy Inefficiency in the US Economy: A New Case for Conservatism.* Research Report RR-89-12. Laxenburg, Austria: International Institute for Applied Systems Analysis.

Ayres, R.U. 1997. Metals Recycling: Economic and Environmental Implications. *Resource Conservation and Recycling* 21: 145–173. (Also in *Proceedings of the 3rd ASM International Conference on the Recycling of Metals*, Barcelona, June 1997.)

Ayres, R.U. and Ayres, L.W. 1996. *Industrial Ecology: Closing the Materials Cycle*, Cheltenham: Edward Elgar.

Ayres, R.U. and Ayres, L.W. 1998. *Accounting for Resources 1: Economy-wide Applications of Mass-Balance Principles to Materials and Waste*, Cheltenham: Edward Elgar.

Battelle–Columbus Laboratories 1975. *Energy Use Patterns in Metallurgical and Nonmetallic Mineral Processing (Phase 4 – Energy Data and Flowsheets, High Priority Commodities)*, Interim Report S0144093-4. Columbus, OH: Battelle–Columbus Laboratories (prepared for United States Bureau of Mines).

Bolch, W.E. Jr. 1980. Solid waste and trace element impacts. In *Coal Burning Issues*, Vol. 12. Green, A.E.S. (ed.). Gainesville, FL: University Presses of Florida, pp. 231–248.

Davis, W.E. 1971. *National Inventory of Sources and Emissions: Mercury, 1968*. APTD-1510. Leawood, KS: W.E. Davis & Associates (for EPA, Research Triangle Park, NC).

Davis, W.E. 1972a. *National Inventory of Sources and Emissions: Barium, Boron, Copper, Selenium and Zinc 1969 – Copper, Section III*. APTD-1129. Leawood, KS: W. E. Davis & Associates (for EPA, Research Triangle Park, NC).

Davis, W.E. May 1972b. *National Inventory of Sources and Emissions: Barium, Boron, Copper, Selenium and Zinc 1969 – Zinc, Section V*. APTD-1139, Leawood, KS: W.E. Davis & Associates (for EPA, Research Triangle Park, NC).

Duchin, F. and Lange, G.-M. 1995. Prospects for the recycling of plastics in the United States. *Structural Change and Economic Dynamics* 9: 335–357.

Forrest, D. and Szekely, J. 1991. Global warming and the primary metals industry. *Journal of Metallurgy* 43: 23–30.

Gadgil, A.J., Rosenfeld, A.H. Arasteh, D. and Ward, E. 1991. Advanced lighting and window technology for reducing electricity consumption and peak demand: overseas manufacturing and marketing opportunities. *Proceedings of the IEA/ENEL Conference on Advanced Technologies for Electric Demand-Side Management*. Paris: OECD/IEA, pp. 83–84.

Gaines, L.L. 1980. *Energy and Material Flows in the Copper Industry*. Technical Memo. Argonne, IL: Argonne National Laboratory (prepared for the United States Department of Energy).

Lovins, A.B. and Sardinsky, R. 1988. *The State of the Art: Lighting*. Competitek Report. Snowmass, CO: Rocky Mountain Institute.

Lowenbach, W.A. and Schlesinger, J.A. 1979. *Arsenic: A Preliminary Materials Balance*. EPA-560/6-79-005. McLean, VA: Lowenbach and Schlesinger Associates, Inc. (for EPA, Washington, DC).

Lübkert, B., Virtanen, Y., Muhlberger, M., Ingham, I., Vallance, B. and Alber, S. 1991. *Life Cycle Analysis: International Database for Ecoprofile Analysis (IDEA)*. Working paper WP-91-30. Laxenburg, Austria: International Institute for Applied Systems Analysis.

McElroy, A.D. and Shobe, F.D. 1980. *Source Category Survey: Secondary Zinc Smelting and Refining Industry*. EPA-450/3-80-012. Kansas City, MO: Midwest Research Institute. (for EPA, Research Triangle Park, NC).

McGannon, H.E. (ed.) 1971. *The Making, Shaping and Treating of Steel*. Pittsburgh: United States Steel Corporation.

Masini, A. and Ayres, R.U. 1996. *An Application of Exergy Accounting to Four Basic Metal Industries*. Working Paper 96/65/EPS. Fontainebleau: INSEAD.

Radian Corporation 1977. *Industrial Process Profiles for Environmental Use: Primary Aluminum Industry*. PB281-491, EPA-600/2-77-023y. Austin: Radian Corporation (for IERL, Cincinnati, OH).

PEDCo-Environmental. 1980a. *Industrial Process Profiles for Environmental Use: Primary Lead Industry*, PB81-110926, EPA-600/2-80-168. Cincinnati: PEDCo-Environmental (for IERL, Cincinnati).

PEDCo-Environmental. 1980b. *Industrial Process Profiles for Environmental Use:* Chapter 28, *Primary Zinc Industry*. PB80-225717, EPA-600/2-80—169. Cincinnati: PEDCo-Environmental (for IERL, Cincinnati).

PEDCo-Environmental. 1980c. *Industrial Process Profiles for Environmental Use:* Chapter 29, *Primary Copper Industry*. PB81-164915, EPA-600/2-80-170. Cincinnati: PEDCo-Environmental (for IERL, Cincinnati).

Raskin, P., Gallopin, G., Gutman, P., Hammond, A. and Swart, R. 1998. *Bending the Curve: Toward Global Sustainability*. Polestar Series Report (8). Stockholm: Stockholm Environment Institute.

Thomas, R. (ed.) 1977. *Operating Handbook of Mineral Processing*, Vols 1 and 2. New York: McGraw-Hill.

United States Bureau of Mines. 1993. *Minerals Yearbook 1993; Vol 1: Metals and Minerals*. Washington, DC: United States Government Printing Office.

Vringer, K. and Kornelis Blok, K. 1995. The direct and indirect energy requirements of households in the Netherlands. *Energy Policy* 23: 893–910.

# 7 Industrial ecology and the built environment

*Iddo K. Wernick*

In its short history, industrial ecology has distinguished itself from other approaches to achieving environmental quality by analyzing the systemic effects of human resource use on environmental burden and by use of an analogy with natural ecosystems. Although there are examples of the application of these principles to consumer products and industrial processes, their application to the built environment presents new challenges. Current recycling levels for many construction materials remain low. Recovered materials and alternative feedstocks can supplant many virgin materials flows but must overcome economic and technical barriers. Property degradation after material use, non-uniform waste streams, and geographically dispersed sources limit the economic feasibility of recovery. Independent of materials use and management, industrial ecology must develop consistent approaches to building design and construction as well as the utilization of land for structures. Finally, the natural ecosystem analogy focuses the question of how industrial ecology should view the built environment: as its own ecology or as part of a larger one. Based on the principles that form its conceptual core, what strategies does industrial ecology offer for smart design and systems thinking about the human built environment?

## Introduction

The human built environment surrounds us. We live and work within it. A growing human population and the rise of the non-agricultural economy in the last centuries, coupled with the expanded use of fossil energy for construction, has dramatically increased the breadth, as well as the height, of the global built environment as we see it today. Matching the growing number of people and enterprises demanding built structures, governments and commercial concerns at all levels continue to have vital interests in their production. An array of demographic, economic, and technical forces contribute to shaping the built environment, and its form reflects the prosperity of society, its aesthetic sense, and the desired functionality of our civilization.

The built environment provides a platform for human impact on Nature by drawing on reserves of renewable and non-renewable resources, and by influencing natural landscapes, biological populations, and the flow of services provided by natural systems. Strategies to reduce the adverse influence of human construction on Nature range from reducing the environmental burden caused by construction materials to designing structures that make better use of sunlight to service their inhabitants. Industrial ecology embraces these and other measures aimed at providing people the services of shelter and comfort using minimal resource inputs and causing the least disturbance to natural systems. It does so by accounting for the materials flows associated with construction activities, analyzing systemic effects of resource use on environmental burden, and by use of an analogy with natural ecosystems. Based on consideration of these core principles, how does industrial ecology address the impact of the built environment on Nature?

Industrial ecology starts by determining the magnitude and complexion of the diverse set of materials that flow to and from the built environment. The largest flows consist of essentially benign rocks and minerals used for structure, some fraction undergoing intensive thermal treatment. Other flows include manufactured building components containing metals and plastics whose production requires more energy and involves more problematic waste streams. To manage these and other resource flows, industrial ecology searches for analytic tools to find opportunities to reduce the impact of flows and investigates technologies and economic arrangements that help to integrate more recovered material and non-traditional feedstocks into primary material flows. Industrial ecology also addresses the functionality of materials and manufactured components embedded in the built environment as they influence the operation of building systems for plumbing, electricity, and climate control and the use of natural resources over the lifetime of a structure.

Land is another natural resource that industrial ecology must consider in trying to reduce the impact of the built environment on Nature. Examining the extent to which the land utilized for structures alters biological activity and other land surface characteristics produces a more accurate picture of the environmental impact of construction. Systematic analysis thus requires that, in addition to tons of materials flow, hectares of land used must be included as input. On the implementation side, industrial ecology must yield plausible options that balance more efficient land use with less disturbance of the land used.

Strict accounting of physical measures of material flow and land use still leaves industrial ecology without the guiding principles provided by the analogy between human-built and natural systems. As an analogy, this assumes a separation that allows for comparison between two distinct entities. The flaw in such separation is twofold. First, despite their qualitatively distinct capacity for self-consciousness and control over their surroundings, humans remain one species among many struggling for survival. Furthermore, many of the environmentally preferred paths for "greening" the built environment require *integration* with Nature by working with it to provide services of shelter and comfort and by avoiding disturbance to natural systems. Thus, in addition to promoting a model of industry *as* ecology, industrial ecology must accommodate the concept of industry *within* ecology. This dual model calls for a built environment that uses fewer resources for construction, integrates structures into the external landscape, and contains structures that work with Nature, not against it, in their operation.

Architects, developers, contractors, regulators, manufacturers, and, of course, consumers all drive the materials flows, land use, and building design that shape the built environment. To effect change, industrial ecology must consider the incentive structure seen by these actors in the construction industry, who are seldom compelled to coordinate their decisions. Innovative environmental practices face economic challenges as well as resistance to changes in proven methods that are supported by an established industrial infrastructure. The incentives offered by the market and by government regulation for enhanced environmental building design and construction practices determine their diffusion in the economy. As is the case for other industries, these incentives must be incorporated into an economy that chronically undervalues environmental assets and liabilities (Heal 1998).

To be useful, industrial ecology should provide analytic tools, design principles, and implementation strategies for improving the environmental character of structures. These elements must all contribute to a consistent approach that recognizes the diversity of materials flows, the many impacts of built structures on Nature, and the far-flung social and commercial interests involved. In addition to providing useful insight into the environmental impact from the built environment, this exercise aims to advance the value of industrial ecology by applying it to new areas of inquiry.

## Material flows

### *Bulk mass flows*

The mass of basic construction materials flowing to the built environment in the USA has increased more than twentyfold over the past century (Figure 7.1) and has significantly exceeded the rate of population growth. Including construction materials used for roads and infrastructure, the built environment consumed six times more material per person in 1995 (approximately 24 kg per person per day) than it did 100 years ago, though much more material moved outside the boundaries of the economic system then and went unrecorded in government-compiled statistics. The relatively stable ratio of tons to dollars of gross domestic product (GDP) over the century shows that material consumption grew in tandem with economic activity, with a surge after World War II owing to US highway construction. Though total tonnage rose dramatically, the US economy grew just as fast, on average, over the century.

Sand and stone dominate the materials considered in Figure 7.1, accounting for nearly 90% of the total tonnage. The bulk of these materials are used in road infrastructure,

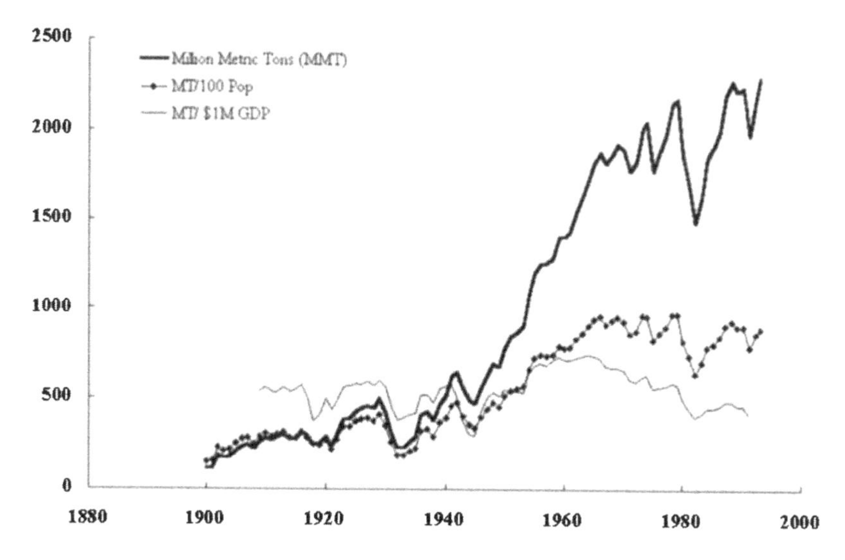


*Figure 7.1* Construction materials consumption, USA, 1900–93. Construction materials include perlite, cement, gypsum, limestone, granite, basalt, sandstone, marble, slate, crushed stone, limestone, dimension stone, sand and gravel as well as lumber and steel. Sources: United States Department of Commerce (1975, 1998), United States Geological Survey (1996), United States Department of Agriculture (1993).

and only a fraction (approximately 20%) is used in buildings. Figure 7.2 looks at a select number of basic construction materials to distinguish trends other than the utilization of sand and stone. Consumption of this more limited group of materials has also increased over the past century, although to a far lesser degree. More efficient use of these materials means that the per capita consumption in the USA is less today than a century ago. Economic growth far outpaced growth in the consumption of these materials. The ratio of consumption to GDP shows that the significance of a ton of steel or cement fell dramatically over the century relative to overall economic activity. Nonetheless, while the economy became more focused on information and value addition, it grew on the base of a built environment whose absolute mass continues to expand, albeit more slowly.

### *Other flow measures*

Tonnage provides only a weak indicator of actual environmental harm. Reductions in aggregate mass *per se* do not necessarily indicate reduced environmental loads. Full life cycle analysis (LCA) of the materials flows to the built environment must include the upstream environmental effects of material consumption as well as the ultimate disposition of those materials in the environment. Energy flows, for instance, depend on high-volume upstream material flows for extraction and processing and end up primarily as carbon dioxide in the atmosphere and ash disposed of on land. Considering the energy content of materials in the built environment by weighting materials according to the energy expended in their production offers additional insight. Using energy instead of mass as the yardstick, the measure of some high-volume materials that generate little direct environmental harm shrinks when compared with the smaller mass flows of cement,

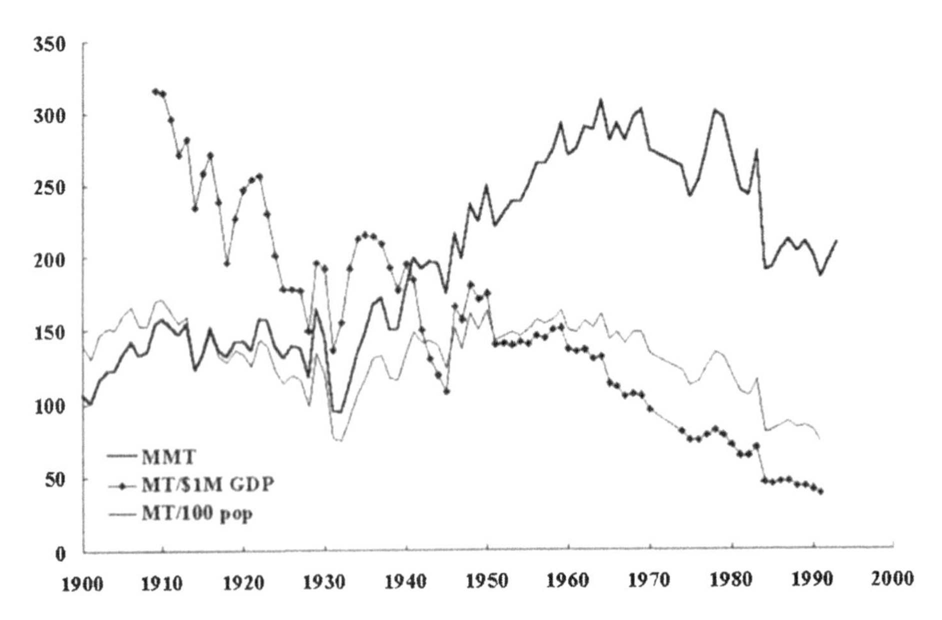


*Figure 7.2* Raw steel production and consumption of gypsum, lumber, and cement, USA 1900–93. Sources: United States Department of Commerce (1975, 1998), United States Geological Survey (1996), United States Department of Agriculture (1993).

metals, and plastics, the processing of which requires more energy. As an illustration, the energy used to produce a single ton of aluminum is equal to that that required to produce 8 tons of steel, 12 tons of glass, 35 tons of cement, or over 50 tons of a typical commodity, polymers (US Congress Office of Technology Assessment 1993). In search of a more ecological energy measure, Odum (see Chapter 2) advocates the use of "emergy" for measuring the environmental impact of a unit of mass of a given substance. Emergy accounts for the total amount of intercepted solar energy involved in producing that mass and thus more accurately reflects the environmental stresses resulting from consumption. To comply fully with the LCA model, toxic flows generated during materials processing must also be taken into account, as well as the durability of materials and their suitability for recovery after use.

### *Strategies*

Strategies to reduce the flow of virgin material inputs begin with efforts to obtain maximum value from existing structures and building components. Once a structure has been demolished, attention turns to the best way to recover value from material waste streams. Reducing the need for new inputs altogether, "dematerialization" offers a further strategy that promotes more efficient design of structures and structural components to reduce resource needs for construction and building operation.

### *Recovery*

Table 7.1 shows consumption and recovery levels for the major construction materials used in buildings in the USA in 1996. Notable are the low recovery levels for most materials, primarily because of the degradation of material properties after use and poor economics. As an example of property degradation, crushed concrete exhibits a broader particle size distribution, more fine particles, lower average density, and different mixing characteristics compared with natural aggregates, which prevents ready acceptance by the industry (Wilburn and Goonan 1998). Additionally, the low value of recycled bulk material and the relative importance of transportation costs continue to favor the economics of virgin sources. Uniform and concentrated deposits of desired materials are more commonly found in natural reservoirs than in heterogeneous waste streams coming from scattered construction sites. Finally, low disposal fees for construction and demolition (C&D) waste and low prices for virgin materials continue to make recycling the more expensive alternative. Taken together, these factors contribute to a paucity in the existing recycling infrastructure for bulk construction materials.

Metals stand out for their high recycling levels. Obsolete structures contain relatively high concentrations of steel and other metals. As for automobiles, favorable economics have led to development of an infrastructure that takes advantage of this source of scrap, particularly for steel. General levels of copper and aluminum recycling[1] were approximately 15% and 22% respectively in 1996 (United States Geological Survey 1997). No data are available regarding the amounts of copper and aluminum recovered specifically from buildings. However, it is likely that the relatively high concentration of these metals found in demolished buildings is one reason for their higher than average recovery levels.

A significant proportion of construction materials is derived from non-renewable organic sources. The use of plastic in construction materials has risen sharply over the

*Table 7.1* Flow of construction materials, USA, 1996

| *Material* | *Apparent consumption in million metric tons (MMT)* | *Recovery level* | *Cascading options* | *Alternatives to traditional feedstocks* |
|---|---|---|---|---|
| Concrete | 378 (Kelly 1998) | 5% from all construction including infrastructure | 85% of recovered concrete to road base | Fly ash, agricultural and metallurgical wastes |
| Wood | 57[a] (United States Department of Commerce 1998) | Insignificant | Mulch, compost, animal bedding, fuel | Wood composites from waste wood |
| Gypsum | 20 (United States Geological Survey 1997) | Small | Animal bedding, soil amendment | FGD, phosphogypsum[b] |
| Steel | 13[c] (United States Geological Survey 1997) | 85%[d] | Depends on scrap quality | Scrap |
| Brick | 16 (United States Geological Survey 1997) | Insignificant | Likely | Likely |
| Asphalt | 8.6 (Kelly 1998) | 75%+ recovery from road infrastructure | Hot asphalt mix for roads | – |
| Glass | 3.4[e] | Insignificant? | Containers, fiberglass, glassphalt | – |
| Carpet | 2–2.5[f] | Insignificant | For nylon and recycled carpet plastic | – |
| Copper | 1 | Old scrap = 15% of apparent consumption | Depends on scrap quality | Scrap |
| Aluminum | 1 | Old scrap = 22% of apparent consumption | Depends on scrap quality | Scrap |

Notes

a Based on US lumber consumption for 1996, excluding lumber used for manufacturing and shipping. Assumes a conversion factor of 0.00236 board feet/cubic meter and an average wood density of 0.5 metric tons/cubic meter.

b FGD is flue gas desulfurization sludge generated as a solid waste by coal-fired power plants. Phosphogypsum is a form of gypsum generated during phosphate rock processing for fertilizer.

c Seventy percent of end uses for steel are reported as steel service centers and "other."

d The 85% (18.2 MMT/21.4 MMT) includes recovery of discarded steel from roads and infrastructure.

e Mass value of apparent consumption based on value ratio of apparent consumption to total production in dollars multiplied by domestic glass production in tons.

f Based on annual carpet discards of 1.4–1.8 MMT and assuming 70% of new carpet is used for replacement.

past half-century (e.g. plumbing, fixtures, carpets), with a concomitant increase in the quantity in the C&D waste stream. However, plastics recycling in the US remains low in general – a little more than 5.4% of the plastics waste generated in 1996 – and is dominated by recovery of standardized beverage containers (United States Environmental Protection Agency 1998a). It is unlikely that plastics recovery from C&D debris exceeds this average value. Asphalt recovery is high, but the figure of 75% mentioned in Table 7.1 represents recovery from discarded road infrastructure. No data are available on the recovery of discarded asphalt from residential roofing, a more dispersed source than the material from road demolition.

Though current recovery levels remain small, the prospect for reducing the need for dimensional lumber and making better use of wood wastes is promising (Wernick *et al.* 1998). Many materials have been used as substitutes for wood in construction over the past century, contributing to a roughly 60% fall in per capita lumber consumption in the USA since 1900 (Wernick *et al.* 1996). The lumber industry has also changed its own mix of products. The past decades have seen an increase in the number of composite panels that use fibers and irregular wood shapes bound with organic adhesives. The manufacture of structural wood products from composites uses fewer individual trees of more species because of the different dimensional requirements and because waste from the forest products industry can be utilized.

### *Cascading materials flows*

Aside from promoting closed-loop recycling of used materials, industrial ecology also seeks to find other opportunities for improved materials management. Cascading materials flows are reused, but the degradation of materials' properties means that they can be exploited only for lower-value uses. Many of the recovered materials listed in Table 7.1 are reused in lower-value products (e.g. crushed concrete for road base) that tolerate significant variability in material properties. Several agricultural uses have similarly low-quality thresholds and absorb some wood and gypsum wastes. High-quality demands also help explain the absence of closed-loop recycling for flat glass in the USA. The markets for waste glass consist entirely of demand for manufacturing containers, fiberglass, and "glassphalt" (i.e. a glass–asphalt mix used for road building), all products more tolerant of deviation in the specifications of feed materials. Notwithstanding the emerging use of wood waste for value-added products mentioned earlier, the non-uniform size distribution of wood waste streams and the presence of contaminants generally restrict recovery to agricultural uses and as fuel.

Carpet also cascades down the value chain. The prospects for closed carpet recycling suffer from poor collection economics and the fact that only a small fraction of the original product (approximately 17%) can be recovered for new carpet facing (Lave *et al.* 1998). Markets for recovered carpet include use as landfill cover after shredding, an extremely low-value use. Alternatively, used carpet can be economically converted to recycled carpet plastic, a polymer that combines the polypropylene and nylon found in carpet to produce a lower-quality polymer with diminished market value. While efforts should be intensified to realize the highest value markets for C&D waste, the use of waste materials for any productive purpose reduces demand for resources and lessens disposal volumes.

### *Non-traditional feedstocks*

Using alternative sources for construction materials provides another avenue for reducing the environmental damage caused by materials flows. For example, additives to cement and concrete come from wastes from the energy (fly ash), agriculture (rice hulls), and metals (steel slag) industries. Suitable material properties and favorable economics play a central role in evaluating the viability of markets for alternatives such as these. As another example, the utility and fertilizer industries generate gypsum wastes several times the amount consumed annually in wallboard production. Although these alternatives to mined natural gypsum require some processing and can contain contaminants that are costly to remove, they are also commonly found in large quantities near central processing facilities, making the economics of collection more favorable.

The integration of non-traditional feedstocks into the construction materials market will require research and development to eliminate the problems preventing greater diffusion of these materials in the economy. While advances in technology may reduce the price of these materials in relation to virgin materials, economic incentives may still be necessary to compensate for added collection costs. The incentive structure for waste handlers and materials processors will benefit from eliminating subsidies to virgin materials producers (United State Environmental Protection Agency 1994), providing favorable financing for non-traditional materials processors, costing primary materials to reflect environmental externalities, and imposing higher disposal fees.

### *Dematerialization*

Better design offers a further strategy for reducing materials flows through improved efficiency, i.e. dematerialization. Structural components that utilize this strategy include lumber-saving wood joists and steels of smaller cross-section to support structural loads. The German cabinet manufacturer, Kambium Furniture Workshop Inc., provides a prime example of reducing environmental stress at the manufacturing stage by directly incorporating renewable sources into its electricity supply, relying on locally available resources, manufacturing durable products, and avoiding synthetic chemicals for treating wood (Liedtke *et al.* 1999). To bypass the problems of carpet recycling, several innovative firms (Interface and Milliken) have developed reusable carpet tile products that are leased by customers and refurbished periodically by the manufacturer. These examples illustrate some of the opportunities for reducing materials flows by changing product and process technologies and instituting new economic arrangements.

### *Accretion*

The accretion of material in the built environment is one more reason why recovery levels fall short of the amounts entering it. Far less material exits the built environment as waste than enters it as new feedstock. A study by Franklin Associates estimated national C&D debris from residential and commercial buildings in the USA to be 135 MMT in 1996 (Franklin Associates 1998).[2] This figure compares with nearly 500 MMT for the list of inputs in Table 7.1, implying a nearly 4:1 ratio of inputs to outputs, or an annual incremental increase of over 350 MMT, not including the amount incorporated into infrastructure. The difference between inputs and outputs of structural material flows shown in Figure 7.3 represents this daily accretion for the average American household.

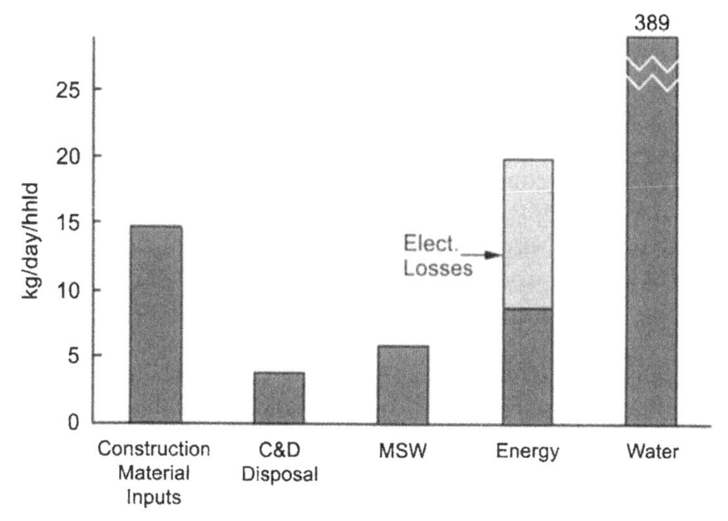


*Figure 7.3* Daily household material flows for construction materials, energy carriers, municipal solid waste, and water, USA, 1996. Sources: United States Department of Commerce (1998), United States Geological Survey (1997), Franklin Associates (1998), United States Environmental Protection Agency (1998a).

By relying on several plausible assumptions, anecdotal information, and data compiled by state agencies, the Franklin study estimates that 20–30% of C&D debris is recovered.

### *Non-structural material flows*

Applying industrial ecology to the material flows arising from building operations is no less important than managing the flow of construction materials. Figure 7.3 shows the flow of energy carriers, municipal solid waste (MSW), and water from the average American household. The mass flow of hydrocarbon fuels (coal, oil, gas) for end uses such as electricity, water and space heating, and cooking exceeds the flow of structural materials. The data make clear that selecting materials to ensure proper thermal insulation and incorporating energy-efficient illumination, appliances, and heating, ventilation and cooling (HVAC) systems can significantly reduce the flow of materials to the built environment over time.

Both consumers and producers influence the size of materials flows, and energy is no exception. Over the last century in the USA, the number of occupants per occupied housing unit has fallen monotonically from 5 in 1890 to a little more than 2.5 in 1996 (United States Department of Commerce 1975, 1998). The trend toward fewer residents in the average individual housing unit acts to counter the improved performance offered by energy-efficient designs. To quantify the relationship between energy use and number of residents, Schipper (1996) suggests that energy use is proportional to the square root of the number of people living in a unit. Thus, for example, an average household of four consumes half as much energy as four individuals living separately. On the industry side, utilities have significant opportunities for reducing primary energy use through technologies such as integrated gas combined cycle turbines, which can offer 60–70% more efficient conversion of thermal to electrical energy than the current average (Hansen 1998).

Municipal solid waste flows from households also exceed C&D debris disposal. Diverting a greater proportion of this waste stream to recycling relies primarily on building residents and managers. Initial building design considerations can ease the process, for instance by designating space for waste sorting in multiunit buildings. Responsible management also offers opportunities for more beneficial use of wastes from landscaping. The United States Environmental Protection Agency (1998b) estimates that a potential 65 MMT of compostable material could be recovered annually from US residences and commercial building sites for beneficial use. The amount of water flowing to and from residences dwarfs the other flows. Opportunities for reducing absolute water flow through efficiency gains include measures such as low-flow plumbing fixtures and recycling gray water for landscaping.

Detailed checklists for environmental building design offer help to contractors and homeowners by systematizing the considerations and opportunities from energy-efficient design to plumbing systems (Lopez Barnett and Browning 1995). The US Green Building Council's "Leadership in Energy and Environmental Design" (LEED) system offers a quantitative scoring and reward system for rating a building's environmental performance (e.g. energy efficiency, use of recycled building materials, indoor air quality, water conservation). LEED promises to appeal to industry's need for clear benchmarks to measure progress (US Green Building Council 1999).

## The land resource

Materials flows provide a valuable organizing principle for analyzing the environmental concerns associated with structures in the built environment. However, in considering the environmental impact of building construction, materials flow analysis alone remains inadequate. Consideration of the impact of the built environment on Nature needs to address the effects of resource extraction and the extent and distribution of the built environment on the surface of the Earth. Mining, harvesting, and processing construction materials can disturb remote landscapes and ecosystems, while buildings themselves alter the landscape and what can live on it. Industrial ecology must offer principles to optimize human use of that finite and precious resource, land.

### *Mining*

Because construction absorbs a significant fraction of the materials extracted from the Earth by humans, reducing demand by society reduces disturbances from extraction activity, which may occur far from the point of use. Examples include surface coal mining, by far the dominant use of land for mining, fuel transportation by rail and pipeline, and power lines to carry generated electricity across continents. Sand and stone may cause disturbances at the point of extraction, yet typically travel only 25–50 miles from local quarries and pits (Barsotti 1994). The disturbances caused by metals mining occur in areas typically thousands of miles from the point of use. Though land used for metal extraction represents less than 0.1% of Earth's terrestrial surface (Barney 1980), exploration and mining activity can affect surrounding ecosystems as a result of infrastructure needs and by dispersing toxic metals and chemicals into the environment (Wernick and Themelis 1998).

The leverage to mitigate environmental impacts from mining most often falls in the hands of mine operators and relies on the economic and regulatory climate in the host country. Ecosystems proximate to newly discovered resource deposits face a particular threat in less developed regions, where the need for foreign exchange overshadows domestic environmental concerns (Hodges 1995). Leading international companies may be motivated to voluntarily adopt the stricter environmental regulations in place in the wealthier nations. However, such action requires a potent mix of altruism and concerns over liability.

### *Area-based measures*

To account comprehensively for the upstream effects of consumption for the built environment, several measures have emerged that relate the area used for procuring resources to consumption, such as the "ecofootprint" model offered by Rees and Wackernagel (1995). In evaluating the environmental stress caused by resource demand, the sustainable process index goes further to consider the land used for resource extraction, processing, and local waste assimilation, and also denominates environmental impact in units of hectares (Narodoslawsky and Krotscheck 1995). These measures allow the area of land disturbance to be used as a generic currency for considering the environmental impact of resource demands for the built environment.

Incorporating area-based measures into the design process emphasizes the value of using land for multiple purposes and using land that has already undergone conversion for human use. Restoring abandoned sites and structures offers the counterpart to materials reuse and recycling for land. For structures, this may involve standard building renovation or reclamation of sites that cannot be used due to past contamination (i.e. "brownfields'). For extraction sites, reclamation of mines and quarries using wastes as fill material offers an option for lessening overall disturbance by using the same land for the material cradle and grave.

### *Infrastructure*

The buildings that form the nucleation points for people and energy in the built environment sit within an extensive infrastructure. Of this vast infrastructure (much of it for automobiles), residential and commercial buildings occupy a small area. "Developed land," defined by the US government as "[L]and that includes urban and built-up areas in units of 10 acres or greater, and rural transportation," occupies about seventeen times the combined square footage of residential (approximately 170 billion sq. ft.) and commercial buildings (approximately 59 billion sq. ft.). This ratio does not take into account the fact that many of these structures are multistorey and thus displace far less land than their measured square footage (United States Department of Commerce 1998). Another salient ratio relates the amount of land covered by infrastructure to population density for a given location. Empirical studies in the USA have found that high population densities are accompanied by a low value of developed land per resident. Figure 7.4 shows that in those states in the USA where population density is lowest, the amount of developed land per person is greatest (Waggoner *et al.* 1996). Conversely, those states with the greatest population density have more people using the same infrastructure. This study focused on built structures for residential and commercial purposes. Future work will assess the environmental impact of the infrastructure that mediates the impact of buildings on the surrounding environment and occupies far more area.

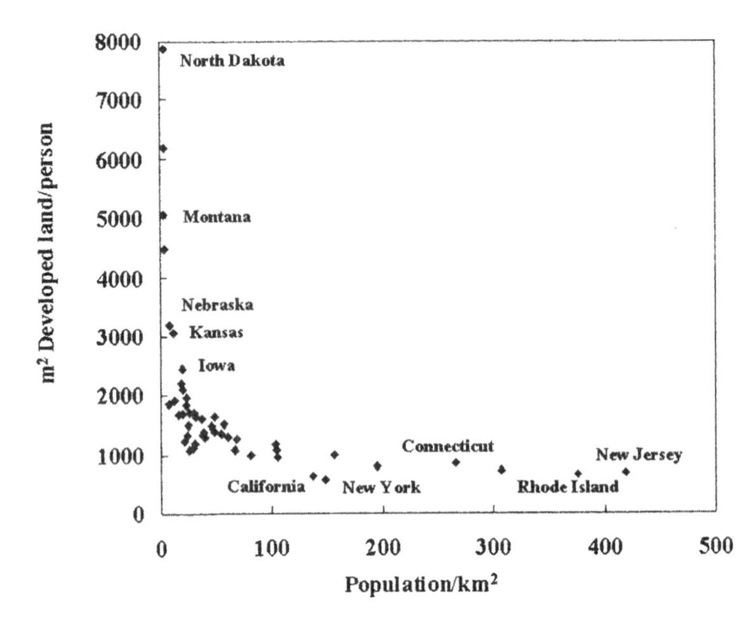


*Figure 7.4* Relation of developed land per person to population density on non-federal land in forty-eight contiguous states of the USA. Population data are from 1990 census and development data are for the year 1987. Sources: United Nations Food and Agricultural Organization (1994), United States Bureau of the Census (1986 and 1991).

## *Land use quality*

Just as analysis of tons is necessary but insufficient for assessing the environmental impact of materials, tabulating hectares of land surface without considering the quality of the land is inadequate. Going beyond simple area measures calls for environmental valuation of the land. Net primary production (NPP) measures the net amount of biomass produced by photosynthesis, a property characteristic of an ecosystem and its typical vegetative cover (Vitousek *et al.* 1986). Considering NPP for prospective construction sites allows planners to determine the impact of altering or eliminating photosynthetic production in a given area as compared with an area of land elsewhere. Consistently assigning productivity values to land also allows for evaluating alternative plans for a site by comparing the amount of biological activity they preserve within the site boundaries.

While environmental impact statements do consider the effects of development on endangered species and hydrology, the cumulative regional effects on habitat fragmentation and hydrologic flow may escape individual assessments. Measures are needed that go beyond looking at local effects on biological populations to consider the regional consequences of development. For instance, considering the contiguity and topology of land affected by development in a region offers a tool for determining whether the land left undisturbed by construction retains its integrity as a habitat for plant and animal species and for providing natural services.

One fundamental environmental value of land stems from the inescapable fact that a finite amount of surface intercepts the solar radiation that supports terrestrial life. The built environment replaces the surface of the natural landscape presented to incoming solar radiation, making the reflective properties of the built environment significant. For structures themselves, surface reflectivity enters into design decisions as it affects the heat stored and released by a structure and thus the comfort of its inhabitants. The surface of the built environment also affects the global reflectivity of solar radiation (i.e. albedo), which contributes to the Earth's radiation budget and influences the global mean temperature. Environmentally responsible practice discourages the use of darker surfaces that absorb almost all of the incident solar radiation in favor of lighter surfaces that reflect more and green (i.e. biologically active) surfaces that use absorbed light for energy to drive photosynthesis.

### *Land use for multiple functions*

Consistent with the principle of resource efficiency, industrial ecology offers a simple principle for reducing land disturbance: *Use the same land for several purposes.* As a rule, using the same land to serve multiple functions leaves more area for nature. For example, designing the places where people live and work as multistory structures in general allows for more undeveloped land to remain undisturbed. Viewed in this light, the rise in the average floor space of US residential homes from 1500 square feet to 2150 over the period from 1970 to 1997 is tempered by the fact that the number of homes with two or more stories grew from 17–49% (United States Department of Commerce 1998). The fraction of multistory commercial buildings is higher still. Of course, site-specific considerations as well as those concerning quality of life and aesthetics qualify this endorsement of building.

Using solar panels on roofs offers another example of using the same surface to serve dual functions, human shelter and energy production. Embedding solar energy collectors directly into roofing materials further combines these functions. In appropriate climates, solar roofing panels can generate a net amount of electricity over their lifetime that exceeds the energy needed for their manufacture and installation (Lewis and Keolian 1997). Even in colder northern climates, large-scale experiments are under way using building-integrated photovoltaics (BIPVs) (Hellmans 1999). Using the sun's thermal energy directly to heat water flowing on roofs, and using skylights to illuminate living and work spaces, also conforms to the same principle of multiple functionality.

Finally, the built environment need not present an entirely lifeless surface. Using the same surface for different functions includes the preservation of biologically active areas on and around structures. Integration of green spaces into cities (i.e. parks) and residences (i.e. gardens) allows for photosynthesis to occur in the midst of the built environment. The surfaces of buildings themselves can support photosynthetic activity in the form of living roofs that can offer superior thermal insulation as well as better solar reflectivity. Underground dwellings whose roofs are flush with the land surface offer the advantage of greater thermal stability indoors as well as leaving the above landscape intact and biologically active. Above-ground buildings, and even road infrastructure, can be topped with living roofs as seen in the designs of the architect Malcolm Wells (Wells 1994) (see Chapter 12).

## The ecological analogy

Industrial ecology draws from models of theoretical ecology that describe how natural systems work, as well as how to better integrate the built environment within them. Nature provides us with a model for design of the built environment and also serves as an essential part of what is being modeled, thus invoking our understanding of industry *within* ecology as well as industry *as* ecology. Kay (1984) provides greater structure to this relationship by identifying a comprehensive framework. In addition to calling for exclusive reliance on renewable resources, he posits that the objective of a more compatible relationship between the built environment and Nature must attend to: (1) the interface with Nature, (2) the model provided by Nature, and (3) the direct use of natural systems.

### *The interface with Nature*

Looking at the interface of the built environment with Nature invites consideration of the direct effect of construction as well as the impacts from consumption of minerals, fuels, and land. To examine the consequences of human resource use, theoretical ecologists draw on examples from Nature. The pulse growth model asserts that, in general, ecosystems initially thrive when offered abundant resources only to falter dramatically or collapse totally later when the available resource base can no longer support existing populations. Other models predict less catastrophic results, with populations stabilizing as they approach saturation levels. Ecologists note that changes in ecosystems can occur suddenly and irreversibly when the threshold of a system's capacity is exceeded for an extended time. The history of several island civilizations offer examples of precipitous collapse, and fears arise that similar consequences will hold for entire regions and for the planet as a whole (Daily 1997). Thus, while the models are suggested for analyzing human impacts on natural ecosystems, they are also indicated for the human populations that rely on them. For a rich discussion of systems ecology and models of ecosystem population dynamics, see Lotka (1924), Odum (1991), and Holling *et al.* (1995).

### *Nature as model*

Using Nature as a model raises a central question regarding the geographic distribution of the built environment. In what direction should planners try to shape the extent and intensity of the built environment in the future? Leaving more of Nature undisturbed suggests concentrating human settlements to minimize land cover, often through greater uniformity in the built environment itself (e.g. high-rise buildings) or for obtaining natural resources (e.g. plantation forestry). Nature itself, however, seems to encourage sparser and more heterogeneous development that lessens the need for concentrated, and vulnerable, industrial activity.

While superior efficiency and productivity may be evident over restricted temporal and spatial scales in ecosystems, longer time scales expose resilience as the guiding principle (Peterson *et al.* 1998). Systems designed for resilience tend to overdesign in order to absorb shocks to the system, such as buildings with walls that can withstand infrequent earthquakes or forests that survive hurricanes. Natural systems also tend to exhibit redundancy, e.g. plants may distribute thousands of seeds and microorganisms divide rapidly, but in both cases each individual has only a slim chance of survival. Thus, the efficiency that industrial ecology seeks to emulate is offset by Nature's drive to maintain resilience, and it is that property that dictates what the system will be optimized for (Figure 7.5).

The example given by Holling *et al.* (1995) brings the trade-off between efficiency and resilience into focus. In this classic example a grassland in its natural state supports both a drought-resistant species of grass and a lusher grass species that exhibits far less resilience in the event of drought. Ranchers, favoring the lusher grass, converted the site

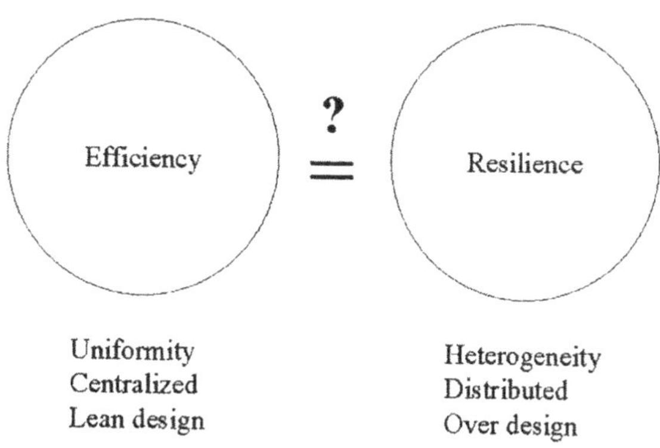


*Figure 7.5* Comparing lessons for design from the ecological analogy.

to support only that species. While the conversion proved more efficient (i.e. more profitable) over the short term, the inevitable onset of drought devastated the lusher grasses, and the eradication of the drought-resilient species left the site barren.

### *Using Nature*

Natural systems can directly provide services to structures in a variety of ways. Strategies for using sunlight range from drawing the curtains to let in light to designing and building passive and active solar collectors that combine building orientation, construction materials, and mechanized devices to maximize the benefit from the sun for internal climate control. In addition to using sunlight, structures can take advantage of the constancy of temperature below the land surface and use the gradient between the surface and below ground to warm air and water in winter and cool them in summer. Replacing expensive, energy- and chemical-intensive waste treatment systems with wetlands that naturally cleanse waste streams provides one more example of using Nature to supplant man-made systems.

Though not identical, the strategies of addressing the interface between humans and Nature, imitating Nature, or using it do complement one another. For example, imitating Nature by making productive use of wastes and closing material loops reduces human stress at the interface with Nature by preserving natural resources and reducing waste disposal volumes. Designing homes to capture useful solar radiation while rejecting unwanted heat reduces reliance on external energy inputs and benefits from imitation of natural design. Responsibly using natural wetlands to treat effluents enhances their growth and at the same time mitigates the environmental stress caused at the interface between the built environment and natural bodies of water. Because imitation is the greatest form of flattery, directly using Nature to provide services provides the best example of using Nature as a model. Taken together, the application of all three strategies promotes building construction that is modeled on Nature and therefore more compatible with it.

## Conclusion

The actual range of landscapes, climates, ecosystems, and available construction materials that shape the built environment challenges industrial ecology to develop analytic tools and implementation strategies that are valuable to practitioners yet retain broad applicability. These include more comprehensive measures of resource use and methods for materials management, improved environmental site assessment and planning, and better design principles for structures. The principles offered must respect the primal human need for shelter and account for the lack of a single central actor in influencing the evolution of the built environment.

Historical analysis of the diverse set of resources flows to and from the built environment offers measures for identifying the impact of technological innovations and government policies on those flows. The analysis shows the gains possible from more efficient material utilization and suggests methods for improving materials management by closing material loops and using industrial wastes for construction materials. LCA studies of the material embedded in the built environment focus attention on diminishing environmental impact from the manufacture of building components and reducing resource use over the lifetime of a structure for internal operations. Using LCA, the environmental assets and liabilities of using a given material can be weighed against its ability to limit other material and energy flows over the life of a structure. From this perspective, efforts to reduce embedded energy and improve recyclability should focus on frequently replaced building components, while materials permanently embedded in the structure should be optimized for reducing resource use for internal operations.

Industrial ecology approaches to land use quality should offer planners a guide on how to disturb less land, and disturb land less, in shaping the built environment. Using the same parcel of the land surface for multiple functions offers one principle for better performance. Designing structures to accommodate local materials, climate, and landscapes offers another avenue to better integrate them into the natural landscape in both form and function.

Effecting systemic change in the construction of the built environment demands identifying and integrating the structural incentives for the actors that shape it. Examples of structural incentives include:

1. developing technologically advanced material recovery systems;
2. improving the economic climate for materials recovery through market and regulatory mechanisms;
3. expanding environmental impact statements to go beyond minimal compliance with existing regulations;
4. instituting local building codes with higher efficiency standards for lighting, thermal insulation, and HVAC systems;
5. providing roadmaps for environmentally responsible practice to architects and building contractors;
6. establishing clear benchmarks for businesses to measure environmental performance.

Improvements in all of these basic items will allow responsible practices to diffuse into the economy and lessen reliance on the actions of a small number of highly dedicated individuals.

From cathedrals to track homes, the built environment both reflects and influences the values of society. Improving the built environment requires accounting for human needs for shelter, comfort, and social function. To be successful, attempts to modify existing designs and practices must fall within the boundaries of what people can accept and must be hearty enough so that the principles used in designing elegant prototypes can diffuse broadly through imitation. While architectural vision must necessarily extend beyond the strictly practical, real change should not rely on massive behavioral shifts or on mass acceptance of a rarefied aesthetics and efficiency as necessary ideals. Achieving broad-based change in the built environment of the future will require greater cooperation between all of us that shape it, a worthwhile goal in itself.

## Acknowledgments

I thank Jesse Ausubel of The Rockefeller University and Robert Pollack of Columbia University for their comments on the text, and Rebecca Johnson of the Columbia Earth Institute for her valuable comments and edits to help prepare this manuscript.

## Notes

1 The figure for aluminum recycling does not include used beverage can recycling.
2 The Franklin study notes that insufficient data as well as differing state definitions of C&D debris make it difficult to distinguish the fraction going to or coming from building construction or to establish a uniform group of materials to consider.

## References

Barney, G.O. 1980. *The Global 2000 Report to the President of the United States*. New York: Pergamon Press.

Barsotti, AF. 1994. *Industrial Minerals and Sustainable Development*. Washington, DC: US Bureau of Mines.

Daily, G.C. 1997. Introduction: What are ecosystem services? In *Nature's Services: Societal Dependence on Natural Ecosystems*. Daily, G.C., Reichert, J.S. and Myers, J.P. (eds). Washington, DC: Island Press, pp. 1–10.

Franklin Associates 1998. *Characterization of Building-Related Construction and Demolition Debris in the United States*. Report No. EPA530-R-98-010.Prepared for the US Environmental Protection Agency, Municipal and Industrial Solid Waste Division, Office of Solid Waste.

Hansen, U. 1998. Technological options for power generation. *The Energy Journal* 19(2): 63–87.

Heal, G.M. 1998. *Valuing the Future: Economic Theory and Sustainability*. New York: Columbia University Press.

Hellmans, A. 1999. Solar homes for the masses. *Science* 285: 679.

Hodges, C.A. 1995. Mineral resources, environmental issues, and land use. *Science* 268: 1305–1312.

Holling, C.S., Schindler, D.W., Walker, B.W. and Roughgarden, J. 1995. Biodiversity in the functioning of ecosystems: an ecological synthesis. In *Biodiversity Loss: Economic and Ecological Issues*. Perrings, C., Maler, K.-G., Folke, C., Holling, C.S. and Jansson, B.-O. (eds). Cambridge: Cambridge University Press, pp. 44–83.

Kelly, T. 1998. *Crushed Cement Concrete Substitution for Construction Aggregates: A Materials Flow Analysis*. United States Geological Survey Circular No. 1177.

Kay, J.J. 1984. *Self-organization in Living Systems*. PhD thesis. University of Waterloo.

Lave, L., Conway-Schempf, N., Harvey, J., Hart, P., Bee, T. and MacCracken, C. 1998. Recycling postconsumer nylon carpet: a case study of the economics and engineering issues associated with recycling postconsumer goods. *Journal of Industrial Ecology* 2(1): 117–126.

Lewis, G.M. and Keolian, G.A. 1997. *Life Cycle Design of Amorphous Silicon Photovoltaic Modules*. Report No. EPA/600/SR-97/081.Cincinnati: USEPA National Risk Management Research Laboratory.

Liedtke, C., Rohn, H., Kuhndt, M. and Nickel, R. 1999. Applying material flow accounting: ecoauditing and resource management at the Kambium Furniture Workshop. *Journal of Industrial Ecology* 2(3): 131–147.

Lopez Barnett, D. and Browning, W.D. 1995. *A Primer on Sustainable Building*. Snowmass, CO: Rocky Mountain Institute.

Lotka, A.J. 1924. *Elements of Physical Biology*. Baltimore: Williams & Wilkins. Reprinted by Dover, New York, 1956.

Narodoslawsky, M. and Krotscheck, C. 1995. The sustainable process index (SPI): evaluating process according to environmental compatibility. *Journal of Hazardous Materials* 14(2–3): 383–397.

Odum, H.T. 1991. Emergy and biogeochemical cycles. In *Ecological Physical Chemistry: Proceedings of an International Workshop, November 1990, Siena, Italy*. Rossi, C. and Tiezzi, E. (eds). Amsterdam: Elsevier Science, pp. 25–56.

Peterson, G., Allen, C.R and Holling, C.S. 1998. Ecological resilience, biodiversity and scale. *Ecosystems* 1: 6–18.

Rees, W. and Wackernagel, M. 1995. *Our Ecological Footprint: Reducing Human Impact on the Earth*. Gabriola Island, BC: New Society Publishers.

Schipper, L. 1996. Life-styles and the environment: the case of energy. In *Technological Trajectories and the Human Environment*. Ausubel, J.H. and Langford, H.D. (eds). Washington, DC: National Academy Press, pp. 89–109.

United Nations Food and Agriculture Organization 1994. *FAO Production Yearbook 47*. Rome: FAO.

United States Bureau of the Census 1986 and 1991. *State and Metropolitan Area Data Book*. Washington, DC: United States Bureau of the Census.

United States Congress, Office of Technology Assessment. 1993. *Industrial Energy Efficiency*. Report No. OTA-E-56.Washington, DC: US Government Printing Office.

United States Department of Commerce 1975. *Historical Statistics of the United States, Colonial Times to 1970*. Washington, DC: United States Bureau of the Census.

United States Department of Commerce 1998. *Statistical Abstract of the United States 1998*. Washington, DC: United States Bureau of the Census.

United States Environmental Protection Agency. 1994. *Federal Disincentives: Study of Federal Tax Subsidies and Other Programs Affecting Virgin Industries and Recycling*. Report No. EPA/230-R-94-005. Washington, DC: USEPA.

United States Environmental Protection Agency 1998a. *Characterization of Municipal Solid Waste in the United States: 1997 Update*. Report No. EPA530-R-98-007. Washington, DC: USEPA Office of Solid Waste.

United States Environmental Protection Agency 1998b. *Organic Materials Management Strategies*. Report No. EPA/530-R-97-003. Washington, DC: USEPA Office of Solid Waste and Emergency Response.

United States Department of Agriculture 1993. *Agricultural Statistics*. Washington, DC: United States Department of Agriculture.

United States Geological Survey 1996. *Materials Database*. Unpublished.

United States Geological Survey 1997. *Mineral Commodity Summaries 1997*. Washington, DC: US Government Printing Office.

US Green Building Council 1999. Leadership in Energy and Environmental Design (LEED$^{TM}$) green building rating system. Available at < http://www.usgbc.org/ >

Vitousek, P.M., Ehrlich, P.R. and Ehrlich, A.H. 1986. Human appropriation of the products of photosynthesis. *BioScience* 36: 368–373.

Waggoner, P., Ausubel, J.H and Wernick, I.K. 1996. Lightening the tread of population on the land: American examples. *Population Development and Review* 22: 531–45.

Wells, M. 1994. *InfraStructures*. Self-published. ISBN 0962187860.

Wernick, I.K. and Themelis, N.J. 1998. Recycling Metals for the Environment. *Annual Review of Energy and Environment* 23: 465–497.

Wernick, I.K., Waggoner, P.E. and Ausubel, J.H. 1998. Searching for leverage to conserve forests: the industrial ecology of wood products in the US. *Journal of Industrial Ecology* 1(3): 125–145.

Wernick, I.K., Herman, R., Govind, S. and Ausubel, J.H. 1996. Materialization and dematerialization: measures and trends. *Daedelus* 125(3): 171–98.

Wilburn, D.R. and Goonan, T.G. 1998. *Aggregates from Natural and Recycled Sources Economic Assessments for Construction Applications: A Materials Flow Analysis*. Circular No. 1176. Washington, DC: United States Geological Survey.

# 8 Construction ecology and metabolism

## Rematerialization and dematerialization

*Stefan Bringezu*

One of the purposes of this volume is to build a bridge between the theoretical requirements for sustainability and some practical measures for planners, engineers, architects, and investors. The central question is how to increase the resource efficiency of construction. It is clear that human society needs to reduce the absolute burden on the environment of all construction activities while at the same time maintaining or even increasing material welfare and services provided to customers. In this chapter, actual policy targets will be presented and strategies for implementation will be compared. Empirical data on the material flows and resource intensity of the construction sector in Germany will be discussed. Examples for practical measures will be presented for the following construction-related activities:

- design of construction products and buildings;
- materials management;
- planning of infrastructure;
- product, facility, and building management.

This chapter will focus on the life cycle-wide material intensity of products and services, the materials intensity per service unit (MIPS) of construction, indicating essential aspects of ecological-economic performance, but also widen the scope to an integrated resource management (IRM) based on a multicriteria analysis considering ecological, economic, and social aspects.

### Requirements for construction ecology

In essence, the general demands for "construction ecology" are identical to those defined for "industrial ecology" (Graedel and Allenby 1995; Ayres and Ayres 1996; Allenby 1999). The anthroposphere is regarded as a metabolic system that receives resource inputs from, and releases waste outputs to, the environment (Ayres 1989; Baccini and Brunner 1991; Ayres and Simonis 1994). The two systems will coexist only if the metabolism of the anthroposphere does not disrupt the life-support functions of the "carrier" or "host" system, the surrounding biogeosphere (Odum 1971). Under sustainable conditions the resource functions are not depleted and the waste absorption functions are not overloaded.

Theoretical arguments have been contributed from a thermodynamics perspective. The path toward sustainability will require the anthroposphere to minimize the losses of useful energy. In other words, the production of entropy should be reduced in order to offset the natural tendency described by the second law of thermodynamics (Ayres 1994).

However, it seems to be very difficult to provide decision makers and engineers with sufficient and practical information on the entropic potential of various processes, products, and infrastructures on a large scale.

Extending the experience on the functioning of natural ecosystems to the anthroposphere, Odum and Odum (1999) advocate that human beings should strive to minimize the use of "empower," i.e. energy per unit of time provided by the sun. Of course, our energy supply will have to be shifted toward naturally renewable energy sources, which in the end all derive from solar input. However, if the empower principle is interpreted in a way that the use of fossil energy (stocked solar energy) is to be maximized, this would contradict international efforts to mitigate global warming and counteract international efforts for climate protection.[1]

Some pragmatic requirements have been formulated from an economic perspective in terms of "management" rules for sustainability (Barbier 1989; Daly 1990). According to the first rule, the use of renewable resources should not exceed the regeneration rate. In order to operationalize this demand, one has to consider that the use of either naturally or technically renewable materials always requires some inputs of non-renewables (e.g. mineral fertilizer to replace nutrients lost as a result of leaching in agriculture and the requirements for materials and energy for recycling processes). As a consequence, the total life cycle of products has to be checked for the use of renewables and non-renewables. The former will have to be distinguished according to criteria on sustainable modes of production in agriculture, forestry, and fishery. For construction purposes, the origin of timber products from sustainable cultivation would be an example.

The second rule states that non-renewable resources may be used only if physical or functional substitutes are provided, e.g. gains from fossil fuel production should be invested in solar energy systems. Here the basic assumption is that man-made capital may be substituted for natural capital ("weak sustainability"). The central requirement from an economic perspective is that the sum of natural and man-made capital is not reduced (e.g. Pearce and Turner 1990). However, from a natural systems perspective it may be argued that there are minimum requirements of Nature that may not be depleted without risk to life support functions.[2] Therefore, man-made capital should not be substituted (permanently) for natural capital ("strong sustainability"). Under this assumption, the second rule would require a minimization of the use of non-renewables.

The third rule states that the release of waste matter should not exceed the absorption capacity of Nature. This can be operationalized by comparing "critical loads" of water, soil, and air compartments with actual levels of emission rates. Once measures have been successfully applied to reduce pollution problems, the "after-end-of-pipe" approach to limit critical loads is also important. The implementation of the third rule is usually based on substance-specific analyses. This approach has some limitations. Generally, it must be acknowledged that we are aware of only the tip of the iceberg with respect to the potential future impacts of all materials and substances released to the environment. Many natural functions react in a non-linear manner. The complex interactions of natural substances such as carbon dioxide, not to mention thousands of synthetic chemicals, cannot be foreseen in total.

From our experience we know that the effects of certain emissions become obvious *after* release and the change in the environment takes place. There is a huge time lag between the scientific finding and public perception and political reaction. Thus, the chances for comprehensive and precautionary materials management are extremely limited. A long-term effective implementation of the third rule should begin before the end-of-pipe and should aim to minimize the environmental impact potential of anthropogenic material flows. This impact potential is generally determined by the volume of the flow times the specific impacts per unit of flow. The second term is unknown for most materials released to the environment. The first term, the volume or weight used or released in a certain time period, can be made available for nearly every material handled. It may be used to indicate a generic environmental impact potential. As long as detailed information on specific impacts is lacking, it may be assumed that the impact potential is growing with the volume of the material flow. The overall volume of outputs from the anthroposphere can only be reduced when the inputs to this system are diminished.[3] This is especially important for construction material flows with a significant retention time within the anthroposphere. Starting from the situation that the assimilation capacity of Nature is overloaded for a variety of known substances, the long-term implementation of the third rule requires a reduction in the resource inputs of the anthroposphere in order to lower the throughput and ultimate output to the environment.

Another rule that has not yet attracted sufficient attention may be derived from the relation of inputs and outputs of the anthroposphere. Currently, the input of resources exceeds the output of wastes and emissions in industrialized as well as developing countries. As a consequence, the economies of these countries are growing physically (in terms of new buildings and infrastructures). The stock of materials in the anthroposphere is therefore increasing. In Germany, the rate of net addition to stock was about 10 tons per capita annually in the mid-1990s. Associated with this accumulation of stock is an increase in built-up land area and a consequent reduction in reproductive and ecologically buffering land. Keeping in mind the limited space on our plant, this development cannot continue infinitely. Thus, a flow equilibrium between input and output must be expected.[4] However, a question naturally arises: When will the economy stop growing physically and to what physical level?

## Strategies and goals for sustaining the metabolism of economies

The focus of traditional environmental policy has been on pollution control. Based on a systems perspective, together with the insight into anthropogenic material flows, modern policy concepts follow a dualistic approach (Figure 8.1).

- On the output side, the release of pollutants is regulated in order to reduce well-known problems. This strategy is directed to specific substance flows (e.g. lead, cadmium, carbon dioxide). Derived policies may be generally characterized as reactive.
- On the input side, the resource requirements need to be diminished in order to lower the impacts of resource extraction as well as the effects of subsequent flows. This strategy is directed to general resource flows (primary material, energy and water) and area requirements. Derived policies may be addressed as proactive.

The two strategies are complementary rather than mutually exclusive (Bringezu 1997a). Pollution control alone cannot restrain increasing resource requirements and a shift in environmental problems as a result of the control of selected substances. A reduction in resource inputs alone may not be sufficient to keep the flows of specific pollutants below critical levels.

Pollution abatement and chemical controls are well-established fields in terms of policy, methods, organization, and technology. However, the reduction of resource inputs is relatively new to the agenda. A proposed goal for industrialized countries is to increase

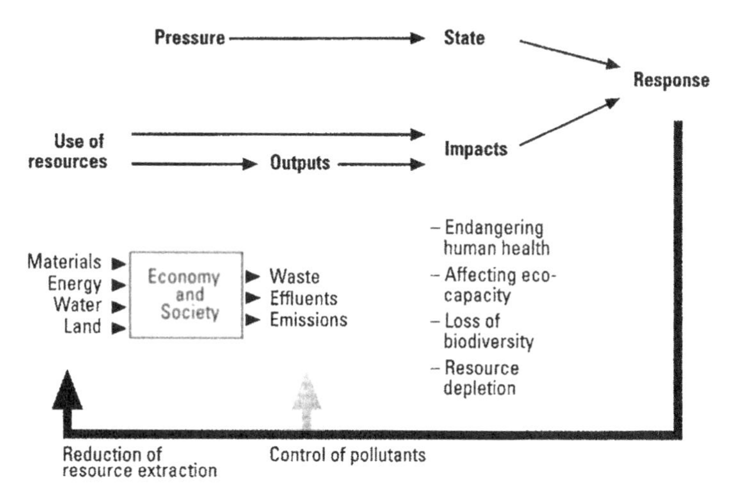


*Figure 8.1* Complementary strategies to sustain the physical exchange of anthroposphere and environment. Corrected after Sachs *et al.* (1998). Copyright Wuppertal Institute, Germany.

the resource efficiency by a factor of 4–10 over the next thirty to fifty years (Schmidt-Bleek 1994; Weizsäcker *et al.* 1995). The increase in resource productivity should guarantee that economic performance and welfare can be increased while the absolute burden to the environment as a consequence of resource extraction is reduced.

On the program level, the factor 4–10 goal has attracted wide attention. It was adopted by the special session of the United Nations (United Nations General Assembly Special Session 1997) and the World Business Council for Sustainable Development (1998). The environmental ministers of Organization for Economic Cooperation and Development (1996) expect progress toward this end. In Japan, a Factor 10 Institute has been founded. Several countries included the aim in political programs (e.g. Austria, Netherlands, Finland, Sweden) (Gardener and Sampat 1998). Even in Germany the draft for an environmental policy program refers to a factor of 2.5 increase in raw materials productivity (1993 to 2020).

The plausibility of these goals has been underpinned by empirical data on the trend and status quo of the resource requirements of industrial countries (Table 8.1). In the four countries studied, the continuous high level of resource flows did not indicate any significant decline over twenty years (1970s to the mid-1990s). Construction flows constitute more than a quarter of total material requirements (TMR). The extraction of construction minerals such as sand and gravel imposes pressure on the local environment through the destruction and long-term change of natural habitats[5] and distortions of the potable water supply. Quarries and the earth movement associated with the build-up of infrastructure significantly contribute to changes in landscape. In the form of buildings and infrastructure, the minerals which are used in construction affect a larger area than the physical area associated with natural deposits. This increase in the built-up area – which is still growing (see below) – is associated with a loss of natural, reproductive, and ecologically buffering land. After use, the construction minerals again burden the environment when deposited in contaminated form or on land that could be used for other reproductive purposes.

The total effects of material flows cannot be predicted in a scientifically satisfactory way. However, the adoption of the per capita resource requirements of industrialized countries by all developing countries – which would result from technology transfer –

*Table 8.1* Resource requirements for construction in selected industrial countries 1991 (Adriaanse *et al.* 1997; Bringezu 1998a)

| | *Construction minerals* | | *Infrastructure excavation* | |
|---|---|---|---|---|
| | *Tons per capita* | *Percent of TMR* | *Tons per capita* | *Percent of TMR* |
| USA | 7.6 | 9 | 13.7 | 16 |
| The Netherlands | 8.9 | 11 | 3.4 | 4 |
| Japan | 9.0 | 19 | 8.9 | 19 |
| Germany | 15.3 | 18 | 3.7 | 5 |
| Average | 9.3 | 13 | 10.5 | 14 |

Note
TMR, total materials requirement.

would probably have severe impacts on a growing number of locations worldwide. The factor 4–10 concept is also intended to reduce the disparity between nations and regions in terms of sharing the use of resources and the burden to the global and regional environment. Starting from the current status quo, the industrialized countries are being challenged to reduce their resource requirements while meeting the demands of their citizens, guaranteeing them adequate material welfare.

There are several strategies that can be implemented to reach these goals and to sustain the physical basis of societies. The main strategies may be described as rematerialization and dematerialization[6] (Figure 8.2). The former aims to close materials cycles by reuse, remanufacturing, and recycling. The reproduction of biomass via agriculture and forestry is also part of this strategy. As a consequence, the requirements for primary resource inputs (and associated hidden or rucksack flows) and final disposal would be reduced while the volume of flows within the anthroposphere would not be diminished.

Dematerialization in a strict sense aims at the absolute reduction of material flows within the anthroposphere (which as a consequence would also reduce the physical exchange with Nature). To this end, industry will have to become more efficient in the use of materials and energy. An essential step will be the delivery of services instead of selling products (see below). The use of materials (and energy) on the consumption side should also become more efficient. The effects of an increase in resource efficiency have so far been compensated by economic growth and rising demand for efficient products. Therefore, an increased sufficiency of final consumption is being regarded as another essential element of dematerialization. Finally, waste management will also have to become more efficient in terms of its own resource requirements as well as in facilitating rematerialization.

## Construction material flows in Germany

In Germany, the absolute volume of material flows for buildings and infrastructure almost quadrupled between 1950 and 1995 on a per capita basis (Figure 8.3). However, the share of the major components remained almost constant, indicating that construction technology in terms of materials choice was not subject to significant change. Nevertheless, the importance of the construction sector for the overall performance of the economy increased dramatically. From 1950 to 1973 – the period of reconstruction after the World War II and the economic boom of the 1960s – this sector was the backbone of economic growth (Figure 8.4). During that time, GDP in Germany increased in parallel with the

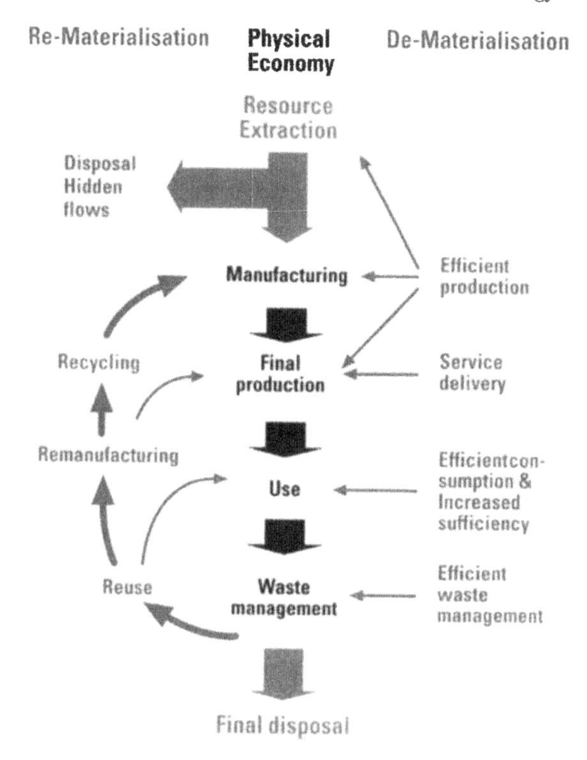


*Figure 8.2* Rematerialization and dematerialization. Copyright Wuppertal Institute, Germany.

increase in the volume of materials used for construction. After the first energy crisis, however, the increase in construction activity ceased although it remained at a constant high level. The economic growth of the national economy continued to be fueled by other sectors (exports, services). Interestingly, the volume of direct material inputs to construction on a per capita basis did not change as a result of reunification. Although there has been significant construction activity in terms of buildings and infrastructure in the eastern part since 1990, the total activity level has remained unchanged since the late 1980s.

The generation of construction waste increased steadily during these recorded periods. However, intensive recycling activities after 1990 had the effect of reducing final disposal significantly (Figure 8.4). In 1995, the physical output of the stock of buildings and infrastructures to the environment[7] was only 1.9% of the inputs. The difference between inputs and outputs – the indicator of physical growth – has always been dominated by the magnitude of the inputs. As a consequence of the effective recycling of construction waste, the physical growth of the technosphere is even more determined by the input of materials to construction (which consist largely of primary materials).

It can be concluded that a steady-state situation between inputs and outputs may only be reached by reducing the inputs to construction. Thus, the successful implementation of a rematerialization strategy increases the need for resource efficiency and effective dematerialization.

The environmental burden associated with construction is not restricted to the materials used directly. There are also upstream flows of resource extraction which are not used for further processing. These are called "rucksack" or "hidden flows". These flows consist of

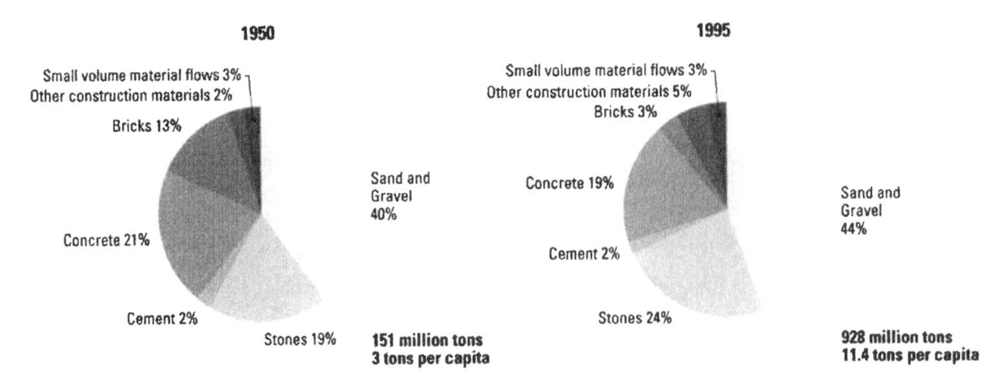


*Figure 8.3* Share of major material flows to direct material inputs to construction in Germany, 1950 and 1995. After Bringezu and Schuetz (1997). Copyright Wuppertal Institute, Germany.

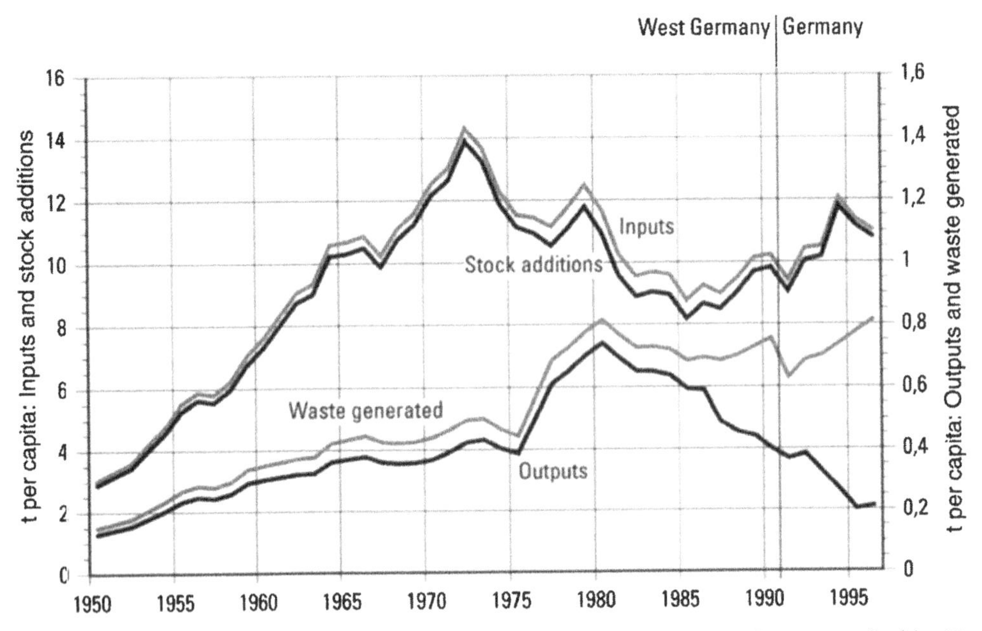


*Figure 8.4* Total volume of direct inputs to and outputs from construction. Data researched by H. Schuetz. Copyright Wuppertal Institute, Germany.

overburden and extraction waste in mining and quarrying activities which are usually associated with impacts to the local environment at the extraction site (devastation of landscape, destruction of habitats, contamination of soil and water, etc.). The magnitude of these flows does not indicate anything about specific impacts [which could be revealed only by an environmental impact assessment (IEA)]. It does, however, indicate the general potential pressure on the environment associated with the demand for products that require these resource flows.

The rucksack of "low-volume flows" (e.g. metals, wood, plastics, glass) per ton of material is higher than the rucksack of "high-volume flows" (e.g. sand and gravel, stones, concrete, bricks) (Figure 8.5). Any assessment of the resource intensity of construction activities will be insufficient if these upstream flows are neglected. A consequence of substituting "low-volume flows" for "high-volume flows" is a possible shift of the environmental burden from domestic quarries and mines to foreign ones. Such a shift must be suspected particularly when some "low-volume flows" increase in absolute terms. This was the case for non-ferrous metals between 1991 and 1996. Materials such as copper and aluminum are associated with an extraordinarily high rucksack per ton.[8]

## MIPS and the method of material intensity analysis

In order to quantify the total material resource requirements associated with certain products and services, the method of material intensity analysis can be applied (Schmidt-Bleek 1994; Bringezu *et al.* 1996; Schmidt-Bleek *et al.* 1998). This method helps to determine the cumulative material requirements (material input) on a life cycle-wide basis and relates it to the service provided. This relation is called the MIPS (material input per service unit). Although the results are condensed as much as possible to support decision making of managers, engineers, designers, architects, planners, and others, MIPS in general is not a single value. The primary materials input is aggregated to five main categories, which are recorded separately: abiotic (= naturally non-renewable) raw materials, biotic (= naturally renewable) raw materials, soil, water, and air (Figure 8.6).

In general, extractions are considered if they are obtained by humans through technological means. All movements of materials are accounted for, irrespective of their economic value. This comprises used materials (e.g. iron ore concentrate) and the rucksack of unused materials (e.g. overburden and refinery waste separated by mining facilities). Biotic raw materials reflect the harvest of green biomass in agriculture or forestry as well as fishery and hunting. Human-induced soil movement is taken to be soil erosion in agricultural fields and forests. For some purposes, the calculation of the plowed soil volume may be appropriate. Water inputs comprise all flows diverted from the natural water path. The air input comprises those atmospheric components that are chemically or physically changed. This is largely the oxygen required for combustion processes. As a consequence, the air input is often proportional to the emission of carbon dioxide. Thus, input-focused material intensity analysis also allows for some comparison of the most important outputs responsible for global warming.

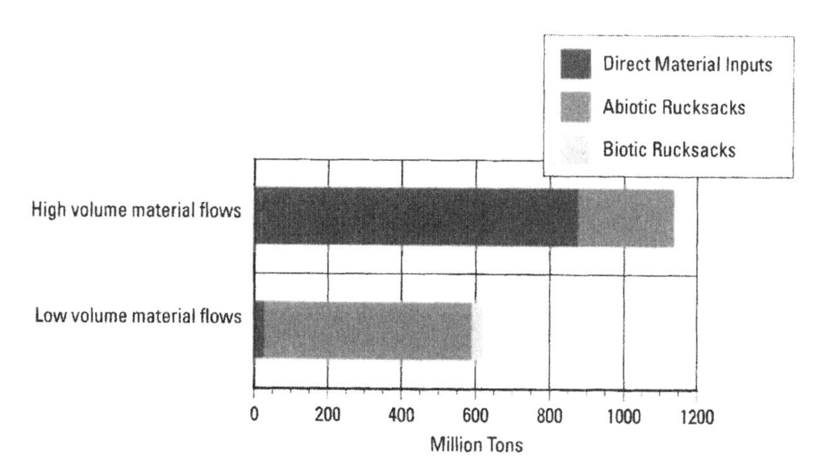


*Figure 8.5* The rucksack flows of "high-volume" and "low-volume" material flows for construction (Bringezu 1998b). Copyright Wuppertal Institute, Germany.

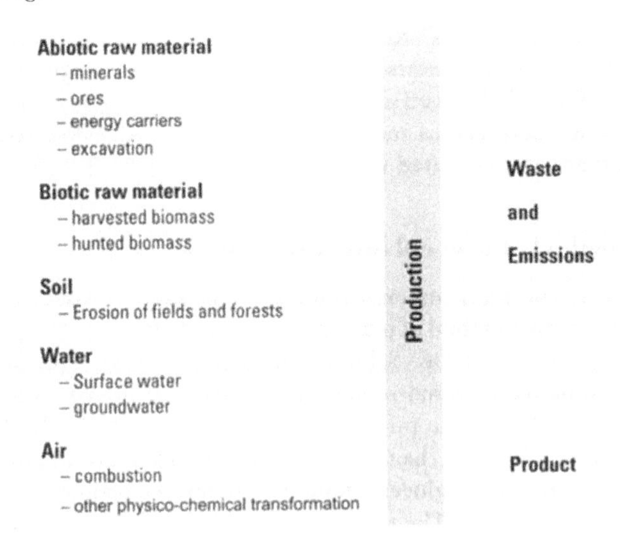


*Figure 8.6* The ecological rucksack of a product comprises those primary material requirements that do not enter the product itself. Copyright Wuppertal Institute, Germany.

Material intensity analysis has been designed to reveal differences in the order of magnitude of material flows. It may be used as a first step of a LCA, especially if the decision making is to contribute to de- and rematerialization and an increase in resource efficiency. The results may be interpreted in a manner similar to an analysis of cumulative energy requirements. The latter does not reveal specific risks, such as those from nuclear power. Nevertheless, it is widely accepted in practice that energy requirements should be minimized while maximizing the efficiency of energy use. By analogy, material intensity analysis provides information on cumulative primary materials requirements which can be used to increase materials efficiency but may not be interpreted in terms of specific impacts per unit of flow. Material intensity analysis also accounts for energy carriers (including the rucksack) on a mass basis, thus allowing comparison of energy carriers and non-energy carriers.

Material intensity analysis can be applied on different levels, from base materials and products to regions and national economies (Bringezu 1997b). This allows the examination of sectors that may be particularly challenged should resource efficiency become an economic or policy requirement. Bringezu *et al.* (1998) ranked the total material requirement of the West German economy with respect to consumer demand and found that the highest demand was for materials for housing, followed by food and leisure. This analysis considered not only construction material flows but also the requirements of energy carriers (including the rucksacks) for construction and maintenance of buildings (especially lighting and heating).

Any comprehensive study of the environmental impact potentials of construction will have to consider not only the construction phase but also the use and after-use phase. For each phase, the upstream (and downstream) processes should be analyzed within the system boundaries definition. The inputs from the environment and the outputs to the environment provide the system boundary for the analysis. If the first step focuses on the input side, the material intensity can be determined as an essential part of a resource intensity analysis (considering also energy and area use separately).

## Design of construction products and buildings

Environmentally benign construction that meets the demands of the users and considers economics starts with the design of buildings, components, and individual products. The "golden rules for ecodesign" (Box 8.1) may give guidance for engineers, architects and planners.

The first golden rule aims to avoid shifting problems between different processes and actors. For instance, if the energy requirements for heating or cooling during the use phase of buildings are not considered during the planning phase, then the best opportunity to achieve energy efficiency will be lost. Similarly, consideration of only the direct material inputs for construction will fail to take account of the environmental burden associated with upstream flows.

The second rule arises from the fact that most building products are not used much of the time. For a considerable part of each day and of each week, homes, offices, and public buildings are essentially unoccupied. Nevertheless, economic and environmental, and probably also social, costs have to be paid for maintenance. Multifunctionality and more flexible models of use may reduce the demand for additional construction and contribute to lower costs for users. The model of car sharing may also be applied to construction. Part-time employees already share the same office. And there is even potential for more efficient building use beyond normal working hours.

The third rule in Box 8.1 goes hand in hand with the factor 4–10 target for material requirements, including energy carriers, and should be applied to the average of products and services. The application of this rule requires more intellectual power to be applied in the search for alternative means of providing the services and functions demanded by users.

As is the case for every industrial sector, the basic question for the construction industry is about the *services to be delivered*. Clearly, this is more than the provision of a box around some number of square meters of ground or about providing single-family houses surrounded by a large garden. Thus, the demand for particular functions – such as brightness, quietness, a healthy climate, facilities for work, rest, and communication – also determines the type and dimensions of the building, the facilities required, and the materials chosen.

The demand for services will lead to new technological answers. For instance, consider the function of cooling, which is provided by refrigerators. This service or function may be described as "the provision of a minimum volume with a maximum temperature, kept dark in the vicinity of the kitchen." In temperate zones, this service can also be delivered by a cooling chamber integrated in the outer wall of buildings (Figure 8.7). Such a device can be significantly less energy and material intensive than a conventional refrigerator

*Box 8.1* The golden rules for ecodesign. After World Business Council for Sustainable Development (1994), Schmidt-Bleek and Tischner (1995)

1. Potential impacts to the environment should be considered on a life cycle-wide basis ("from cradle to grave").
2. Intensity of use of processes, products, and services should be maximized.
3. Intensity of resource use (material, energy, and land) should be minimized.
4. Hazardous substances should be eliminated.
5. Resource input should be shifted toward renewables.

(Tischner and Schmidt-Bleek 1993). But it will only be realized by a new approach to engineering, planning, and construction.

Different technological options that provide the same service can be compared in terms of resource intensity. For example different types of houses (wooden post-and-beam structure versus concrete-brick construction) or major components of buildings (e.g. outer walls with different structure and composition) as well as individual products (e.g. water pipes) can be compared in terms of intensity of material and energy use.[9] While the summary of those comparisons can be used for different management and sourcing decisions, the detailed data allow different construction items to be optimized with respect to material efficiency.

The following parameters have a major influence on the material and energy intensity of construction:

- Type of construction. This determines the volume and – together with the choice of materials – the mass consumed for the building.

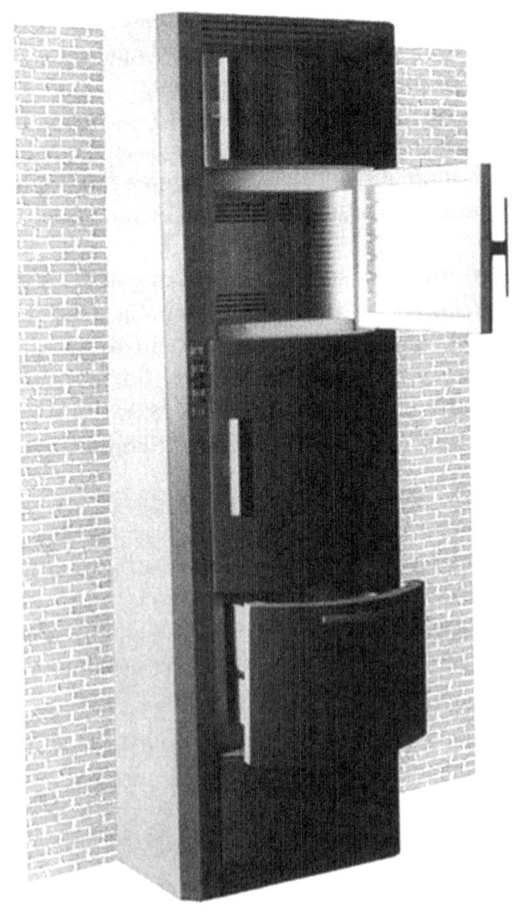

*Figure 8.7* Services may be provided by other technologies which require fewer resources (Tischner and Schmidt-Bleek 1993). Copyright Wuppertal Institute, Germany.

- Materials chosen. These are associated with different direct inputs and various rucksack flows.
- Durability. A long life will reduce the need for replacement and new resource requirements.
- Repairability. This could significantly contribute to an increased life.
- Dismantlability: This is a prerequisite for repair, reuse, and recycling. The current practice of using composite materials and complex systems is a severe impediment.

If the mass of a new construction could be reduced by one-third, if the choice of materials could reduce the material requirement, again by one-third, if the durability was, on average, extended by one-third, and if the repairability and dismantlability contributed the same one-third factor, then the overall effect would, theoretically, be a factor of 5 reduction in resources. Thus, the factor 4–10 goal is not just a vision but could be realized by an effective combination of measures. Further practical examples are given by Weizsäcker *et al.* (1995).

## Materials management

Whereas resource management considers all options in seeking to maximize the sustainable use of resources – including efficiency gains through dematerialization – materials management in a strict sense concentrates on the possibilities for rematerialization as well as detoxification. The latter aims to replace toxic and ecotoxic chemicals as well as to eliminate those substances that may interfere with recycling. The former will be addressed in more detail.

In response to governmental demands for recycling in 1996, the German construction industry committed itself to increasing the recycling rate of demolition waste by 50% by 2005 (base year 1991). Unfortunately, the validity of data on waste generation and recycling is generally rather weak. There is a discrepancy between official statistical data and specific studies (Enquetekommission 1998). In 1996, the Environmental Ministry came to the conclusion that the industry had already achieved this goal (Kohler 1997).

However, one should be careful with the interpretation of recycling rates. Most so-called recycling of construction waste *de facto* results in cascading use, i.e. "down-cycling". Recycled demolition waste from buildings is used solely for road construction, for which lower-quality performance is required. Higher quality standards, especially for concrete used in buildings, are currently being revisited in order to allow the use of secondary raw materials while guaranteeing adequate safety.

Sustainable materials management would require an increase in high-level recycling. This would mean that waste from various construction activities is primarily used again for the same purpose.

The use of secondary raw materials from recycling or cascading use can significantly contribute to reducing requirements for primary resources. However, recycling processes *per se* also need a certain amount of materials and energy. Some recycling technologies require even more primary resources than the original primary route of processing (Figure 8.8). Thus, the construction and recycling industries will have to examine whether recycling makes economic and environmental sense.

*Box 8.2* Status quo versus future demolition material recycling to eliminate down-cycling

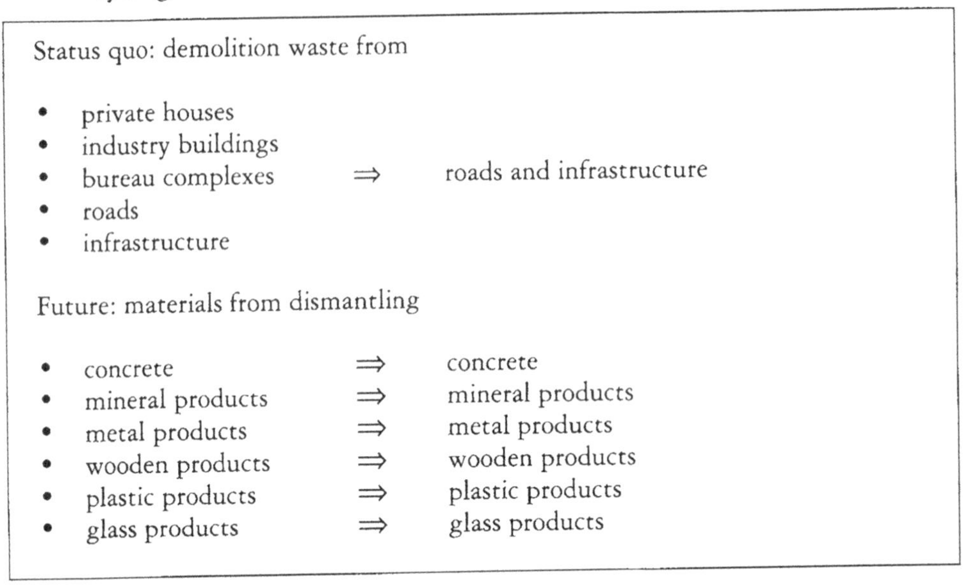

Status quo: demolition waste from

- private houses
- industry buildings
- bureau complexes ⇒ roads and infrastructure
- roads
- infrastructure

Future: materials from dismantling

- concrete ⇒ concrete
- mineral products ⇒ mineral products
- metal products ⇒ metal products
- wooden products ⇒ wooden products
- plastic products ⇒ plastic products
- glass products ⇒ glass products

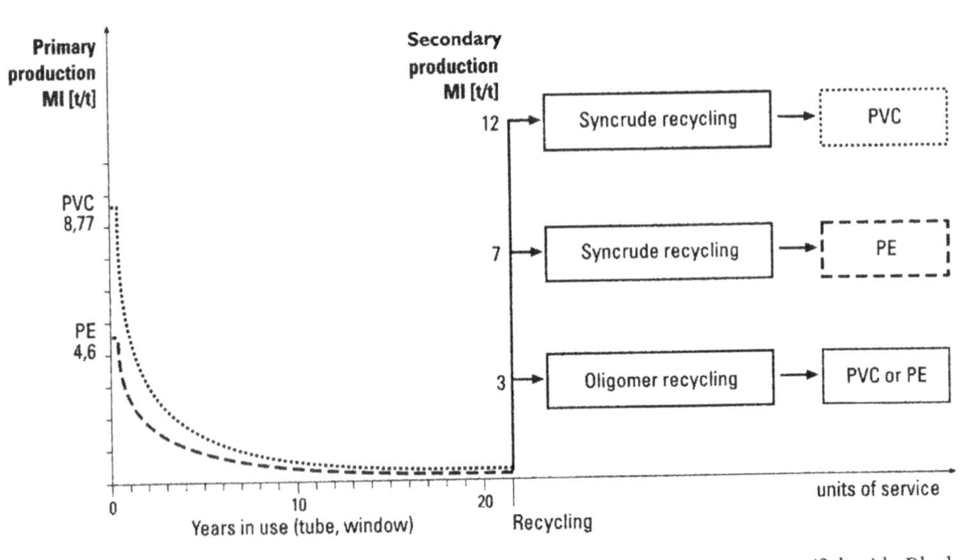


*Figure 8.8* Recycling technologies require different amounts of primary resources (Schmidt-Bleek and Liedtke 1995). Copyright Wuppertal Institute, Germany.

## Planning of infrastructure

Buildings are not the only interim part of a construction materials flow system between raw material extraction and final waste disposal. Homes, workshops, and offices are also interwoven within a network of supply and waste management flows for the provision of general services. Material flows for energy supply constitute a major part of societal metabolism. The flow of water represents the highest throughput volume and is interlinked with critical substance flows such as flows of nutrients and pollutants. Additionally, supply and waste management systems may have different resource requirements.

Planning for energy supply should:

1 determine the demand for electric and thermal energy of the planned construction;
2 check the possibilities of reducing requirements for non-renewable energy;
3 select those alternatives with the lowest resource requirements on a life cycle-wide basis.

The first step is normal practice for engineers. The second step requires extended know-how about possibilities for increasing the energy efficiency of buildings as well as the use of renewables. Solar architecture, in particular, is an art that deserves increased attention. The third step may be assisted by material intensity analyses.

For instance, the analysis of some wind converters used in Germany demonstrated that the material requirements for construction and maintenance are far lower than for the provision of the electricity from the grid (Figure 8.9). On the one hand, it is obvious that the use of renewable energy requires the use of non-renewable materials. On the other hand, this example shows that some portion of energy demand can be met with minimal resource inputs.

The results of a material intensity analysis may also be used for the further development of certain technologies. Modern processing of photovoltaic cells has been associated with

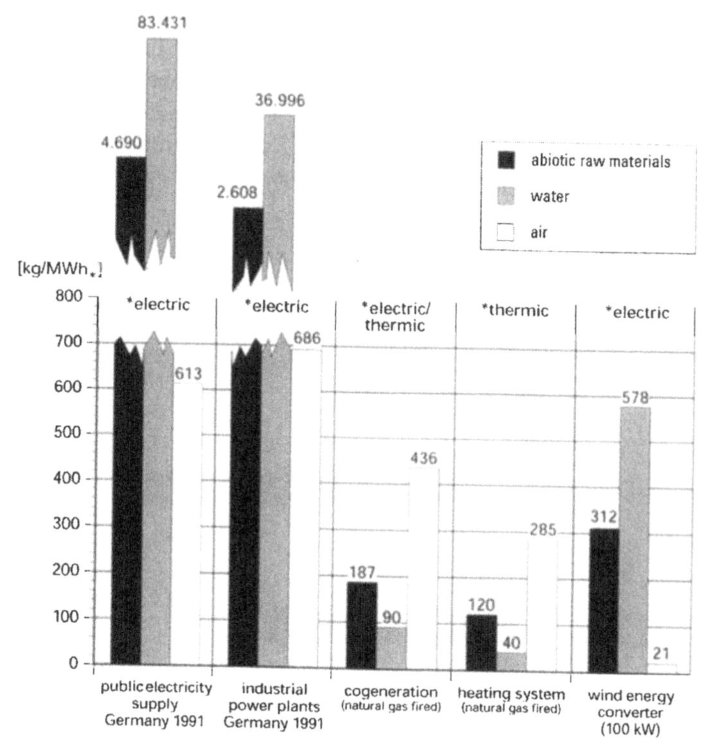


*Figure 8.9* Material intensity of different energy supply systems used in a small-to-medium sized company in Germany. Data calculated by Manstein after Manstein (1995). Copyright Wuppertal Institute, Germany.

resource requirements for abiotic raw materials that are of the order of magnitude of hard coal mining, and the emission intensity for carbon dioxide exceeds that of nuclear power plants by a factor of 10 (Spies 1997). The results indicate that photovoltaic technology must be significantly improved in order to contribute significantly to the factor 4–10 goal.

Similar to energy planning, increased efficiency in use should be considered before the provision of new water supply systems (Bringezu 1999). For example, in Germany the installation of water-saving technology in households allows a reduction in water consumption per capita per day to 100 liters. This technology is cost-effective and is the state of the art for new construction. Collecting rainwater to supply toilets and washing machines and to irrigate gardens may further reduce the demand on the potable water supply (Figure 8.10). However, this approach is associated with additional requirements for abiotic raw materials and somewhat increased carbon dioxide emissions (which are proportional to air inputs). Therefore, the installation of additional rainwater collection facilities can only be recommended in regions with a significant scarcity of potable water.

Wastewater management in Western countries relies on a flushing system that uses precious drinking water to transport feces out of homes and cities. Although this system resulted in a significant improvement in hygiene and health, especially in cities, during the last century, it has shifted the problem to rivers and the sea, which have been loaded with nutrients. Centralized sewage treatment has been used to reduce the eutrophication of rivers, and it has become a widespread belief that sewage problems will be solved when all households are connected to municipal sewage treatment plants.

However, the cleaning efficiency of modern municipal sewage treatment plants is poor for a variety of chemical elements (Raach *et al.* 1999). Even with state-of-the art denitrification technology about 25% of the nitrogen remains in the water. Even so, a huge amount of financial and natural resources is consumed in this type of sewage treatment. Municipal sewage treatment accounts for approximately 3–4% of the total material requirement of Germany (Reckerzügl and Bringezu 1998). There are alternatives

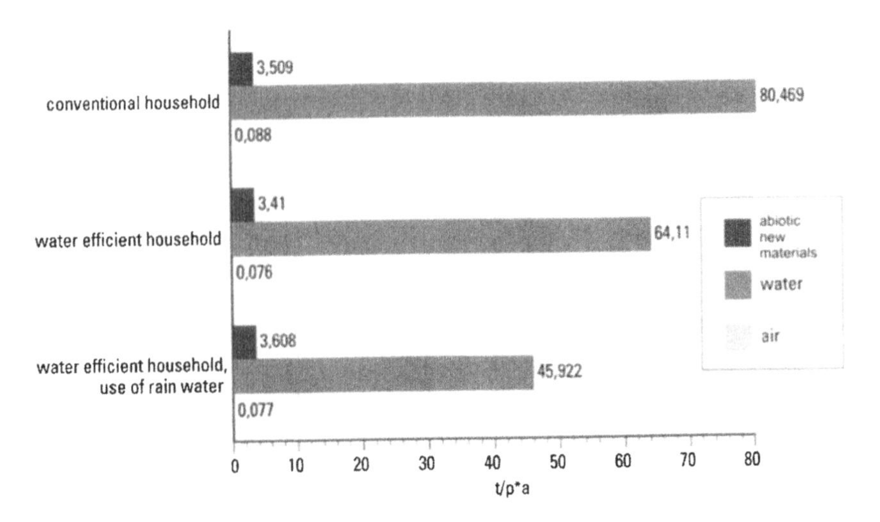


*Figure 8.10* Material intensity of rainwater use facilities with and without water-efficient technologies in households (Boermans-Schwarz 1998). Copyright Wuppertal Institute, Germany.

for rural areas as well as also for the suburbs of bigger cities. Comparative analyses have been performed for an area with 15,000 inhabitants.

Two semicentralized systems have been studied, in which rainwater is conducted underground and gray water is treated in constructed wetlands. The feces are also diverted separately. In the first system, the fermentation system, the feces are collected in vacuum toilets (like those in airplanes and ships) and transported via vacuum pipes to a fermentation container (Figure 8.11). This container also receives biowastes from households. Anaerobic processes result in the production of methane, which can be used for electricity production. The nutrient solution is transported regularly to farming areas, where it is either stored temporarily or used directly for fertilization of the fields. This also contributes to increased nutrient cycling. In the second alternative studied, the feces are collected by compost toilets and a container in the basement of houses (normally no more than two toilets per container).

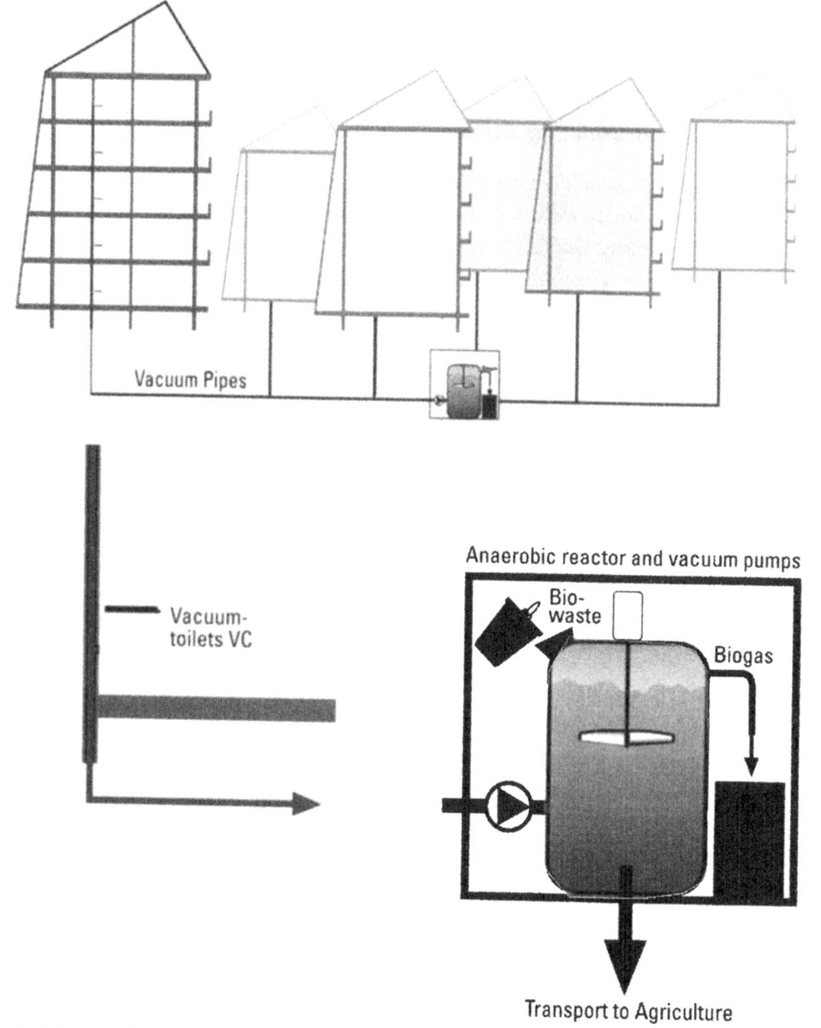


*Figure 8.11* Scheme of a semicentralized sewage treatment system with vacuum transport of feces and fermentation tank. After Otterpohl *et al.* (1997). Copyright Wuppertal Institute, Germany.

A decentralized alternative to the municipal system was also included in the study. An individual sewage treatment plant is used in many rural places. Feces and gray water are led into a septic tank. The overflow is treated by a constructed wetland. Stormwater is drained separately.

The results of the comparison indicate that the semicentralized alternatives, especially the fermentation system, are associated with lower material intensity (including water) and lower carbon dioxide emissions than the municipal system (Figure 8.12). The individual sewage treatment plant also has clear advantages. However, water consumption is hardly changed and the cleaning performance, especially with respect to ammonium and nitrogen, meets the standards but remains rather low.

Initial calculations of the costs of these different systems does not indicate any significant difference. Thus, semicentralized systems for sewage treatment should be considered by planners and engineers when new infrastructure is to be built in areas of low to medium population density (resulting in at least 6 m of sewage canal per person for the municipal centralized system). Given an area of 2.5 $m^2$ per person for the constructed wetland, these alternatives can also contribute to the aesthetic value of the settlement.

## Product, facility, and building management

In order to make profit, products are sold and the ownership is transferred to the customer together with the responsibility for the further fate of the product. The interest of the producer is to sell as many products as possible and to promote any opportunity to do so. If the ownership of the product were to remain with the producer or provider, these actors would have an increased interest in long-term durability, good repairability, and low maintenance costs. As a consequence, one would expect that the requirements for natural resources would be significantly reduced. For instance, carpets, lighting equipment, and heating facilities can be rented. In principle, this could also be applied to windows, flexible walls, and other modular components of buildings. An increasing number of companies are embarking on the road to selling service contracts rather than hardware.

Facilities management companies usually dedicate their services to the maintenance of infrastructure installations such as heating devices, cooling equipment, and lighting systems. The next step will be to guarantee the services provided by the machinery, for example a convenient and constant temperature and humidity in rooms and adequate brightness at different places within buildings. The facility management firm could be an intermediary between the hardware provider and the user of the building, with the producer of the equipment renting its products to the facility manager. In case of breakdown there will be an incentive to repair before replacing parts. Easy dismantling and low final disposal costs, together with good recycling potential, with the prospect of earning money, even from waste parts, will be the consequence of this type of product and facility management.

In the case of whole buildings, it is debatable whether ownership is profitable not only in economic, but also in environmental terms. Today, those who can afford to are buying houses. In the future, customers may prefer to rent rather than be bothered with the maintenance requirements of ownership. Construction firms may find that the shift toward providing service will mean that they have to become real estate companies, owning and managing buildings for different purposes.

In the end, a change of product management may significantly contribute to changing a distorted incentive structure that impedes resource efficiency and dematerialization.

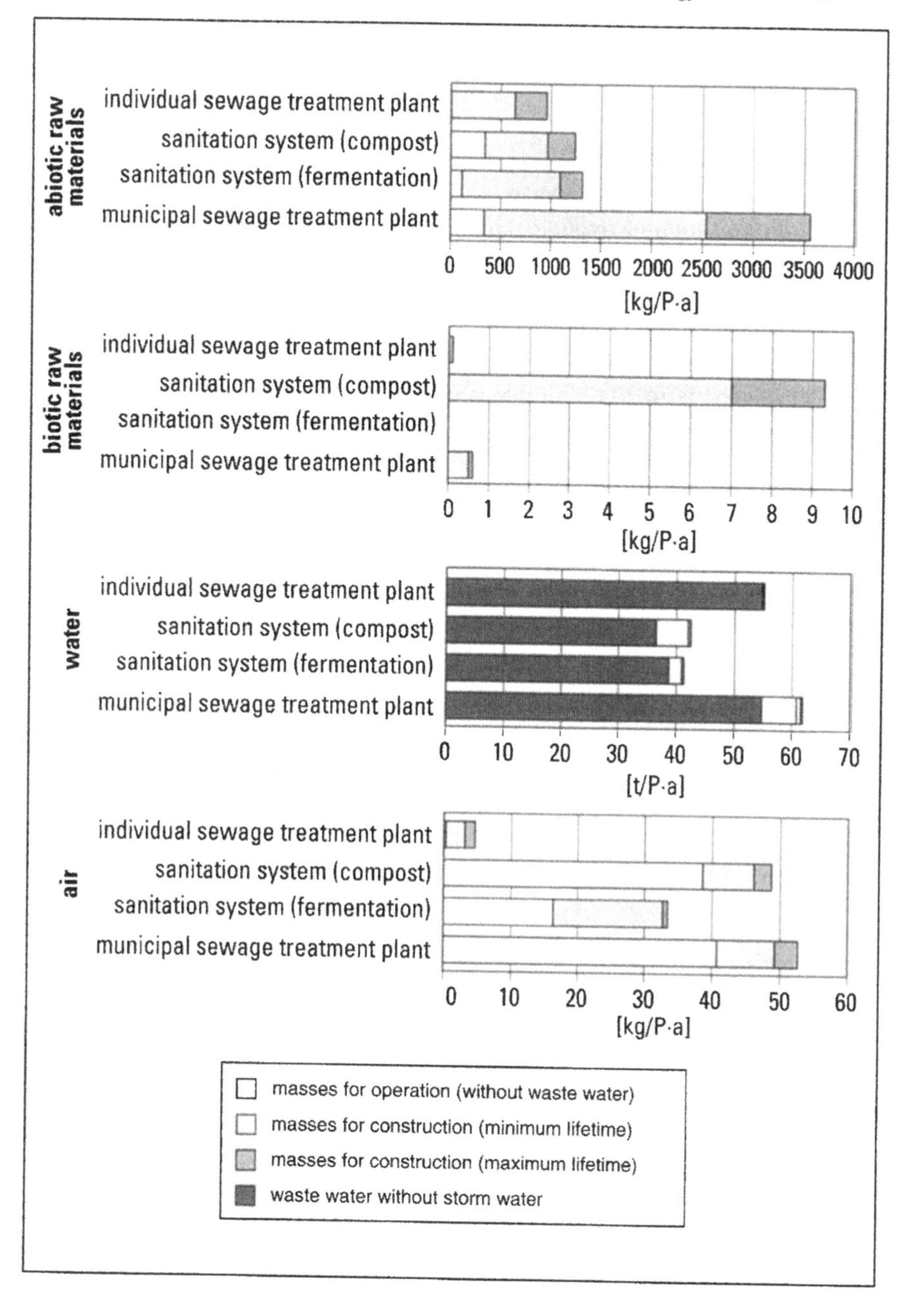


*Figure 8.12* Results of the material intensity analysis of different sewage treatment systems (Reckerzügl and Bringezu 1998). Copyright Wuppertal Institute, Germany.

However, other obstacles remain. For instance, architectural fees should be based on avoided costs for energy and materials instead of the volume or weight of buildings. Construction standards will have to be changed in order to allow for the use of secondary raw materials and to minimize the use of materials while guaranteeing safety conditions.

Depreciation rates should be revisited in order to consider the real condition of materials stocks and to provide an incentive for a prolonged use of buildings and infrastructure.

### *Integrated resource management*

Progress in sustainable development requires integrated resource management (IRM), comprising minimization of resource requirements and the control of critical emissions together with optimal use of financial and social resources. IRM requires comprehensive and decision-oriented information for the different actors and its implementation is strongly dependent on cooperation across sectors.

For instance, the increased cycling of nutrients from settlements with vacuum systems and semicentralized fermentation tanks requires contracts with farmers. The incentive for farmers to use the fermented sewage will depend on the price structure for mineral fertilizer. Thus, governments will have a role to play in promoting efficient use of fertilization by agriculture (to minimize nutrient loads on groundwater and rivers) by imposing levies on mineral fertilizer, and as a consequence must also support the further development of infrastructures for sustainable wastewater management.

Decision makers in construction need information that is easy to understand and sufficiently comprehensive to guide the decision in the right direction. For this purpose, the relevant parameters must be aggregated. As a consequence, detailed information is lost in favor of condensed information. In order to provide managers of construction

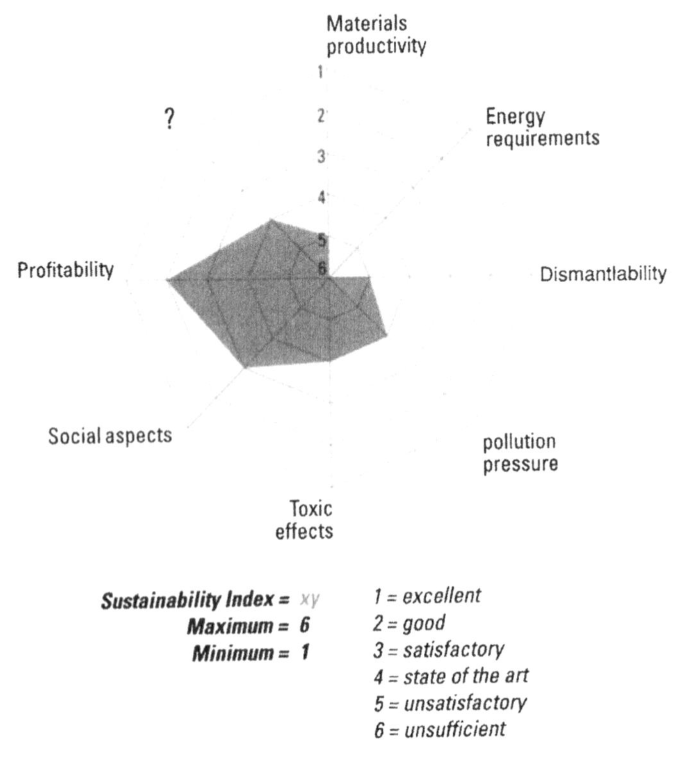


*Figure 8.13* Example of benchmarking buildings or construction components using to multiple criteria. After Wallbaum and Liedtke (1999). Copyright Wuppertal Institute, Germany.

companies with information on the ecological, economic, and social performance of the types of buildings they offer, Liedtke and Wallbaum have proposed a multcriteria index system. The three dimensions may be assigned equal or different weight. Figure 8.13 presents an example that can be used to benchmark construction products according to different indicators. Input-oriented indicators are considered (materials productivity and energy requirements) as well as output-oriented parameters (pollution pressure and toxic effects). Dismantlability has been included as a key concern for construction.

Materials productivity is equal to the inverse of MIPS and considers the sum of abiotic and biotic raw materials in relation to the services provided. Energy requirements – in this example – may focus on the energy consumption during the use phase of the buildings.[10] Pollution pressure may consist of the emissions regarded as most relevant at the national level ($CO_2$, $SO_2$, $NO_x$, etc.). In their work, Wallbaum and Liedtke (1999) considered construction activity to be "state of the art" if firms complied with emission standards. The operationalization of the parameters "toxic effects" and "social aspects" is still under development. The former is directed to the elimination of hazardous substances in construction products (e.g. carcinogenic substances shall be excluded). The latter is related to the acceptance by the customer of the building product.

Another example is the comparison of construction materials, as shown in Figure 8.14. Two input-oriented environmental parameters, MIPS and CER (cumulative energy requirements), and two economic parameters, price and value added, are considered. Two easily available social factors, labor intensity and average wages level, are also included and are considered positive indicators because they reduce unemployment problems and provide benefit for workers.

These examples demonstrate different possibilities for multicriteria-based decision making. The development of a method to operationalize social aspects of sustainability and to aggregate output-related environmental information is ongoing. The selection of different parameters as well as the form of aggregation will always be debatable. From an analytical systems perspective it seems necessary that:

1 Input- and output-based indicators should be used to reflect sufficiently and comprehensively the life cycle-wide performance of firms and products with respect to sustaining industrial metabolism. Whether the main outputs, such as carbon dioxide, are accounted for directly or indirectly via the cumulative energy requirements seems to be of secondary importance.
2 Ecological information should be included with economic and social aspects. Here the concentration on key parameters reflecting core concerns may be legitimate.

## Conclusions

- Conventional construction is associated with resource requirements that are in conflict with the conditions necessary for sustainable development. The status quo is characterized by physical growth of the technosphere that depends largely on non-renewable inputs.
- In order to reconcile the construction metabolism of the anthroposphere with the reproductive and assimilative capacity of the biogeosphere, and with regard to the demand for material welfare, an increase in resource efficiency is a key strategy.
- To this end, measures for rematerialization and dematerialization are required. Empirical data for Germany show that recycling of construction waste can be very

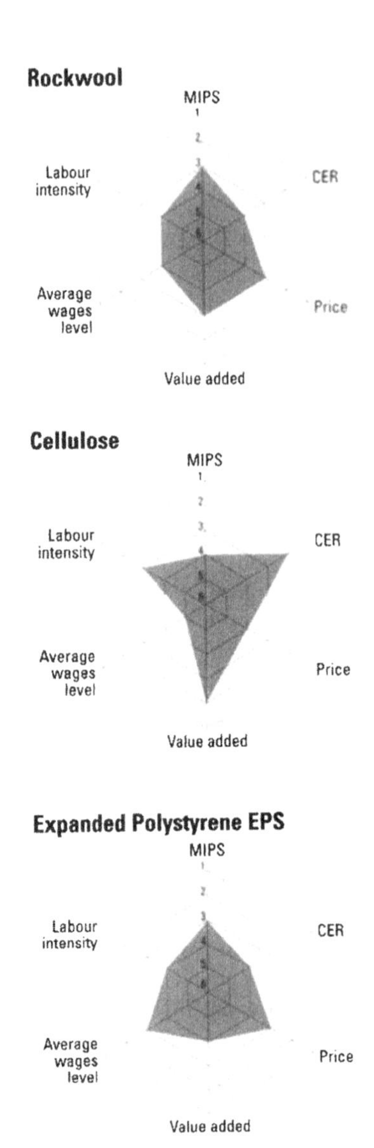


*Figure 8.14* Example of benchmarking insulation materials according to ecological, economic, and social criteria (Wallbaum and Liedtke 1999). Copyright Wuppertal Institute, Germany.

effective. As a consequence, however, the need for dematerialization and for a reduction of the resource inputs has become more important.

- Design, materials management, infrastructure planning, and product management may all contribute significantly to an increased resource efficiency of construction.
- The analysis of life cycle-wide material intensity can be used to provide practical information for managers, engineers, architects and planners. It may be included in a multicriteria assessment that focuses on using parameters essential for sustainable development.

## Notes

1 H.T. Odum's conclusions are based on the insight of Lotka that self-organization within ecosystems tends to develop those parts, processes, and relationships that draw in the most energy and use it with the best efficiency. This is obviously true for natural systems, and it seems to be the tendency for the anthroposphere, too, under an uncontrolled "business as usual" scenario. However, the question arises as to whether human beings are able to steer development in such a way that energy consumption is optimized (rather than maximized) in order to avoid catastrophic events (which are also characteristic of natural systems).
2 Although environmental scientists can hardly define absolute thresholds which can be related to (avoided) specific impacts (also to living conditions, welfare, etc.) on the one hand and to specific actors who should adapt their activities on the other hand.
3 This is a consequence of the first law of thermodynamics concerning the necessary mass balance between inputs and outputs of processes.
4 This does not imply any prognosis regarding the future dynamics of the material stock of the technosphere or anthroposphere or the length of the time required for flow equilibrium. If the stock becomes too high for some reason, physical decline may follow.
5 In abandoned quarries, increased species diversity often occurs after some years. Similarly, if humans are excluded from an area for an extended period of time, a similar scenario (with other species) would be expected.
6 In other texts the former has been included within the latter.
7 Materials found on disposal sites are regarded as being outside of the anthroposphere because, in the long run, their composition cannot be controlled and they are no longer in productive use.
8 Data for a variety of base materials produced in Germany can be found at http://www.wupperinst.org/Projekte/mipsonline/download/download.html
9 A PhD thesis considering each of these levels is being prepared by H. Wallbaum of the Wuppertal Institute.
10 They may also include cumulative energy requirements. The consideration of materials intensity (in which energy carriers are already included) and energy intensity represents a double weight to the energy-related flows. However, one may also use energy requirements to represent output-related pressures subsequent to energy consumption.
11 For information on ConAccount see http://www.conaccount

## References

Adriaanse, A. *et al.* 1997. *Resource Flows – The Material Basis of Industrial Economies.* World Resources Institute, Wuppertal Institute, Netherlands Ministry of Housing, Spatial Planning and Environment, Japanese Institute for Environmental Studies. Washington, DC: World Resources Institute Report.
Allenby, B.R. 1999. *Industrial Ecology. Policy Framework and Implementation.* Upper Saddle River, NJ: Prentice Hall.
Ayres, R.U. 1989. Industrial metabolism. In *Technology and Environment.* Ausubel, J. and Sladovich, H. (eds). Washington, DC.
Ayres, R.U. 1994. Waste potential entropy: the ultimate ecotoxic. In *International Symposium – Models of Sustainable Development. Exclusive or Complementary Approaches of Sustainability?*, Paris, 16–18 March 1994, vol. I, pp. 171–190.
Ayres, R.U. and Ayres, L.W. 1996. *Industrial Ecology: Towards closing the Materials Cycle.* Cheltenham: Elgar Publishing.
Ayres, R.U. and Martinas, K. 1994. Waste potential entropy: the ultimate ecotoxic. *International Symposium – Models of Sustainable Development. Exclusive or Complementary Approaches of Sustainability?*, University of Paris 1 Pantheon-Sorbonne, AFCET, Paris, 16–18 March 1994, Vol.1, pp. 171–90.
Ayres, R.U. and Simonis, U.E. 1994. *Industrial Metabolism: Restructuring for Sustainable Development.* Tokyo: United Nations University Press.

Baccini, P. and Brunner, P.H. 1991. *Metabolism of the Anthroposphere*. Heidelberg: Springer Verlag.

Barbier, E.B. 1989. *Economics, Natural Resource Scarcity and Development. Conventional and Alternative Views*. London: Earthscan.

Boermans-Schwarz, T. 1998. *Materialintensitäts-Analyse von Anlagen zur Nutzung von Regenwasser im Haushalt im Kontext einer nachhaltigen Wasserwirtschaft*. Wuppertal: Diplomarbeit am Fachbereich Sicherheitstechnik der Bergischen Universität Gesamthochschule.

Bringezu, S. 1997a. *Environmental Policy. Fundamentals, Strategies and Starting Points for an Environmentally Sustainable Economy* (in German). Munich: Oldenbourg Verlag.

Bringezu, S. 1997b. *From Quantity to Quality: Materials Flow Analysis*. In *From Paradigm to Practice of Sustainability*. Proceedings of the ConAccount Workshop, Leiden, 21–23 January 1997. Wuppertal Special 4. Bringezu, S. Fischer-Kowalski, M., Kleijn, R. and Palm, V. (eds). Wuppertal Institute, Germany, pp. 43–57.

Bringezu, S. 1998a. Comparison of the material basis of industrial economies. In *Analysis for Action: Support for Policy towards Sustainability by Material Flow Accounting*. Bringezu, S., Fischer-Kowalski, M., Kleijn, R. and Palm, V. (eds). Proceedings of the ConAccount Conference, 11–12 September 1997. Wuppertal Special 6. Wuppertal Institute, Germany, pp. 57–66.[11]

Bringezu, S. 1998b. *Material Flow Analyses for Sustainable Development of Regions* (in German). Habilitation thesis. Technical University Berlin. Published as *Resource Use of Economic Regions* (in German). Heidelberg: Springer Verlag.

Bringezu, S. 1999. Material flow analyses supporting technological change and integrated resource management. In *Ecologizing Societal Metabolism – Designing Scenarios for Sustainable Materials Management*. Proceedings of the ConAccount workshop, Amsterdam, November 1998. Kleijn, R., Bringezu, S., Fischer-Kowalski, M. and Palm, V. (eds). CML report 148, Section Substances and Products. Leiden: Leiden University.

Bringezu, S. and Schuetz, H. 1997. *Material Flow Accounts*. Part II. *Construction Materials, Packagings, Indicators*. Statistical Office of the European Communities, Doc. MFS/97/7, 87 pp. Available at http://www.wupperinst.org/download/index.html.

Bringezu, S., Stiller, H. and Schmidt-Bleek, F. 1996. *Material Intensity Analysis – A Screening Step For LCA*. Proceedings of the Second International Conference on EcoBalance, Tsukuba, Japan, 18–20 November 1996, pp. 147–152.

Bringezu, S, Behrensmeier, R. and Schütz, H. 1998. Material flow accounts indicating environmental pressure from economic sectors. In *Environmental Accounting in Theory and Practice*. Uno, K. and Bartelmus, P. (eds). Dordrecht: Kluwer Academic Publishers, pp. 213–27.

Daly, H.E. 1990. Towards some operational principles of sustainable development. *Ecological Economics* 2: 1–6.

Enquetekommission 1998. *Schutz des Menschen und der Umwelt des Deutschen Bundestages*. Bonn: Konzept Nachhaltigkeit. Vom Leitbild zur Umsetzung. Abschlußbericht.

Gardener, G. and Sampat, P. 1998. *Mind over Matter. Recasting the Role of Materials in our Lives*. World Watch Paper 144. Washington, DC: Worldwatch Institute.

Graedel, T.E. and Allenby, B.R. 1995. *Industrial Ecology*. Engelwood Cliffs, NJ: Prentice Hall.

Kohler, G. (ed.) 1997. *Recyclingpraxis Baustoffe*. 3. erw. Aufl. Cologne: Der Abfallberater für Industrie, Handel und Kommunen.

Manstein, C. 1995. *Quantifizierung und Zurechnung anthropogener Stoffströme im Energiebereich*. Diplomarbeit an der Bergischen, Universtitat Gesamthochschule Wuppertal, Fachbereich Sicherheitstechnik, Wuppertal.

Odum, E. 1971. *The Fundamentals of Ecology*. Philadelphia: WB Saunders.

Odum, H.T. and Odum, E.C. 1999. The ways of energy and materials in all systems. *Manuscript for discussion in the workshop reader of the Rinker Eminent Scholar Workshop on Construction Ecology and Metabolism: The Materials Cycles*. Gainesville, FL: University of Florida, Chapter 4.

Organization for Economic Co-operation and Development 1996. Meeting of OECD environment policy committee at ministerial level, 19–20 February 1996, Paris. Paris: OECD Communications Division.

Otterpohl, R., Grottker, M. and Lange J. 1997. Sustainable water and waste water management in urban areas. *Water Science and Technology* 35: 121–133.

Pearce, D.W. and Turner, R.K. 1990. *Economics of Natural Resources and the Environment.* Baltimore: Johns Hopkins University Press.

Sachs, W., *et al.* 1998. *Greening the North.* London: Zed Books.

Raach, C., Wiggering, H., Bringezu, S. 1999. Stofffflußanalyse Abwasser – eine Abschätzung der Substanzflüsse deutscher Kläranlagen. In *Vom Wasser*, Vol. 92. Weinheim: Wiley-VHC, pp. 11–35.

Reckerzügl, T. and Bringezu, S. 1998. Vergleichende Materialintensitäts-Analyse verschiedener Abwasserbehandlungssysteme. *Gas- und Wasserfach (gwf) Wasser + Abwasser* 11: 706–13.

Schmidt-Bleek, F. 1994. *Wieviel Umwelt braucht der Mensch ? MIPS – Das Maß für ökologisches Wirtschaften.* Berlin: Birkhäuser Verlag.

Schmidt-Bleek, F. and Liedtke, C.1995. *Kunststoffe – Ökologische Werkstoffe der Zukunft?* Symposium Kunststoffe+Umwelt, Frankfurt/Main, 27–28 June 1995, Frankfurt: Verband Kunststofferzeugender Industrie, pp. 38–45.

Schmidt-Bleek, F. and Tischner, U. 1995. *Produktentwicklung. Nutzen gestalten – Natur schonen.* Vienna: Schriftenreihe des Wirtschaftsförderunginstituts 270.

Schmidt-Bleek, F., Bringezu, S., Hinterberger, F., Liedtke, C., Stiller, H., Spangenberg, J. and Welfens, M.-J. 1998. MAIA. *Einführung in die Material Intensitäts Analyse nach dem MIPS-Konzept.* Wuppertal Texte. Berlin: Birkhauser Verlag.

Spies, H. 1997. *Vergleichende Materialintensitäts-Analyse von Photovoltaikanlagen nach dem MIPS-Konzept sowie ein Vergleich mit anderen Energieträgern.* Diplomarbeit vom Fachbereich Sicherheitstechnik der Bergischen. Universität Gesamthochschule Wuppertal.

Tischner, U. and Schmidt-Bleek, F. 1993. Designing goods with MIPS. *Fresenius Environmental Bulletin* 2: 479–84.

United Nations General Assembly Special Session 1997. *Programme for the Further Implementation of Agenda 21.* Adopted by the Special Session of the General Assembly, 23–27 June 1997, Paragraph 28. New York: United Nations.

Wallbaum, H. and Liedtke, C. 1999. MIPS und COMPASS – Methoden und Instrumente zum ressourcenschonenden Bauen. Dokumentation des Kolloquiums "Ressourcenschonendes Bauen" der Universität Stuttgart, Fakultät Architektur und Stadtplanung, April 1999, available only at Universität Stuttgart.

Weizsäcker, E.U., Lovins, A.B. and Lovins, L.H. 1995. *Factor Four. Doubling Wealth Halving Resource Use.* London: Earthscan.

World Business Council for Sustainable Development 1994. *Getting Eco-Efficient – How Can Business Contribute to Sustainable Development?* Report on the Eco-Efficiency Workshop, Antwerp, 1993. Geneva: World Business Council for Sustainable Development.

World Business Council for Sustainable Development 1998. *WBCSD Project on Eco-Efficiency Metrics and Reporting.* State-of-Play Report by M. Lehni. Geneva: World Business Council for Sustainable Development.

# 9 Construction ecology

## An environmental management viewpoint

*Fritz Balkau*

### Introduction

The construction industry is characterized by a high level of resource consumption and significant generation of pollution and waste compared with other service sectors. Numerous environmental programs to address individual environmental impacts have already been established by national authorities, however a more systematic and holistic approach still remains to be developed by the industry itself. Industrial ecology has the potential to provide a framework for such wider coordination, and to address the issues in a more comprehensive way.

Industrial ecology can be regarded as a study of the material and energy flows, population dynamics, operational rules, and interrelationships between the various components of the entire production system. Industrial ecology is not an attempt to copy Nature; it simply refers to a way of looking at an industrial system as a complete dynamic entity rather than focusing on individual components.

In order to implement these ideas in industry, two challenges have to be met:

1. how to present the "ecology" concept as a complete model that addresses all environmental policy areas; and
2. how to select an effective combination of management instruments that can apply the model in a real situation. This chapter will try to focus particularly on the second item.

### The concept of industrial ecology

Some excellent perspectives on industrial ecology have been published by various authors. Among the main elements commonly described are:

- industrial metabolism;
- industrial ecosystems (associations);
- materials cycles in Nature and in industry;
- evolution of industrial technologies.

Underlying the above, several subsidiary ideas are relevant to the practical implementation of sustainability concepts in industry, in particular:

- the precautionary principle;
- the prevention principle – cleaner production and ecoefficiency;

- life cycle management;
- the zero emissions concept;
- dematerialization – the factor 10 concept;
- integrated environmental management systems;

The best-known application of industrial ecology is in "industrial symbiosis", as described in places such as Kalundburg in Denmark, in Tsumeb in Namibia, and in some of the industrial complexes in Russia. Here the materials loop has been closed by linking several different production units that use each other's waste as secondary raw material. The synergies of such an approach can be considerable, leading to both economic and environmental benefits. The recently established "Valuepark" in Germany is an example of the strategy of creating synergies between aspects of corporate operation such as transport, supplies, services, and linked markets. Unlike Kalundburg, which evolved over time, the Valuepark was planned from the outset to maximize synergies, for example by favoring certain types of companies. However, although the idea of industrial symbiosis is attractive, the interdependence can also create problems when one of the linked units fails or wants to change technologies. And at a collective or macrolevel it requires a degree of industrial cooperation that rarely occurs spontaneously in a competitive market.

Industrial symbiosis concentrates on optimizing material and energy flows, but this is insufficient if we are aiming at a more sustainable pattern of development worldwide. Additional environmental factors such as land use and biodiversity, for example, also need to be optimized.

## Implementing industrial ecology models

The study of industrial processes does not by itself lead to changes in industrial planning. Development is often driven by political, commercial, and personal decisions that have nothing to do with material things. We need therefore to consider the relationships of all the partners and players that influence or depend on construction activities, from building firms and their suppliers to the clients and users, including also the rule-making authorities and institutions. In particular, we need to study the management mechanisms and instruments by which they implement their organizational and personal objectives.

As yet, no mature industrial ecosystem has consciously applied the relevant management elements to create an artificial "ecology". Instead, there are a number of situations in which the separate components can be seen in action. Such restricted application of individual elements of an ecological management system is unlikely to give optimal results overall, but it allows us to see how the dynamics work in practice, and to identify the main instruments that we can eventually combine into a larger model.

Some current examples of such management elements include:

- Improved individual corporate decisions on sustainability. Major international companies are now re-examining their basic products. For example, the oil company BP is moving into renewable energy, and some mining companies, such as RioTinto, are abandoning lead. The company Interface leases rather than sells carpets, thus recovering the primary resources at the end of their useful lifespan and avoiding waste disposal burden on the environment.
- Adoption of formal environmental management systems (EMS) by individual enterprises, and more recently by industrial estates in France and South-East Asia.

The use of EMS makes it easier to consider more thoroughly the life cycle aspects of an operation or an industrial estate, although many company and especially estate EMS still focus on only a limited part of a production or service process.

- There is increasing application of the practice of supply chain management and extended producer responsibility, whereby linked liability and organizational dependence is created along the product chain. Such linking allows greater synergy and collective action. Major international companies, such as Nike, Philips and IBM, are already using SCM to influence their suppliers in developing countries, thus exerting an influence worldwide.
- Central or collective infrastructure management of a number of environmental functions by the owners of large industrial estates. The most common services are related to utilities, emergency response, and waste treatment; however, there are many untapped opportunities for expanding the service into other environmental areas, including cleaner production. Some estates also have a common building code.
- Cooperative environmental programs in large industrial estates in France, where the developer, individual companies, service providers, and the local community agree on an environmental charter and action plan for an industrial estate. The focus is on total environmental performance rather than simply waste exchange, although waste disposal services are included in the charter.
- Government industrial development policy to create an appropriate framework for corporate decision making, as for example the National Cleaner Production Strategy of the Australian government. The Netherlands also has well-developed policy positions that are driven by integrated life cycle thinking.

The above include instruments that are company specific as well as example of collective action and government frameworks.

Taken together, the above could lead to a more ecological approach to industrial development if they are combined intelligently in a systematic way. The question is whether we take a "hands-off" approach and let the ecological system evolve naturally in a Darwinian fashion (the Kalundburg example), or if we wish to accelerate the evolution by intervening in the process of industrial planning and operations (the Valuepark example). In the latter case we would need a clearer view of the planning process to identify where intervention is most effective, and the creation of mechanisms to bring the relevant partners together. It is unlikely in our current political system that one single player could make the change alone. It also brings us back to the role of government regulation to create the right framework, and in taking strong implementation action. Overall, natural evolution will not give us useful results in the time span dictated by our current environmental problems.

To explore the effectiveness of various instruments when combined into a more complete model, the United Nations Environmental Program (UNEP) has been examining the environmental dynamics in large industrial estates. Working with industrial estates is useful because it involves a relatively straightforward administrative structure that lies between that of individual companies (limited effectiveness on their own) and that of national government (slow and complicated). UNEP's work has involved the development of technical guidelines for estates, followed by a number of regional training workshops to discuss key issues. This work has identified a number of possibilities for estate managers to provide a more comprehensive framework – i.e. an industrial ecology – for companies to reduce their individual environmental burden, as well as identifying synergies in services and operation. It is clear, however, that the management instruments developed for individual companies must be further adapted before they are effective at the estate level.

### *The environmental agenda revisited*

Problems regularly arise in industry when we have only an incomplete understanding of the environmental agenda. A group of companies may, for example, plan collective treatment facilities only to be told that waste reduction is now a higher priority. Solving problems one at a time rather than in an integrated fashion runs the risk of merely moving the problem around. This approach also quickly loses sight of the evolution of the entire environmental agenda, with consequent attempts to apply old-fashioned environmental solutions. Industrial ecology, similarly, must also integrate new environmental and social issues that come up, and stay on top of the changes that occur in environmental policy.

For the construction sector it is useful to recall some of the main environmental issues that need to be considered:

- local pollution and waste issues around construction and building sites;
- contributions to global pollution impacts, e.g. ozone layer, climate change;
- exposure to hazardous chemicals and contaminants;
- local amenity and visual impacts, during construction and in the long-term;
- land use conflicts;
- habitat modification and impact on biodiversity;
- modification of water regimes (surface and subsurface), and perhaps wind.

If we take a life cycle approach to environmental management, the above factors need to be considered during the building use and demolition stages as well as during the construction phase. Some recent national studies have focused strongly on "sustainable buildings" which consume less energy and incorporate environmentally friendly materials and whose components are able to be recycled after demolition. The design of energy-efficient buildings is probably the most widespread application of this approach, followed by more efficient water management. In some counties, the recycling of demolition waste is becoming more widespread. But it is also important not to lose sight of emerging issues. The reduction of ozone-depleting substances in hotels, for example, is becoming a more important issue, and UNEP has responded with a recent technical guide on this subject.

A life cycle approach also looks at the secondary impacts away from the construction site. Mining of construction materials, energy use in transport, and impacts during the manufacture of building components thus all become a prominent part of the equation. The consequences that buildings have on the capacity of municipal waste treatment and disposal sites should also be given proper attention. Taking this line even further, we can consider, for example, the impact on local amenity and transport congestion, both during construction and during later use. Compared with sustainable buildings, there are fewer studies that deal with these off-site environmental impacts, which may nevertheless be a major part of the life cycle burden of buildings on the environment.

Looking even further into the future, if we apply the new concept of "sustainable consumption" to the construction industry, we could predict that the construction industry will eventually no longer be a simple supplier of a building to a client, but rather will ensure that the "function" required by the client is provided in the most environmentally sound fashion across the entire supply chain. Functionality of buildings has of course been an architectural objective for many years, but sustainable consumption goes further by also looking at how to influence user behavior itself in order to reduce the impact.

While the application of the above ideas is not yet widespread, we are already seeing a start in this direction in the practices of some component suppliers who, for example, lease components such as carpets rather than sell them. Management innovations are also being seen, as for example in the tendering practices of clients such as the Australian Olympic Commission.

## Construction ecology

Using the foregoing discussion as a guide, we can start to list some of the more influential management instruments available to the construction sector to make the change to a more sustainable future in the framework of a "construction ecology" model:

- clear and comprehensive environmental standards, and regulations on construction in protected areas;
- national targets and goals on, for example, energy, waste, and water;
- national (environmental) standards for building components, e.g. on energy efficiency, chemical content, etc.
- building codes and permits applied by local and national authorities;
- tendering procedures that make explicit reference to low environmental impact;
- financing conditions that include environmental criteria;
- voluntary codes of environmental conduct by key sections of the building profession;
- operational codes for environment protection during construction, operation, and demolition;
- environmental management systems by construction companies and building owners;
- wider application of supply chain management, extended producer responsibility, and green purchasing;
- greater use of assessment tools such as EIA and LCA and predictive models;
- greater use of environmental audits and performance reports.

These instruments are applied by quite different bodies at different times in the building cycle. It would clearly be of enormous help if there were a common understanding of their use and some coordination in their application. As it stands, it is quite possible that different instruments are applied in inconsistent and perhaps even contradictory ways.

Coherent action can only occur if the environmental goals are clearly defined. Here it has to be admitted that, although environmental quality standards are now almost universal (although not universally comprehensive), performance standards for other parameters such as energy efficiency, chemical contamination, and habitat modification are still rare. In effect, the government framework is still incomplete.

Accordingly, it would help if the industry itself developed a common view of the environmental agenda, and addressed the national and global priorities in a coherent way. This could be done, for example, by organizing a regular, ongoing, and comprehensive dialog on environment with all internal and external stakeholders, with the aim of eventually forming a voluntary code on the environment similar to what we see in some other sectors.

Such dialog often needs to be stimulated by an outside agent, and it is interesting to ask what guidance is available from international bodies such as the UNEP.

### *Some UNEP activities relevant to construction ecology*

UNEP's Division of Technology, Industry and Economics is undertaking a number of program activities that can be directly applied to the construction industry:

- The Cleaner Production program applies the "prevention" approach to industrial activity by taking action along the entire life cycle of a process or product. Applied to the construction industry, this means the design of low-impact buildings, selection of building components and materials with a small environmental footprint, the use of construction techniques that minimize the amount of produced waste, and the reuse of as much of the waste on-site as possible.
- The Industrial Pollution Management program focuses on sound recycling, treatment, and disposal of residues that cannot be avoided in the application of the "prevention" techniques above.
- The activities of the Sustainable Consumption program focus on creating understanding and commitment by all stakeholders in achieving a different pattern of consumption – both personal and industrial - that is in line with the Earth's long-term carrying capacity;
- The Responsible Entrepreneurship activities focus on promoting the wider use of corporate environmental management systems and tools such as corporate environmental reports, supply chain management, and environmental accounting;

The above elements can come together in different ways, as for example in UNEP's initiative to bring environmental management concepts into industrial estates worldwide. Buildings and construction are a significant factor in estate management, and the environmental aspects are treated at some length in the UNEP program. Because estates provide a useful decision-making framework, the way in which the physical facilities in such estates can be developed along more ecological lines constitutes an important element of UNEP's work.

Another area where specific focus on construction has occurred is in the hotel industry, and the UNEP has published several technical guides that include consideration of building design and operation.

## Conclusion

While energy-efficient buildings and waste management at construction sites are now well understood, the environmental agenda for the construction industry also includes other issues that require a more integrated approach. Life cycle approaches to the industry will require closer consideration of the role of both upstream (suppliers) and downstream (occupiers) stakeholders in addressing environmental issues systematically. A comprehensive "ecology" model of the construction industry that is amenable to implementation has to incorporate all these factors.

The complex relationships between major stakeholders in the construction industry make it necessary to identify more clearly some of the management mechanisms that will push the construction sector to incorporate ecological principles in its development plans. These mechanisms include instruments applied by corporations, governments, and by developers themselves.

Industrial estates provide a convenient context within which these management mechanisms can be examined and tested.

## Part 3

# The architects

*G. Bradley Guy*

In the days of agrarian societies, economic and cultural systems were dependent upon local resources and flows of energy, such as the rain and sun that fell directly upon the land where settlements were located. Buildings of those times were based upon designs that expended the least amount of energy and used the materials closest to hand. The architecture of the Industrial Age has not fitted Nature as well. Witness the effects of fossil fuel use on climate change. Witness that most buildings are uninhabitable without continuous, high-intensity infusions of energy. Each of the architects in this section has found an architectural expression that is his own profound attempt to return environmental balance to the creation of buildings.

The three chapters on ecological design take different approaches, but with many similarities, to describing how modern society can design buildings to realize the patterns of bioregional "fit." These perspectives range from the literal "planted" architecture of Malcolm Wells to the abstracted ecology of Jürgen Bisch's urban office buildings, which seek to minimize resource consumption while using glass, steel, and concrete as the predominant materials. On the other hand, Sim Van der Ryn and Rob Peña speak less of a style than an approach that may be said to fit somewhere between the first two approaches, using both low-technology, biological materials such as straw bale, and high-technology products such as photovoltaics, while shaping projects to operate in tune with the climatic rhythms of their sites. The Real Goods project they describe is in a rural area, whereas the office buildings of Jürgen Bisch are located on highly urbanized sites, and Malcolm Wells' buildings seek to disappear altogether, buried in the ground.

Sim Van der Ryn and Rob Peña (Chapter 10) discuss how we must learn from natural ecological systems and use them as an analog for the built environment. They believe that constructed metabolism, or the intense use and conversion of energy through building systems, has replaced/displaced form as a means to adapt to natural forces. They simultaneously point to the fundamental problem in trying to make buildings more like Nature, which is that buildings are fixed in time, whereas natural forces are in constant cycles, day and night, and season to season. We see a static condition, a building, yet all around it are incredible forces impacting upon it, from the microscopic to the cosmic, with motion from the very rapid to the very slow. In the modern age, buildings are typically designed neither to evolve and be adaptable nor to last. Somewhere in between these extremes lies the answer.

Van der Ryn and Peña imply that ecological design is physiological, i.e. is about the structure of the building, the organs, and their interrelationships, both internally and to the external environment. An analog of natural systems might be a device like skin that would bring internal moisture to the surface of the body to allow evaporative cooling to take place. In the building's case, which exists to provide shelter for people, the building's users have a role to play as the "nerves" of the building, acting as sensors that tell the building when and how to act. In this way, users are change agents for a building, from simply rearranging furniture, lowering blinds, or opening windows, to designing and adapting the building as their needs warrant. Over time, as technological or functional needs change, the users will make more intrusive alterations, such as rearranging walls and systems.

No matter how well the designer creates the initial building to meet known needs, it is inevitable that the occupants will need to make the building their own in some manner. This interaction between the occupants and a building will have a benefit beyond physical comfort, imbuing the building with the dynamic qualities of life. And as is the case with the human nervous system, which over time learns functional patterns, training the body to become more skillful at survival, so too must designers learn from the users of buildings.

Ecological design is a design of place, the place of the users, the climate, the topography, and the local culture. As in Nature, where species from similar genetic strains will evolve into subspecies when faced with different bioregional forces, so will the generic "modern" architecture building either fail in its location (environmentally) or adapt. The glass box that is identical on all four sides will inevitably require tremendous amounts of energy to operate, no matter where it is. By its very homogeneity it may seem to fit anywhere and, at the same time, fit nowhere.

According to Van der Ryn and Peña, an ecological design will have a multiplicity of parts such that, as much as possible, specialization is eliminated. If the parts have multiple functions, even though each part might not be the most precise and singularly efficient component, when it is all put together, the building will be a collection of resilient and interdependent components that can match the changing needs of both user and Nature. A resilient and user-adaptable building is able to take the path of least resistance, just as is the case with energy use by living organisms. The simple process of using materials as close to their source as possible and with the least amount of processing and additives will reduce the metabolic flows that not only consume energy, but also result in waste energy and materials. An example of this is the use of solar energy directly at the building, gathering it from the roof for heating water and internal spaces, using the least possible amount of transmission energy. By allowing convective forces to move air and water, the source of energy, the sun, provides its own "power line."

Optimizing a building's relationship to the forces which impact it is accomplished by considering its form, orientation, and the thermal properties of its materials. The result will be a reduction in the metabolic requirements of the building itself. The building becomes a processor rather than a consumer. In Van der Ryn's view, each design starts with "what is here, what Nature will permit, what Nature will help us do."

Bisch (Chapter 11) has a similar technical approach to Van der Ryn and Peña, but his approach is "leaner," more refined, less enamored of the resiliency and redundancy that Van der Ryn and Peña believe are fundamental to ecological design. It is helpful to see the application of Bisch's ecological design principles in the setting of the modern office building, an application where, no matter the context, the ecology of the building remains the same. It is composed of spaces, an envelope, and systems that feed the building light, air, water, heating, cooling, and energy, either indirectly through external sources or directly through the skin of the building itself using local sources of energy.

Both Van der Ryn/Peña and Bisch agree that good design is an investigation and an understanding of site, materials, and functional needs that require careful consideration, not easy in the world of corporate architecture. Bisch provides a simple illustration of optimization in using cantilevered structures to reduce materials and open up the span of an interior space and allow for the separation of A materials (long-lived) and B materials (short-lived). All agree that making buildings and components able to be removed, remanufactured, reused, and recycled is an important concept. This is the basis of ecological materials flows whereby each material is constantly becoming "food" for another process.

An interesting aspect to Bisch's attempt to create dematerialized and integrated structures, such as using hollow core concrete slabs for combining air distribution and wiring, is that this is somewhat in conflict with the use of demountable structures and separation of A and B systems. Although a reduced floor to floor height will save on the total materials needed to build the same number of floors and total area, it can reduce other more subjective aspects of a building such as the fenestration that creates daylight penetration into a space. Using radiant floor systems is an efficient means for combining the heating distribution into the very materials of the floor itself, but this entangles the water or air distribution with the concrete slab. While no doubt leaner and less materials intensive, this refinement potentially makes the building less flexible. Nonetheless, this too follows a pattern in Nature of evolution, whereby species adapt geographically to their place and within the cooperative ecosystem of which they are a part. An office building is one organism in the complex weave of an urban community, comprising housing, commercial, and civic structures. By each optimizing its place within the system, each building type is less competitive with the others for both appropriate expression and resource consumption.

Malcolm Wells (Chapter 12) suggests a simple and profound concept, that buildings should be either buried underground or have the earth and plants spread on top of them after construction. The reasons for this are simple: plants are the one true producer and able to absorb and ameliorate the effects of air or soil pollution. Malcolm Wells' book *Gentle Architecture* is an apt expression of this philosophy, which is a wake-up call to respond to the brutality and hubris of the international style of architecture practiced so widely in the middle half of this century and then the styles of post-modernism and deconstructivism in the last quarter of the twentieth century. His call to architecture is for a retreat from paving over of the natural environment and for an acknowledgement of Nature's "green" architecture of plants. The question to be asked in this regard is whether there is such a thing as a universal "style" that is so inherently ecologically appropriate that it can be applied anywhere with equal success. Wells apparently believes this to be the case. Certainly, there is a lesson in the fact that the first dwellings were caves, using the protection and stabilizing effects of the Earth in order to provide shelter.

The chapters is this section provide a perspective based on ecologic analogs and human beings as change agents in developing an architecture that mimics Nature, allowing for redundancy and resiliency, a perspective based on rigorous analysis and dematerialization or "lean" architecture, and an architecture that is barely about buildings at all, but about creating shelter that is subsumed by the earth, built in mass and stability.

Each of these approaches suggests key questions about construction ecology. Should buildings be "loose" and open to adaptation and operation by their human inhabitants and highly evolutionary ? Should they be ecologic machines, rigorously analyzed and sleek, combining many elements in a finely tuned system ? Or should they be built "like a rock," immutable, stable, highly tempered, and universal, needing little in the face of Nature's cycles ?

# 10 Ecologic analogues and architecture

*Sim Van der Ryn and Rob Peña*

"Ecologic design" uses the lessons learned from the study of natural systems ecology to design the built environment, including architecture, engineering, and community planning. Natural systems ecology runs the gamut from the findings of microbiology on the organization of molecules within cells to the findings of astrophysics on the cosmic scale of universes beyond our own. Ecology imbues the physical and biological sciences which, taken together, encompass physical scales from $10^{-22}$ meters to $10^{24}$ meters; from the nanosecond subatomic time scale of an electron particle to the light years time scale of galaxies. In this sense, ecology is "Nature writ large". A generation ago, we drew our inspiration from an understanding of ecology as presented in such classic college texts as Eugene Odum's *Fundamentals of Ecology* and the work of such masters as Aldo Leopold, Rachel Carson, and Gregory Bateson. Today, the ecologic ethos is growing wider and deeper as science at all scales and complex systems theory link small patterns into larger ones, continually revealing more of what Bateson called, "the pattern which connects."

Designers looking to natural systems to discover analogs useful in the design of the built environment have an extremely large field of potential information to absorb. We are only at the beginning of the learning curve, shifting the guiding metaphor in architecture from thinking of buildings as static machines or works of sculpture to conceptualizing them as dynamical living systems that are the very nature of Nature. In the nineteenth century, the English Romantic writer John Ruskin declared that architecture was like frozen music, and the American architect Louis Sullivan prescribed that form follows function. Twenty-first century architectural design may be guided more by the ideas that *architecture is music* and *form follows flow*.

As we look forward to a design paradigm responsive to the flow of energy, matter, and human activity, we might also take a look back to a time when adaptation to place in all its dynamic complexity was largely a function of form, location, and organization. Prior to the availability of fossil-fueled means of heating, cooling, and lighting, the connection between indoor and outdoor environments was a fundamental source of form and order in architecture. Such buildings were as characteristic of the cultures that built them as they were particular to their place. Today, however, *metabolism*, the conversion of energy, has displaced form and location as the primary means of adapting to the climatic forces of place and satisfying the physical needs of people.

The essence of ecological thinking is that virtually everything we perceive as fixed is actually in a state of flow and change. Everything is engaged in an intricate dance that we cannot see because the elements of the dance are either too large or too small, and the music being danced to is either too fast or too slow for us to take notice.

Nature's design is adaptive design. At a long time scale, design for adaptation equals evolution; at a short time scale adaptation is *ad hoc* innovation and change. Ecologic building design takes into account a wide time scale of adaptive strategies and scenarios involving *place, people, and pulse*. Examples include changing landscapes and global warming; the shift from command and control hierarchies to problem-solving networks; and the move from centralized energy sources to distributed sources such as fuel cells and photovoltaics.

Ecologically based design is not an ideology, fashion, or style, like post-modernist or deconstructivist architecture. Ecologic design is no more ideological than global warming. It is an extension of scientific thought to the solution of design problems. As a method, it strives for the same open inquiry process, rigorous analysis, and critical thinking which we expect of science. Each building we design is built of potentially testable hypotheses about place, people, and pulse. Lacking agreement about the essential nature of design as applied scientific process, there is no constituency that demands and participates in the testing of design hypotheses. Architects consider themselves lucky if their clients are happy at the completion of a project, a critic writes something nice about their building, or it wins an award.

Ecologic design as a process does not follow the rules of the scientific method but is increasingly informed by progress in mapping the intricate architecture of living systems. The goal in ecological design is to create buildings and environments that are "ecomorphic," i.e. their internal structure mimics and integrates with the natural systems within which they are embedded and connected. Ecomorphism means something different than an architectural form derived directly from Nature, such as the structure of a bridge resembling the structure of a bird wing, or a house looking like a nautilus shell. These are examples of "biomorphism" – forms taken directly from Nature. Ecomorphism goes deeper, implying architectural processes and forms at many scales adapted to Nature.

We provide examples of an ecological approach to design that follows a very simple observation: *architecture is a dynamic adaptation to place, people, and pulse.* This simple dictum proposes that architecture respond to these three key shapers of form that are largely ignored by today's architects. Most contemporary buildings are shaped by the abstract short-run economic programs of corporate and institutional clients and by the fashion dictates of their architects. People, the eventual users and occupants of a building, enter most building programs only as a quantitative factor, not as a qualitative co-creator, inhabitant, and change agent of built form. Short-run, narrow-focus economics dominates the design program and design process.

Buildings are not fixed entities but are constantly changing as a result of the throughput of energy, materials, information, and context. But the input that creates the most change is the reason that buildings exist: people. In *How Buildings Learn: What Happens After They're Built* Stewart Brand (1994) argues that people will always find a way to change buildings that their designers thought of as fixed and immutable. An ecologic design process invites "non-professionals," the building's users and occupants, to be active *participants* in the design process. An ecologic building is designed to adapt to changing human needs, wants, preferences, and aversions.

In the era of globalization and cheap and abundant energy and materials, *place* seems not to matter. Place has dematerialized: place doesn't matter. One place seems interchangeable with another. If we found ourselves in the downtown area of any large city in the world, we would know we were downtown somewhere, but where? Architects and their clients have created a "geography of nowhere," disregarding the physical and cultural context that makes one place unique from another. "Community" is a much overused and imprecise word. *Place*, on the other hand, implies all those ecological connections, flows, cycles, and networks, cultural as well as physical, that give a place its qualities and character. A biome is a particular community and so is an ecosystem. The heart of ecologically intelligent design is to design with and for *place*.

## Place: form follows flow

> Design can be defined as the intentional shaping of matter, energy, and process to meet a perceived need or desire. It is the hinge that inevitably connects culture and nature through exchanges of materials, flows of energy, and choices of land use. In many ways the environmental crisis is a design crisis. It is a consequence of how things are made, buildings are constructed and landscapes are used.
>
> Van der Ryn and Cowan (1996)

The wisdom of Nature's design and the common sense of local genius may be our best sources for design knowledge. Having evolved adaptive mechanisms to thrive within the particulars of place, plant and animal communities provide all sorts of clues about how to build our own communities more responsibly. "Nature is more than a repository of resources to draw on: it is the best model we have for all the design problems we face" (Van der Ryn and Cowan 1996).

All living things use specialized strategies for adapting to the dynamic conditions of their environment. They consistently locate, form, and convert energy in a manner characteristic of the creatures they are and the places in which they reside. For example, desert plants and animals have evolved specialized strategies for adapting to extremes of temperatures, sunlight, and moisture. The waxy green trunk, branches, and spiny leaves of the palos verde uses every exposed surface to capture and convert sunlight, optimizing surfaces that gain energy but lose moisture. The common western jackrabbit has evolved large ears that not only improve its ability to detect predators, but also help its body maintain thermal equilibrium, radiating excess body heat to the environment. Taking cues from these creatures, an ecologically designed desert structure might apply analogous adaptations, using the building's skin to convert sunlight into electricity or perhaps integrating radiators and cooling towers with the building's communication systems. The idea is to design as Nature does: optimizing systems by finding multiple functions for each part.

Humans also employ specialized adaptive strategies for environmental adaptation. We are able to survive, for at least a short time, in all but the most extreme environments on Earth with only rudimentary clothing, shelter, and fire. However, to thrive as the communal creatures we are, we need more massive, permanent structures. Such structures have evolved through a process of incremental steps aimed at reducing the stresses caused by environmental forces. Much like biologic evolution, this process of adaptation is small in scale, incremental, and responsive to the dynamic flows of wind, energy, sunlight, and precipitation. Each adaptive step along the way is discrete, directed at a specific purpose, and congruent with available materials and technology.

Over time, form emerges and becomes interwoven with human purposes and the cultural context. The emergence of form, like a stone in a stream, is shaped by the flows of human activity, cultural evolution, and environmental forces. Over time, building form becomes inseparable from purpose, culture, and place.

As converting energy has always been a costly technological proposition, design strategies involving building location, orientation, juxtaposition, shape, and surface treatment have always been the dominant means of adapting to the environment and optimizing comfort and function.

For example, the Pueblo peoples of the desert south-west of what is now the USA built massive stone, earth, and timber villages organized to optimize their exposure to beneficial sun, wind, and light, while limiting their susceptibility to environmental stress. Pueblo Bonito, an Anasazi city located in present-day north-west New Mexico, is perhaps the best example of such a place. In section, the stair-stepped apartments and storage spaces are arranged to optimize their exposure to winter sun while minimizing summer gain through their better-insulated roof decks. These multilevel apartments are arranged in a large south-facing D-shaped plan enclosing two main plazas. It is easy to imagine how the daily and seasonal passage of the sun choreographed the rhythm of daily life. Knowles (1981) demonstrated that the buildings of Chaco Canyon, Mesa Verde, and Acoma Pueblo, were designed and built to optimize available sunlight and minimize environmental exposure by virtue of their particular uses of location, orientation, juxtaposition, and form. In doing so, they were able to limit reliance on costly energy conversion with wood fires, wood being a precious resource in a desert landscape. Like biological systems, successful adaptations are achieved through optimizing location and form, thereby reducing metabolic demands.

Industrialization and the technology for generating heat and light from fossil-fueled machinery has fundamentally reorganized the way buildings are designed. Sun, wind, and light have traditionally been the fundamental sources of form, order, and organization in the generation of buildings. The ability to produce heating, cooling and lighting by mechanical means has turned the tables on the way today's buildings are designed. *Metabolism* has displaced form and location as the primary means of adapting to the climatic conditions. One consequence is that buildings are now responsible for over a third of all the fuels burned worldwide. Half the electricity generated in this country is used just to light our buildings, many of which are now either windowless or have light-rejecting skins. Burning fossil fuels to heat, cool, and light our poorly adapted buildings adds a heavy burden of carbon and greenhouse gasses into the atmosphere.

Better design begins by seeking answers to three fundamental questions about place. Good design begins, as Berry (1987) stated, by asking, "What is here? What will Nature permit us to do here? What will Nature help us do here?" The assets and liabilities of a place must be understood before appropriate strategies can be identified. This requires careful observation, thoughtful questioning, and a measure of local genius and common sense. The answers to these questions will be reflected in buildings that are rich in regional integrity and character, built environments where people can live well in place.

### *Real Goods*

The Real Goods Solar Living Center (Figure 10.1) is the flagship outlet for a company retailing renewable energy products. Located in a temperate inland-Pacific climate 100 miles north of San Francisco, the building and landscape are intended to demonstrate how a building can operate wholly within the flow of site resources. Real Goods wanted a building that "takes less from the Earth and gives back more to people" (Schaeffer *et al.* 1997). The design brief was to create a building that uses no outside energy sources, uses locally available, environmentally benign materials, recycles wastes generated by its

*Figure 10.1* Real Goods Solar Living Center, Hopland, California. ©Charles C. Benton.

occupants, and restores the badly damaged site to full biological productivity. To accomplish these goals, place-specific adaptive strategies were employed to direct the flow of available sun, wind, light, and water through the site and building for beneficial use, eliminating the need to convert outside sources of energy for heating, cooling, and lighting.

Analysis of the microclimate led to several design hypotheses. Winter days are sunny and mild with some morning fog. The building is occupied only during the day, suggesting that direct solar gain may satisfy most of the heating and lighting needs. Summers are sunny, with many days exceeding 100°F, so the need for summer shading is critical. Low relative humidity suggests the potential for evaporative cooling. Day to night temperature swings are large, so night-time ventilation of thermal mass is a potential cooling strategy.

Orientation and form of this building emerged in direct response to these conditions. Beginning as a narrow, east–west elongated form, the building is wrapped in shallow arc along the north edge of a large circular courtyard. The building remains narrow to provide full interior daylight penetration as well as a strong connection to the courtyard, which serves as an extension of the indoor space. South-facing insulated glass is shaded by a combination of fixed horizontal overhangs and trellises planted in Japanese grape. This grape breaks into leaf in late spring and holds its foliage into the fall, matching the seasonal lag of high summer, which occurs one to two months after the sun traces its highest path through the sky on 21 June. The deep shade of the oasis environment fronting the building provides a cooling microclimate, and includes a drip-ring structure supporting broad-leaved gray poplars and a spiraling watercourse that cools the air sweetened with the smells of thriving plants.

Wall and roof systems were selected specifically for their climate-responsive performance. Most of the north-facing non-glass portions of the building envelope are constructed of straw-bale walls with 3–4 inches of concrete gunite and soil–cement finish to achieve an insulative value of R-65. The roof contains 12 inches of cellulose insulation, a radiant barrier, and a 1½-inch air space to ventilate heat gains, and a white reflective roof membrane. The curved shape of the sloping roof optimizes stack ventilation, which is reinforced by low pressure across the front of the building as prevailing north-west winds flow across the top. Operable view windows along the front and back, as well as clerestory windows along the stepped roof, take advantage of these same principles to facilitate cross-ventilation and stack ventilation. The result is a thermally tight, "switch-rich" envelope that can be easily adjusted and "rigged" to "sail" comfortably through a broad range of weather conditions.

Inside, insulating light shelves help scatter and balance incoming daylight and solar gain. These are lifted against the south glass wall on cool winter nights to help retain the day's warmth stored in the mass of the floor, walls and columns. The gently arching ceiling helps further scatter and balance the daylight from front to back. The segmented roof rises in steps to the tallest portion, which faces south-south-east, optimizing exposure to winter morning sun while eliminating its exposure to mid-afternoon sun in the summer. On busy days, when there is a constant flow of people into and out of the building, inside temperatures remain well below summertime highs and comfortably warm in the winter, thanks to the stabilizing effect of the building's thermal mass. During a large portion of the year the building can operate refreshingly with the doors and windows wide open.

Relying on location and form as the principal adaptive strategies keeps the metabolic demands of this building low. Back-up cooling is provided by modest evaporative cooling units located on the north side of the building. While the air-handling function of these coolers is sometimes used to flush the building with cool night air in the summer, they are rarely used in evaporator mode. Likewise, wood stoves installed and displayed in the showroom are used more for ambience than for need. Substantial thermal mass inside a well-insulated "switch-rich" envelope, summer-shaded south glazing, and climate-responsive form have resulted in an indoor environment that is comfortably stable without the need for imported sources of energy.

Most of these strategies are simply commonsense approaches resulting from attention to the dynamic conditions and qualities of *place*. They would hardly merit description were not thoughtful adaptation to the climate, respect for the land, and attention to people's preferences and aversions all but forgotten in the design of today's buildings. We have become accustomed to spiritless buildings in which meaningful connection to the environment is mostly accidental, and the ability to open a window, control a thermostat, or turn off a light is denied.

We have only recently become an indoor species, having spent most of the last 10,000 years outside in daylight. So it comes as little surprise that retail sales are increased by 40% in skylit stores compared with identical stores with only electric lighting, or that children taught in daylit classrooms learn as much as 26% more than their contemporaries taught in nearly windowless spaces (Heschong 1999a). We have always sought places to reside and ways to build that improve our prospects as the creatures we are and which reinforce our essential connection with the world of which we are bone and sinew. The interior spaces we have created are artificially maintained at great peril to the life-giving systems of the planet. They also reinforce our disconnection with the sources of our good life.

Berry (1987) suggests that to practice *good work*, we must also practice good *home economics*, attending first to the ways in which our lives connect through food, shelter, and livelihood to our immediate place. Good work requires responsibility over one's livelihood, which requires that we narrow the reach of our consumption. If we extend our perceptions of time and flow, apply a little local genius to the challenge of environmental adaptation, and build in ways that honor the flow of matter and energy, we are more likely to make uniquely habitable places, rich in qualitative detail, and good for living well in place.

## People: every voice matters

Organizational development practice made big strides in the 1970s and 1980s, prompted by the alienation of many workers from their jobs; the feeling of being a small cog in a big machine; and the inability of large "command and control" hierarchical bureaucracies to respond to rapid changes in the business and technological environment. The result was the adoption of training methodologies that focused on breaking through bureaucratic modes of communication, building teamwork, and participation in organizational mission.

Out of this yeasty period of organizational group experimentation came a score of innovative participative technologies, such as "skunkworks" (flexible innovative special project teams); team-building exercises offering physical challenges aimed at interpersonal growth, such as Outward Bound; and design or problem-solving workshop or retreats in which a graphic facilitator produces a real-time visual record of the meeting process and results. An alternative to the old single-scenario five-year plan is planning sessions in which the group produces sets of alternative scenarios or plans.

Both of the authors have participated in the development and application of these new forms of "organizational ecology" and have adapted them for use in institutional and corporate facilities planning, an endeavor that too often suffers from "organizational arteriosclerosis," which shuts off blood to the brain and incapacitates the organization.

To understand the problem, you need to understand how most buildings built for large organizations, whether they are public, private, or non-profit sector, are designed. Someone is in charge of "facilities," meaning buildings. They generally are charged with getting the most square footage of space for the least money. The requirements for the new project are generally communicated to the architect in a "building program" and a budget. The building program is generally a thick book listing all the required spaces the building must contain and other standards that need to be met in the design. The budget sets a dollar number for the project cost. Significantly, it does not contain a budget for continuing operating and maintenance cost.

Usually the program omits any specific goals for the building's performance in either material or human terms. In the typical organization, personnel represent 92% of total costs; buildings, including first and operating costs, only 8% (Romm 1999). Yet few organizations attempt to connect the design of a facility to the desired outcomes in terms of the building's use. Does one type of school facility produce better results than another type? Does one office facility result in lower absenteeism than another office? How does the design of a health care facility enhance or deter recuperation from illness? How do particular environments promote health and well-being and others promote ill health and unease? One would expect that the large investments in facilities would be guided by the answers to questions that relate people's actual behavior to a particular designed environment. For example, is the typical windowless office "cubicle farm" – the subject of a thousand Dilbert cartoons – the most cost effective corporate work environment or simply the result of lazy, uninformed design? In The Netherlands and Germany, based on health and productivity studies, labor law requires that a workstation may be no more than 7 meters from an operable window. Recent studies of school and retail environments document sales increases in daylight spaces and improvement in learning (Heschong 1999a,b).

The relatively new field called "building science" is beginning to document physical building performance, particularly in relation to flows of sun, wind, and light. Thirty years ago, psychologists and architects, including the senior author, who wrote a first such study (Van der Ryn and Silverstein, 1967), pioneered the post-occupancy evaluation of buildings in an effort to build an objective reliable database on how designed environments affect occupants' health, well-being, and performance. The architectural profession resisted this effort at objective evaluation as an infringement on professional prerogatives, and institutional clients discouraged outside evaluation as a potential source of criticism of their decisions.

Lacking a scientific basis to plan for people's needs, preferences, and aversions in buildings, the next best thing is to have people participate directly in planning their buildings and environments through intensive interactive workshops known as "charettes." The charette roughly approximates a project's human ecology by including a sample of all the stakeholders in the group. We give two examples of recent charettes we participated in.

### *La arbolera de vida*

For many years, the people of the Sawmill Neighborhood, a small, culturally diverse working-class neighborhood on the edge Old Town Albuquerque, New Mexico, lived beside industrial neighbors who dumped high volumes of toxins into their air, water, and land. By the mid-1980s they had had enough, and a small group of activist neighbors got organized. Ten years later they had succeeded in shutting down the polluters, reclaiming the land adjacent to their community, and establishing an unusual land trust to steward future development. The mission of the development entity created through this process is to regenerate the land, provide housing, and create sustainable livelihood opportunities for families of the Sawmill Neighborhood.

In the summer of 1997, we were brought in as members of a multidisciplinary design team to develop a master plan for the community, in collaboration with the people of the neighborhood. The success of this charette was built upon years of community work prior to our intensive week of actually articulating their vision through design. In addition to having a well-functioning development entity established to carry out all the details of financing, permitting, and constructing their vision, the Sawmill Advisory Council has empowered the community to make decisions about its own future. The community's principal goal for this masterplan is a design that will facilitate its ability to steward the health of the land for the benefit of future generations.

We worked in a large warehouse space near the site, facilitating a constant flow-through of community members, their ideas, recommendations, and feedback. Each day ended with a community open house at which the day's observations were discussed and the emerging masterplans reviewed and critiqued. The results became the starting point for the following day.

A vision that emerged from these conversations with the community was to restore a fundamental element of the neighborhood's past: the irrigation canals, or *acequias*, that once watered the neighborhood's gardens and orchards. Thus, the *arbolera de vida*, or tree of life, became the central organizing theme for the masterplan. The hypothesis is that orchards and gardens will provide stewardship opportunities for the community and a chance for *local genius* to be passed along from generation to generation. Gardens and orchards will also create habitat, restore soils, and help regenerate this badly damaged land, as well as temper the microclimate. The *acequias*, once the central organizing feature of community life, will reconnect this desert neighborhood to its lifeblood: water. The acequias will also provide the refreshing sensations of running water.

On day 2 of our charette a workshop was held for the junior high and high school members of the neighborhood. Part of the day was spent exploring their vision for the neighborhood, and part of the day was spent planning the next day's workshop for the younger children. This process yielded a wealth of detailed information about the present conditions of the neighborhood, as well as the hopes and visions for its future from the next generation. It also yielded an important yardstick for measuring the design: if it's good for children, it's probably good for everyone; good neighborhoods are pedestrian friendly, small in scale, and rich in qualitative detail.

### *The De Anza College workshops*

De Anza Community College is located in the heart of Silicon Valley, California (Figure 10.2). A group of De Anza faculty, staff, and students spent nine years envisioning a new educational center combining interdisciplinary studies, energy management, and distance learning. It will also be a gathering place for community and business functions. The group succeeded in securing a site on the campus master plan, which is located on derelict paving adjacent to the campus native plant demonstration oasis. The new project is to be a model of practical climate-responsive design, including photovoltaics for on-site electrical production.

*Figure 10.2* De Anza College Environmental Studies Building, Cupertino, California. ©Van der Ryn Architects.

When the group interviewed us as the prospective architects, we said that we felt they weren't quite ready yet for a building design. We proposed first a series of open workshops to explore and document what the project's goals were and to provide them graphic materials for fund raising. The group jumped at the idea. They made the connection that the living building they dreamed of needed a living design process.

We proposed four workshops. The group publicized them widely, both on the campus and on the Internet. As De Anza is a commuter campus, the extra efforts to encourage participation were especially important. Our goal was to assemble everyone interested and provide a forum for all ideas and critiques to be heard.

The first workshop was an open forum for listening to and recording ideas. Everyone had a chance to be heard. The core group members provided site constraint information and programming wish lists. Students, faculty, and staff poured out ideas.

This session was particularly productive. We organized it in a formal brainstorming format, listing a series of topics on big pads of newsprint. As the ideas flew we recorded every one – there was no judgment. The power of this technique is that inhibitions dissolve, and a wonderful collection of phases emerged before us, ranging from the goals of the project to specific building techniques. Most importantly, everyone attending felt heard and respected.

At the next session everyone met at the site. We observed and recorded important site relationships and their connections to the flow of human activities. Most importantly, we all felt a growing sense of common purpose. Back at the meeting room we poured over the lists from the first session. Together we built consensus for the most important features of the project. By the end of the session we were sketching the first ruminations of the building's form with the group.

For the third session we presented a set of rough concept drawings to prime the well of discussion, but *we didn't start with those drawings*. We started with a *site plan* showing the constraints and opportunities of the specific site. Then, on layer after layer of tracing paper, we sketched what the group had asked for. It was obvious to everyone that the emerging design grew directly out of *their* criteria. The atmosphere in the room was electric with excitement. A student representative exclaimed, "this makes all the meetings in my years of student government worth it!" We marked the group's ideas for changes directly on the drawings. Later the group meet on its own and sent additional ideas for changes.

The final session included our engineers and cost estimator. It is unusual to include these functions so soon, but their perspectives are essential for getting a project off to a solid start, and the group appreciated their inclusion. We presented the updated design drawings as we did earlier: sketching the group's requests and showing how the design grew out of those requests. At the end of the session the group enthusiastically told us to prepare a booklet summarizing their criteria along with the design drawings. They are now busy fundraising, knowing that the design concept is solidly based on their own ideas.

Poorly designed buildings not only threaten the health of the environment, they erode human health and happiness. The physical and psychological needs of people have been reduced to a narrow band of physical parameters, and performance is measured by the bottom line. The question most often raised regarding green buildings is whether the incrementally higher first cost of a building incorporating ecologic principles is worth it in terms of tangible human benefits, particularly as measured in enhanced performance. We have forgotten that the places where we live and work influence the flow of ideas, the quality of learning, and our relationships with each other.

The process of design is an opportunity for a community to deliberate over the ideas and ideals it wishes to express and how these are rendered into architectural form. What do we want our buildings and communities to say about us? What do we want them to say about our ecological prospects?

## Pulse: metabolism and flow

We start with a little-known interesting fact: regardless of size, every mammal lives for about $1.5 \times 10^9$ heartbeats. However, a shrew's heartbeat rate, or pulse, is many times that of a horse or a whale. Per unit of weight, it uses up many times more energy than larger mammals, i.e. it has a higher *metabolism*. We use "pulse" as a synonym for "metabolism" – the physical and chemical flows and cycles within an organism that maintain life. All materials, systems, and also cultures, are entrained in complex temporal and spatial pulses. This suggests several principles in designing buildings that fit their environment. First, think of buildings as human-designed ecological systems. Second, depict diagrammatically and, if possible, measure their metabolism – the input, conversion, and output of energy and materials. Third, optimize their "ecological footprint" – the impacts and interactions of their metabolism on other systems.

The key strategy in the ecologic design of buildings is to consider the building as an analog of a natural system. Ecologists study such systems by tracing all the energy and material input and output flows through the organism or system: its *metabolism*. The idea of a human-designed system as an analog of a natural system is a relatively new idea. In the past, the only input/output measure that designers and their clients used was dollars: cost and projected economic benefit.

Thinking about human-designed systems as analogous to living systems in terms of metabolism is a radically new form of thinking, yet it is critical in a world in which our decisions about the form and shape, or *morphology,* of human-designed objects and systems, has profound consequences for the living systems in which they are embedded.

Let us take a familiar example: the typical home. Before the home was built its site may have been a forest, grassland, farmland, or wetland, each with its own metabolic flows converting solar energy to biomass, absorbing carbon dioxide and producing oxygen, and providing habitat for a myriad of small creatures each with their own metabolic cycles. All of these are altered by the act of building. To construct the house, trees are cut in far-off locations, metals refined, plastics manufactured. Each building material has its own metabolic cycle that interacts with natural systems. When the house is completed and occupied, a new set of cycles comes into play. Gas, oil, or electricity is used to heat, cool, and light the house. The output is waste heat and carbon dioxide dumped into the surrounding atmosphere. The occupants travel from and to the house by car, again burning fuel. Food grown in far-off locations is purchased and consumed and the wastes disposed of in landfills. Clean water is brought in and discharged together with human wastes and other water-borne debris. Multiply the single home by tens of millions and we find that the metabolic flows arising from human design decisions and the living patterns they produce have a huge effect on the metabolism of natural systems at a planetary scale.

Ecologic design connects specific design decisions to their impacts on natural systems, and through that process generates very different design decisions and very different design solutions. Consider the simple example of the new home that is ecologically designed. It will not be built of virgin wood, but of materials reclaimed and remanufactured from the waste stream. Its energy demand will be reduced through climate-responsive design and energy-efficient equipment. It may produce part of its energy from the ambient environment. It will recycle its water and wastes, and its occupants may eat lower off the food chain or grow some of its own foods. The neighborhood and community pattern will be designed to reduce automobile use.

In this example, we have considered different classes of metabolic cycles. First, the design and production of the home; second, its use and operation; and, third, form and metabolism in the larger context of community patterns and infrastructure. The different players in this design drama – land developers, local government agencies, material producers, builders, and consumers – are probably unaware of the metabolic chaos their decisions unleash. It is not part of their decision or design process. They do not know or share the concepts and language that would allow them to behave differently. And so what the ecologists Harden and Baden (1977) called "the tragedy of the commons" unfolds step by unwitting step.

The process we have described in these few paragraphs involves a set of complex economic, industrial, governmental, and cultural ecologies that have yet to be mapped. The first task is to map the metabolic flows described above, giving people a graphic picture of how the designed and natural worlds interact. We are only in the first stages of being able to translate ecologic thinking into design tools that allow us to trace metabolic effects through the system. Today there are few incentives for doing so, as government regulations and our economic system tend to ignore the relationship between self-interest and common interest, the relationships among three types of capital: *financial, human,* and *natural* capital.

Although the first two terms – financial and human capital – are familiar to most people, the concept of "natural capital" is not. Natural capital refers to the services provided by ecosystems. This means the provision of not only resources, but key services such as climate stabilization, soil fertility, air with the right oxygen content, fresh water and its purification – all the processes that keep the biosphere in balance and which conventional economics takes for granted as "free" services. It is the multiplying effects of design decisions uninformed by ecologic considerations that are putting at risk our planet's natural capital, threatening the integrity of both human and financial capital. Intelligent design decision making needs to take into account the complex interaction of all three forms of capital. The three are not commensurable, and they operate on different time and space scales. Here science can help us. Faced with complex phenomena, science turns to reductionism, analyzing small pieces for clues as to how the whole interacts.

This leads us to an emerging conceptual tool in the development of ecologic design practice: *ecological footprint*. An ecological footprint is an accounting of the resource flow and ecosystem services required to bring a particular design product or system into being, whether it be square yard of carpet, a genetically engineered tomato, a new home, office building, or the sum total consumption of a median American lifestyle. A yard of synthetic carpet requires electrical or heat energy in its manufacture. It requires water, and it requires hydrocarbon feedstocks from virgin or recycled sources. Its manufacture produces heat, carbon dioxide, wastewater, and fiber waste. Computed per yard, these various inputs and outputs equal the carpet's metabolism, or footprint.

Wackernagel and Rees (1996) have calculated the ecological footprint of per capita of selected countries in terms of the amount of forest land, cropland, and grassland required to sequester carbon, grow food and fiber, and sequester wastes. This provides a baseline to determine how many people the planet can support at a particular level of affluence and technology. The footprint concept offers a promising concept to translate the complexities of design metabolism into a resource and ecosystem language and currency that people can understand and use in evaluating design and consumption choices.

In the generation since sustainability first became a design criterion, the tendency has been to focus on easier to measure energy and material flows that result from resource extraction. "Picking the low-hanging fruit" was an effective initial strategy. Now we have to shift our focus to the far more difficult task of accounting for natural capital, for which no technological substitutes exist, and which is far more difficult to measure.

We return to considering metabolism as a function of pulse or time. In the carpet example above, metabolic flows are accounted per spatial unit (square yard). But metabolism is a "flow" concept measured by a unit of time. It makes a big difference if the life of the carpet is three years or thirty. It also makes a difference if it can be easily recycled and remanufactured.

The British architect Frank Duffy first noted an important insight regarding buildings and pulse (Duffy 1990). He observed that modern buildings are composed of at least five layered systems – site, structure, services, skin, and stuff – each having progressively more rapid life cycle and metabolic rates. The pieces wear out at different rates. Their different pulses are affected by technological and cultural change (new fashions and inventions), the effects of environment and weather (oxidation and UV exposure), geotechnical and ecosystemic effects (earthquakes, floods, decline and renewal of urban districts). Site presumably changes only in geologic time. This may be true in the sense that a fixed spatial location remains fixed although its cultural and ecological context may change every aspect of its context except its spatial location.

The other four S's seem to fall nicely into line from slow to fast pulse rates. Building structure is very fixed. When the structural systems need to be replaced because of material failure or possibly new information, such as produced by observing structural behavior during earthquakes, the consequences can be drastic, the construction equivalent of a heart transplant.

HVAC, electrical, and communication systems are nowadays relatively fixed and difficult to change. Advances in technology such as shallow underfloor plenums rather than overhead fixed duct systems may change this equation. Modern building skins such as modular curtain walls tend to have a still shorter life, and "stuff," such as interior partitions, furniture and equipment, an even more rapid pulse.

This brings into focus the question of durability and life cycle cost. Before the modern era, buildings tended to be "single layered," i.e. structure, systems, and skin were one and the same. Durability was a function of the material properties of the single layer. Stone castles might last indefinitely whereas thatch huts would have to be renewed every generation. Today, buildings are composites of different systems with different pulses and different useful lives. What is the role of durability? How do designers minimize the metabolic rate or pulse of the building as a whole?

Brand (1994) observes that, "slow processes rule fast processes," or slow pulse rules fast pulse. This is obvious to anyone who reads trend lines such as the stock market. If you look at the hourly averages on a day that the market advances without seeing the monthly or yearly trend line, you could make a bad decision. And of course that is what we do every day, unless we are aware of the behavior of the more relevant larger, slower pulses. In dry years, people build on flood plains. Eventually the pulse of the fifty- or hundred-year flood manifests itself. If we build with the short term in mind, our buildings are likely to be swept away by the slower, less immediately visible pulses of change. Sustainability implies the ability to survive across many cycles of pulses, slow and fast.

When we begin to design buildings from the point of view of metabolism and pulse, we move to three important strategies: integrated life cycle costing, decarbonization, and dematerialization. Integrated life cycle costing establishes the value of the building over time both as a whole and also for its particular components. Replacing moveable furnishings does not seriously interrupt a building's use, whereas replacing a HVAC system does. We have only one of most of the key body organs such as heart, brain, and liver. If one of these malfunctions or wears out, we may die. But we have a redundancy of other organs – limbs, eyes, kidneys, and lungs – that ensure our durability. Integrating mechanical systems with natural systems – such as daylighting plus artificial lighting, or natural ventilation plus mechanical ventilation – may be one way to design redundancy in our buildings, extending useful life and reducing metabolism.

Reducing the throughput of carbon in buildings is critical to coping with the growing problem of global warming. The obvious measures include energy efficiency and climate-responsive design. The latter, if taken seriously, would have the effect of outlawing our current "big box" building footprints, which cannot function without massive, energy-intensive HVAC and lighting systems. Less obvious and more intriguing is to design buildings with built-in carbon sinks such as a second skin of living materials that absorb carbon dioxide and other toxins. As thirty- and forty-year-old glass and metal walls wear out, they can be replaced with a double skin of energy-producing, heat-reflective, high-performance glazing and an outer skin of carbon dioxide-absorbing plants. As parts of the urban fabric wear-out, they could be replaced with forests. Decarbonizing strategies such as these would give true meaning to "greening the city".

*Dematerialization* – doing more with less by substituting design intelligence for brute force and stuff – has been around a long time. One need only read a book such as *Undaunted Courage*, Stephen Ambrose's account of the 1803 Lewis and Clark expedition, to comprehend how far we have come (Ambrose 1996). As Ambrose points out, at the time of the expedition, the speed of human travel had not advanced in thousands of years since the domestication of the horse. The Pony Express took weeks to carry a message across country, and at a great material cost. Now travel and communication have not only speeded up almost infinitely, they require vastly fewer resources per unit of service. Ambrose tells us of the thousand pounds' worth of equipment required by the explorers to locate their geographic position in latitude and longitude. Now we can pinpoint our location through the use of global positioning system (GPS) handheld equipment weighing a few ounces. The weight required to produce a gigabyte of computing power has fallen from tons in thirty-year-old mainframe computers with vacuum tube circuits to fractions of an ounce in today's miniaturized silicon microchip circuits. These are examples of dematerialization and miniaturization through design.

The building sector has had its share of visionaries preaching design through dematerialization, ephemeralization, and miniaturization. Among the most important was Buckminster Fuller, who sixty years ago was thinking about buildings as ecological systems. To some extent, intelligent design may reduce the initial input of materials in the building, but the key to dematerialization in buildings is likely to revolve around design for reuse and remanufacture. Modern materials – plastics, aluminum, steel, and composites – tend to have high embodied energy. If they are deliberately designed for reuse and remanufacture, their initial metabolism and footprint is extended over many lives.

## Conclusion: breaking through the barriers

The building and construction sector uses more energy, materials, and land than any other sector of human activity. If we are going to reverse the present alarming decline in the health of the planetary environment, changes in how, what, and why we build play a key role in determining the human and planetary future. What barriers exist in implementing the kinds of changes in design process and product we have discussed?

We would be remiss if we did not discuss one barrier that we are most familiar with: the education of architects. Basically, the culture of architectural education has changed little since the founding of the first architectural schools in the middle of the nineteenth century. For the most part, it is a profoundly antiecological culture in its approach to both people and the environment. The persistent image that is cultivated in the schools – particularly the prestigious ones who have celebrity architects on their faculty – is that of the architect as an isolated artist battling to impose his personal gifts on a crude, uncaring world. This is an image first popularized by Ayn Rand in her novel *The Fountainhead*, and it endures in the schools, modeled by the design studio that emphasizes personal expression above all other values. Architecture is an art, but in a culture in which art is synonymous with celebrity and commodity the art of architecture is losing meaning.

The emphasis on the architect as lone creator is at odds with how design happens in the real world. Design is a collaborative team activity. Personal vision and creativity are important, but not more important than the ability to listen to, understand, and work with others. These are skills that are valued in the real world but largely undervalued in the design studio, which has followed other academic departments down the dead end of post-modern deconstructionism in which there are no facts, and no reality other than the one you choose at any given moment. It is a strange blend of nihilism: a world without meaning and narcissism; a person with no context other than self-reference.

While the design studio is unique to architectural education, many of its other features were adopted from engineering education that arose in the French and German polytechnic institutes in the nineteenth century. In these curriculums, applied mathematics and mechanics are the core; the natural environment is simply a challenging but inert landscape on which the works of engineering are played out. This is still reflected in today's architectural curriculum, which seldom have required courses in the natural and environmental sciences. Thus, architecture students are often faced with the worst of two worlds: a culture of "art" at a time when art has no content or meaning except as celebrity and commodity; and a lineage in applied science and engineering that is not rooted in an understanding of living systems.

There have been significant attempts at reform. The German Bauhaus movement, which migrated to North America at the beginning of World War II, was rooted in cross-disciplinary collaboration, social purpose, and learning the substance of design by actually working with the materials processes. In the 1960s and 1970s, there was renewed interest in the social and environmental context of design, but this is once again muted in most schools.

In the 1950s, the cultural historian Jean Gebser described the era we are living in as "the late stage of mentally dominated consciousness; a world above the given world of nature." He and others have traced the 10,000-year history of human culture through distinct phases from the magical world of our hunter–gatherer tribal forebears, through the development of complex agricultural societies grounded in myth, ritual, hierarchy, and war, and moving into the current stage, which begins with the written word and mathematics. Symbols come to represent events in the real world, and the human mind and the culture it has created allows us to manipulate almost all elements in the living world.

Culture is expressed in what and how we build and use the land, water, and other resources. Today, the sights and the trends are not pretty. Satellite images reveal the blotchy, rapidly growing gray patches of megacities – sprawling, polluted prototypes of an urban future. The same dumb building design templates are replicated globally. Good farmland and soils continue to be depleted at accelerating rates; grasslands turn into deserts; forests into wastelands; rivers and coastal waters into stinking cesspools. Species are disappearing at a rate unparalleled since the catastrophic collision of meteors and Earth ended the age of the dinosaurs sixty million years ago. And everywhere on Earth, complex human communities and natural ecologies are destroyed and processed into economic commodities.

Much has been gained so far in our human journey as a species, but we also sense the shadows, the losses, and the inner hollowness that comes with the hubris of seeming control and power over people and all other forms of life. Our evolution as a species is not complete. The design of the human brain suggests that we are capable of greater good and greater wisdom if we can evolve the collective cultural forms that encourage all of us to realize our full potential as humans who are part of the larger flow of life. That is our hope.

The heart of ecologic design is not efficiency or sustainability. It is the embodiment of an animating spirit, the soul of the living world. It is the spirit of the magical preindustrial world wanting to be reborn in the post-industrial ecologic world.

The concept of efficiency as applied to work, whether human labor or thermodynamic, is an industrial concept. It is morally and spiritually neutral. The expression of an animated spirit – the soul of a living being – is not neutral. It is the essence of art and of craft before art became artifact and craft became commodity.

To build in a way that expresses this hope and continuity is our purpose. The barriers exist primarily in the flawed mechanistic mental models that still dominate political and corporate organization. Still largely unchallenged is the dangerous hegemony of conventional economics and its fatally flawed accounting system, which fails to account for the "natural capital" of ecosystem services that cannot be replaced by any technological fix.

The cautionary tale is that of Biosphere II, a $250 million effort employing the best science and technology to recreate in the Arizona desert a miniature closed system including all the major terrestrial and aquatic ecosystems on Earth. When it was fully functioning, eight scientists were sealed inside to live in and off these systems. Within a very short time, the natural systems could no longer maintain oxygen levels to support life, and most ecosystems collapsed.

What does this say about economics and design? First, $250 million could not buy enough oxygen for eight people, yet the planet provides oxygen for six billion people "for free." Stanford economists have calculated that the value of natural systems that never enter the calculations of conventional economics, and for which no technologies can substitute for, is many times the present value of the measurable world economy.

Biosphere II is also a cautionary tale for designers. We cannot recreate four billion years of evolution. But we can work to slow the rate at which things get worse, and we can create environments in which people feel their connection to the pulse of Nature's nurturing life forces. In so doing, we can help bring into being a consciousness that reveals to us the truth of our collective past and present, guiding us to a new ecologic future.

## References


Ambrose, S. 1996. *Undaunted Courage: Meriwether Lewis, Thomas Jefferson and the Opening of the American West*. New York: Simon & Schuster.

Berry, W. 1987. *Home Economics*. San Francisco: North Point Press.

Brand, S. 1994. *How Buildings Learn: What Happens after They're Built*. New York: Viking Penguin.

Duffy, F. 1990. Measuring building performance. *Facilities* 8(5): 17.

Gebser, J. 1986. *The Ever-present Origin*. Athens, OH: Ohio University Press.

Harden, G. and Baden, J. 1977. *Managing the Commons*. San Francisco: W. H. Freeman.

Heschong, L. and the Heschong Mahone Group 1999a. *Daylighting in Schools: An Investigation into the Relationship Between Daylighting and Human Performance*. Pacific Gas and Electric Company.

Heschong, L. and the Heschong Mahone Group. 1999b. *Skylighting and Retail Sales: An Investigation into the Relationship Between Daylighting and Human Performance*. San Francisco: Pacific Gas and Electric Company.

Knowles, R.L. 1981. *Sun, Rhythm, Form*. Cambridge, MA: MIT Press.

Rand, A. 1943. *The Fountainhead*. New York: Bobbs-Merrill.

Romm, J.J. 1999. *Cool Companies: How the Best Businesses Boost Profits and Productivity by Cutting Greenhouse Gas Emissions*. Washington, DC: Island Press.

Schaeffer, J., Arkin, D., Heusley, N., Jackaway, A. *et al.* 1997. *A Place in the Sun: The Evolution of the Real Goods Solar Living Center*. White River Junction, VT: Chelsea Green.

Van der Ryn, S. and Cowan, S. 1996. *Ecological Design*. Washington, DC: Island Press.

Van der Ryn, S. and Silverstein, M. 1967. *Dorms at Berkeley: an environmental analysis*. Berkeley: The Center for Planning and Development Research.

Wackernagel, M. and Rees, W. 1996. *Our Ecological Footprint: Reducing Human Impact on the Earth*. Gabriola Island, BC: New Society Publishers.

# 11 Natural metabolism as the basis for "intelligent" architecture

*Jürgen Bisch*

Many would characterize this millennium as an "Age of Progress" as they reflect on all the opportunities that have come from the explosive growth in the use of energy and materials. No other world view is as pervasive as progress, a myth that looks beneficial even when people recognize that such growth cannot continue for much longer without resulting in the collapse of the world's supporting ecosystems. Like many others in the second half of the last century, I have been influenced by a rising tide of doubts and opposition to this tradition of progress and growth. And my curiosity has followed these doubts back to the roots of my profession: architecture. Many great journeys start almost innocently, like steps that continue down the road long after the reason for leaving has faded. I would like to share such a walk with you.

Years before I finished my education as an architect, I began to appreciate the limits of this wonderful, blue planet. In 1973 Dennis and Donella Meadows' seminal book, *The Limits to Growth*, made a whole generation aware of the global limits to space and resources and our responsibility to act on this. My search for awareness evolved from perceptions of resource limits to a sense of responsibility that grew as I reflected on history. This responsibility transcends the beauty that emerges from one's love of creating – the original meaning of the word "amateur." I learned that "responsible" or significant acts are inspired by a "cause" or "idea" that resonates through our cultural history.

The responsibility to act is personal. In each creative act I choose which spring to draw from: love or money. With time a more mature responsibility emerged: could I make love "practical" and combine the two? But my search back into history showed me that architecture involves far more than love or money. In fact, architecture is far wider than any one discipline, such as designing "beautiful" structures. Architecture emerges from philosophy, sociology, psychology, urbanism, history, physics, chemistry, and even ecology. Architecture, understood in its fundamental, non-materialist meaning, is an "applied art," aimed strongly at all that moves the individual human.

If one seeks their foundations in classical Greek antiquity, one finds that the arts, philosophy, and architecture all strive for the "harmony of the spheres," in their search to understand. Classical literature from Pythagoras[1] and Plato[2] to the Roman Vitruvius[3] leads us to the view that space, Earth, Nature, and humanity are "one." Philosophy originally was a "Nature-philosophy," which also formed the fundamental "Nature-science" of the West. The rediscovery of this unity during the Renaissance coincided with Europe's first forays to discover and conquer the rest of the world. This conquest exported European "culture" worldwide to all climates. The science driving this revolution increased in its power to predict and to change the world, but was restrained by Christian clerics until the beginning of the seventeenth century, when Descartes' *Discours de la Méthode*[4] led the world into the Age of Rationalism. Descartes' leading idea, "the radical doubt," began the disintegration of "Nature-science', and science started to separate from metaphysics.

## The rise of science and industry

The "post-metaphysical" (Habermas 1988) split between science and philosophy released a flood of ideas that was mirrored in surges of power as bond after bond of the old world were split: minerals from the earth, trees from the forest, people from the countryside. The ascendance of material thinking appeared inexorable as the Industrial Revolution's deliberate exploitation of continents of resources and people gave its captains the sense of titans ruling the Earth. For almost two centuries the mounting stream of energy and materials made resources appear limitless, and industry was celebrated by engineers in massive gleaming machines and by architects in tremendous industrial palaces crowned with smoke stacks. The architectural design axis was no longer centered on the bedroom of Ludwig IV but on the industrial boardroom.

The forces behind the industrial upheaval of the ancient agricultural world may not have been apparent to many at the time, but a century later Gideon (1948) documented how, by the second half of the nineteenth century, mechanization of functions sprang from earlier advances of science. Watt's early triumphs with steam power led to Linde's cooling technology.[5] Just as was the case with the outward blossoming of culture during the Renaissance, only in a more complete and perfect way, architecture followed industry out from its European origins, elegantly dressing the modernizing world in glass and steel. With the dawn of the twentieth century the pace quickened as the "century of information" spread these ideas like a fountain, with architecture recasting cities in this "international style."

While the revelations of the Enlightenment had shifted human faith from God toward science, the mechanized application of science had conquered the popular human imagination. "Trust in progress" was converted to "trust in production", and the glamour of world expositions was a stylish cover for the grease, smog, and dirt of the industrializing modern world. In the twentieth century Ford shifted the intention of nineteenth century mechanized production to close the loop between production and consumption.[6] Ford priced his products such that his own factory workers could afford them. With this step, machines claimed center stage, with man subordinated to designing, engineering, cleaning, and feeding them and finally consuming their products (Ford 1922). Marketing evolved to drive overproduction by nurturing a perpetually changing set of "requirements" or "needs" for modern humans, and as the lifespan of these changing fashions declined the volume of waste increased (Burghardt 1985; Steffen 1996).

## Mechanization and society

The human consequences of this product utopia have been frequently noted and critiqued since Marx. Friedman (1975), a Hungarian architect and professor, continued this refrain in his discussion of social utopias, when he proposed that the unquantifiable loss of human potential far outweighed measurable gains in productivity. He writes: "… the social and intellectual growth of the citizen is much more important than to double the production of goods every ten years." This critique attempts an ecological extension of our view of man to encompass the natural world that supports society. "What we need," Friedman wrote, "… is an Operations Manual for the spaceship Earth." In one chapter he points out that the most fundamental of all the "basic illusions" of our civilization is that, despite the fact that we recognize that energy and raw materials are not unlimited resources, we still believe that such limitations are problems to be confronted by future generations.

Jungk (1986), a German future scientist, in his book *And Water Breaks the Stone* continued this effort to identify the flawed thinking that underlies our social–ecological dilemma. He stressed that unless thinkers, be they architects or scientists, squarely face their philosophical responsibility, they can succumb to the superficial views of our age that allow people to be reduced to "consumers" by economists and "voters" by political scientists. And he noted that people fail to rally and mobilize their inherent power to think when they are too distracted by an age of consumerism to realize the threat.

The concept of philosophical responsibility involves extending one's view beyond one's profession or discipline and to integrate other views. For example, C.P Wolf identified the leading wave of science as not solely in technical innovation but in "social impact assessment".[7] Alvin Toffler expresses a similar outlook in *The Third Wave* (Toffler 1980) but with more urgency, "the fat years are passed and the active years do follow."

And so, Christopher Freeman from Sussex University, introduced his five-point list for a technical revolution![8] Why not innovation, as Robert U. Ayres (1984) showed us in *The Next Industrial Revolution*. If we can address this responsibility to think in more holistic and integrative ways, technical innovation might develop new opportunities for sustainability (Ayres and Simonis 1994). Perhaps our confrontations with this philosophical responsibility will redefine the values, the sense of "intelligence" that needs to be applied to steer our development of technology. Why shouldn't this intelligence be derived from the technology and the architecture inherent in natural metabolism and the patterns and interactions of natural processes?

## Applying natural metabolism to architecture in Europe

Some architects and engineers in Europe have already begun to learn from Nature and thereby redefine the aims of our work. Our clients' enthusiastic acceptance of our first experiments shows us that there is an opportunity for change. These successes are sweetened by the fascination gained in our attempts to learn from Nature and its metabolism. One develops a healthy respect when viewing the results of millions of years of natural experimentation. Our market-driven society forces much shorter time horizons for experiments, so why shouldn't we take advantage of these planet-wide experiments that have continued for eons longer than even our imagination can comprehend? This natural intelligence is a "wheel" that we do not have the time or money to reinvent, so we should open our inquiry to integrated, interdisciplinary investigation of the intelligence of natural systems. As Aristotle, the great Greek philosopher, said: "For to think philosophically, logically we need time." Nature has had this time!

What are the basic ideas that we have gained in our first applications of Nature's metabolism to architecture in Europe? They are streamlining, material cycling, and minimizing the use of raw resources and energy. For an engineer or an architect, streamlining – designing the necessary and reducing the building's material impacts – becomes a fundamental law in the design process. Aiming for more natural metabolic cycles has meant the use of building materials that are easy to recycle, either directly into another building or for other uses. Finally, the energy principle of natural metabolism is to adapt construction and design to the regional climate such that the building has the lowest energy use during the operational phase. Just as is the case with energy, we strive to minimize the use of high-quality resources, such as drinking water, during the building's operational life.

Recognizing that few people have worked as we have to gain our perspective on natural metabolism, we have sought to open the door to these ideas through "innovative pricing." This is more than "green" marketing of the same "quality" at lower prices or higher ecological "quality" at higher prices. The attractive opening gambit we offer is based on *why* our prices are lower, for both construction and operation. When customers see that we achieve higher "living quality" standards at lower costs *because* of the "intelligence" of natural metabolism, then the way is clearly open for them to follow our lead in re-examining their assumptions. It is easier for both of us to share a new appreciation for older ideas that lead further than we first imagined. "Less is more", an old thesis in architecture, takes on a new meaning. Exactly the opposite of the "economy of scale" of the past century, designing for less energy and materials gives more freedom to a building's users. Lowering these inputs also lowers expenses, freeing money for other purposes and increasing business efficiency.

These immediate paybacks are sweetened with the gratification that we are leaving more for our children. Further, the quick and obviously practical advantages of applying the ideas of natural metabolism help to point toward a less materialistic humanism that could boost the social quality of life. Society could traverse the intimidating desert of "limits to growth" if it had a vision rich in social quality as a guide. Many in society understand intuitively how empty the "quality" of our lives is when it is based on quantity. Quantity by itself is so unfulfilling that it has now been replaced with growth in quantity. Increasing the flow of goods is now an index of quality. Virilio (1993), a sharp diagnostician of our technocratic, high-speed society, described the cruel emptiness of this cornucopian life that we are living in as a "furious motionlessness." We are beginning to develop the conceptual tools to see and measure how efficiency works. The concept of an *ecological rucksack*[9] gives us a tool to measure the impact of resource use (Schmidt-Bleek 1998). The *factor 4* concept directly links quantity and quality with the idea that increasing efficiency by a factor of 2 will double wealth. Looking at our society through the intelligence of natural metabolism yields a delightful irony: quality lies not at the end of this stream of consumption but under it. Intelligent and innovative architecture and engineering can create the physical examples that will house us more efficiently but also will feed our imagination and help us find the "red thread" (the enduring theme) that leads to a higher-quality, more ecological future.

## Metabolism – streamlining the design

What building design is, what building design incorporates, depends not only on its elementary and structural function, but also on the personality of one's client and his or her image of what the building has to represent. It is the client's view that pervades, concerning formal standards, technical comfort, implemented forms and the sense of quality in the architecture. But beyond all this, an intelligent building design must first be efficient in its economic qualities. To streamline the building design is almost directly coupled to streamlining the building budget. Exactly at this point, we start discussing the idea of streamlined design with the clients, the users.

The best way to protect the environment is, of course, not to build, not to use raw materials. The second best way is to use as little raw material as possible, through intelligent design of long-lasting construction. The classic qualities of timeless architecture arise from durable construction designed using forms of simple elegance that are adaptable to fickle changes of fashion. Timeless form achieved with great flexibility in using and cycling raw materials can reduce the building price and guarantees long-lasting use. Architecture can help lead the path away from runaway demand for cheap industrial resources to adept use of raw materials, the value and timelessness added through intelligent labor. For example, building design becomes long-lasting when its form remains useable no matter what social or economic changes occur. An architect should design for future demand by increasing the economy and flexibility with which different functions are realized. Then buildings will remain useful for decades with little increase in expense or material use due to reconstruction.

A simple, reduced picture (Figure 11.1) shows an example of our approach to streamlined design. Conventional design of a standard office building would employ a structural system composed of precast beams and poured-in-place concrete flooring, an effective but heavy and materials-intensive approach (top row). By employing an additional column to reduce the span, the beam can be removed from the design, allowing the use of a simple concrete floor slab system (middle row). Additionally, by pulling in the exterior columns, one column can be eliminated, thereby greatly dematerializing the design of the structural system and using the minimal quantity of materials in construction (bottom row). A similar approach can be applied to a wide variety of building systems.

Figures 11.1 and 11.2 illustrate how innovative techniques can reduce materials use and construction volume in two different parts of a building's structure. First, we coarsely divide the elements of building construction into durability classes: long-life (A materials) and medium-life parts (B materials). Long-life elements include the load-bearing structure and the facade. Technical items (electricity, plumbing, network technology, etc.) of the building infrastructure, light (non-load-bearing) walls, carpets and suspended ceilings are classified as medium-life parts or B materials. Medium-life materials and construction are usually changed during a building's lifetime and therefore make up a large percentage of the construction waste volume. The best chances for swift, dramatic, and economic benefits come from any efforts to reduce the use of or increase the recycling quality of these materials.

The successful conclusion of this streamlining design process has resulted, so far, in the elimination of double floors and suspended ceilings. By eliminating unnecessary B materials, we can reduce building volume and, ultimately, the building's skin and skeleton

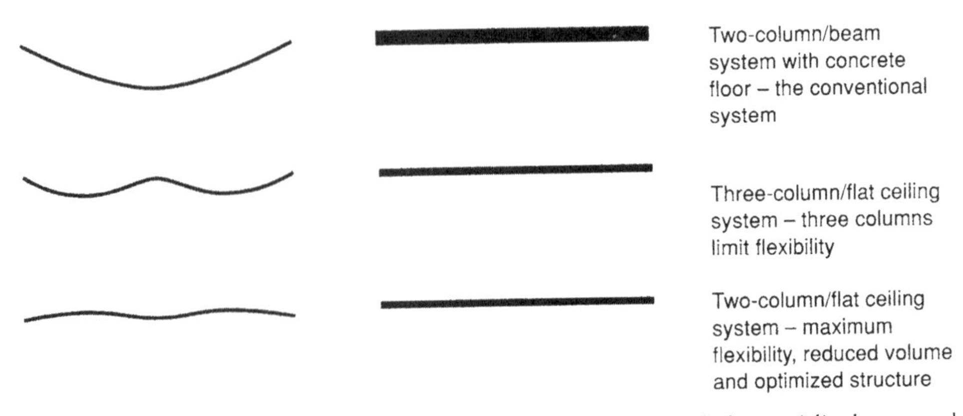


*Figure 11.1* Evolution of a conventional structure design to an optimized, dematerialized structural system.

*Figure 11.2* Combining structural and technical (HVAC, plumbing, electrical, communications) elements of the building can result in a smaller building volume with the same useful building space.

or A materials. The result is a reduction in the volume of those items most costly to install and replace – building structure and façade elements. We have developed new techniques to refine our design and construction of A materials so as to incorporate the technical functions of the B materials we replace. Innovations in the design and construction of concrete ceilings have replaced the need for suspended ceilings and double floors with no loss in functional capability or our freedom of design.[10]

To evaluate the building design quality in this aspect we found that the percentage of structure volume in relation to the total building volume should be smaller than 9%.[11] The structure volume in this evaluation has to integrate the non-room airspace between ceiling and suspended ceilings and the space of a double floor.

One of our construction projects in Regensburg, Germany, shows a possible result of this streamlining process. The building is a research and development complex for Siemens-Germany with space for two thousand engineers working on electronic development for the automotive industries. By drawing our clients into close consultation, they assumed part "ownership" of a streamlining process that changed the office from an expensive "high house" (in the sense of an overbuilt structure following the traditional German building codes) to a conventional, easier to build yet cheaper design. The key innovation was to simply change the "traditional" electronic infrastructure system to an "activated ceiling system."[12] The results of implementing this technology are listed in Table 11.1 and the principal vertical design elements are illustrated in Figures 11.3 and 11.4.

These reductions are dramatic enough to gain the attention of and win sales with even traditionally inclined clients. But far more important is that the promise of the streamlining process is only just beginning to be realized. These savings were achieved by implementing a very simple duct system, that works by integrating conventional infrastructure systems into the building's concrete construction ceilings. I expect that similar reductions will be found as this process is repeatedly applied in design.

## Applying simple physics to activate the function of ceilings

Our "two column–flat ceiling" system follows the principles that only load-bearing elements such as building structure (A materials) are massive, and that they have dual functions. For example, massive walls are mostly prefabricated concrete elements that

*Table 11.1* Savings realized through the streamlining process

| *No.* | *Description of type of reduction (units)* | *Quantity* |
|---|---|---|
| 1 | Building height (meters) | 3.5 |
| 2 | Area of type B building materials (square meters) | |
| 3 | Area of suspended ceilings (square meters) | 36,000 |
| 4 | Area of double floors (square meters) | 28,000 |
| 5 | Volume of type A materials – alum/glass façade (square meters) | 3,010 |
| 6 | Concrete (cubic meters) | 230 |
| 7 | Fire prevention costs – fewer fire escapes (percent) | 2 |
| | Total cost reductions as percent of building budget (percent) | 8 |

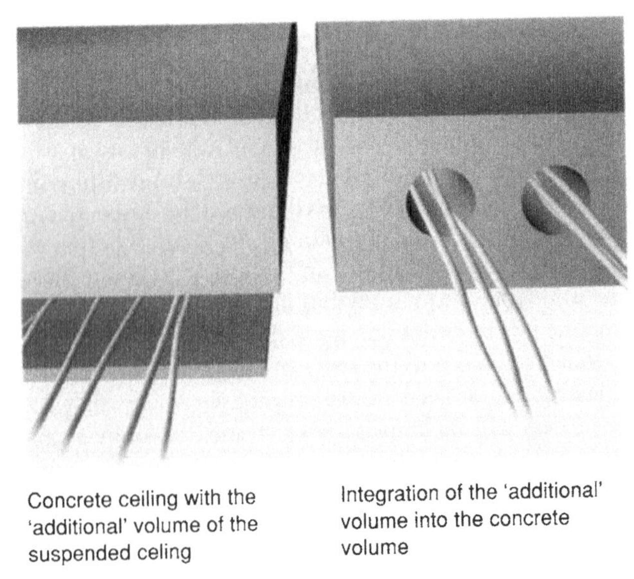


*Figure 11.3* Example of integrating technical volume (electrical wiring) into the structure of the building.

*Figure 11.4* Both buildings have the same office space and identical room volume. The building on the right is dematerialized by integrating technical functions into the structural elements.

serve as stair houses and as fire prevention structures. Ceilings serve as support for people and property and as conduits for air, water, and electricity. Even the simple physics of gravity ("what goes up must come down") comes into play. Removal of double floors and suspended ceilings brings the air of the living/working space into direct contact with the massive ceilings and floors. This contact facilitates energy transfer from the massive energy-storing floors to the air for transport and circulation about the work space, like the cooling effect of adding ice to water (Figures 11.5 and 11.6a and b). This allows us to manipulate the energy gradient cheaply and efficiently and then ride it toward the desired room temperature. We can reduce or lift the room air energy level by about 40 $W/h/m^2$. We can further assist the end users' temperature comfort by dampening temperature peaks with "intelligent" building façade and ventilation technologies.

Our design, aided by an advanced computer simulation program, Transys, has in practice limited the building's summer/winter temperature range to 20–27°C. This software system allows real-time simulation of energy flux in a form-finding process based on a continuous energy-optimizing feedback loop. We employ this iterative feedback loop to examine the energy consequences of design changes during the entire architectural design process, evaluating even the first design sketches. All our design work starts with simple, small-scale paper models to explore the context of site quality, urban orientation, and our own basic design ideas. These models are then transferred into the Transys simulation software to see how our first designs adapt themselves to regional climate conditions during an entire simulated year.

We also use simulation to examine how the building's internal temperature production, affected by biological (people) and electronic (computers) sources of heat in a room, influences the overall energy budget of the design. We focus on the air temperature of several reference zones in the building as indicators of energy-efficient design. These reference zones are always rooms that we expect to experience the most extreme climate situations, such as rooms with southern exposure or which will contain large electronic

*Figure 11.5* Entropy, Nature's technique for "balance."

*Figure 11.6* (a) Conventional approach to building climate control. (b) Advanced approach to building climate control.

equipment, etc. When this basic approach is finished we add specific technologies, such as "ceiling activation," shading system or ventilation, to evaluate the building climate reaction on the computer model.

Figures 11.7 and 11.8 show the influence of the so-called "ceiling activation" on the temperature level of a south-facing office room during a warm period in the central European summer.

Figure 11.9 shows the installation of the piping for a thermal ceiling activation system

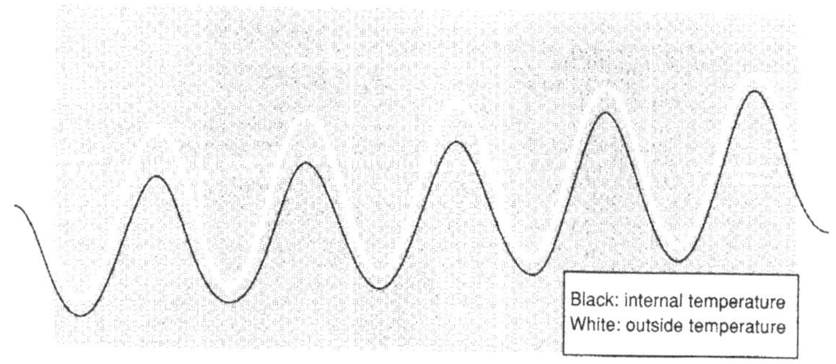


*Figure 11.7* A conventional room without ceiling activation – internal room temperature follows the outside temperature amplitude. The example is for a five-day "hot" summer period in central Europe.

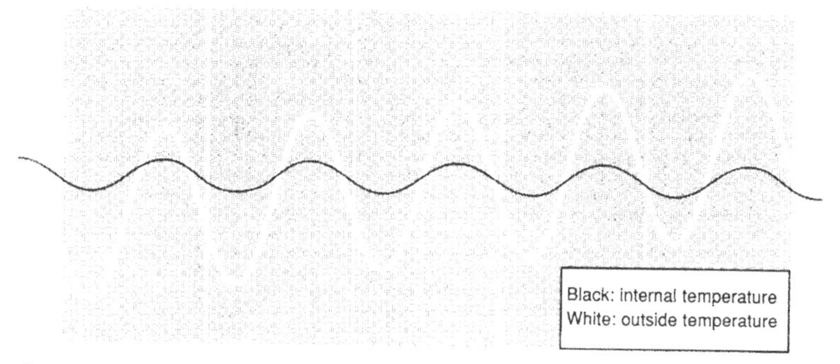


*Figure 11.8* The same room with "ceiling activation" for dampening the internal room temperature.

into a ceiling slab. Groundwater is circulated through the piping in the ceiling slabs to create a radiant heating or cooling system and take advantage of the thermal storage capacity of concrete. Figure 11.10 illustrates the installation of electrical/electronic systems conduit in the ceiling slab. The use of the ceiling slab for the installation of piping and conduit and for heating/cooling can result in the elimination of dropped ceilings and raised floors, thus significantly reducing the materials requirements and the costs of the building.

## Design using the efficiencies of past and future centuries

The key to energy efficiency in ecologically intelligent design is to eliminate the need to adjust temperatures with fossil fuels or electricity. We avoid this by allowing Nature to do the work of creating the desired temperature. Our job is to create designs that exploit temperatures or temperature differences that naturally occur in various media such as air, earth, and even water. First, we couple interior air temperature to the building's thermal mass through our "activated ceiling" design, and then we link the thermal mass to the desirable temperature where it naturally occurs (Figure 11.11). With this link we adjust a building's temperature by driving it up or down with a temperature differential between the working space and the ceiling (mass). We set the ceiling's temperature by a circulating water system that includes a heat exchanger, which we can use to exploit seasonal temperature gradients. In the summer, we use the temperature gradient (25 versus 16°C) between the air and the groundwater. During the winter, a three-way mixer

*Figure 11.9* Thermal ceiling activation system under construction (Gerling Insurance Company Building, Frankfurt am Main, Germany).

*Figure 11.10* Thermal ceiling activation system under construction with additional electrical/electronic infrastructure (Gerling Insurance Company Building, Frankfurt am Main, Germany).

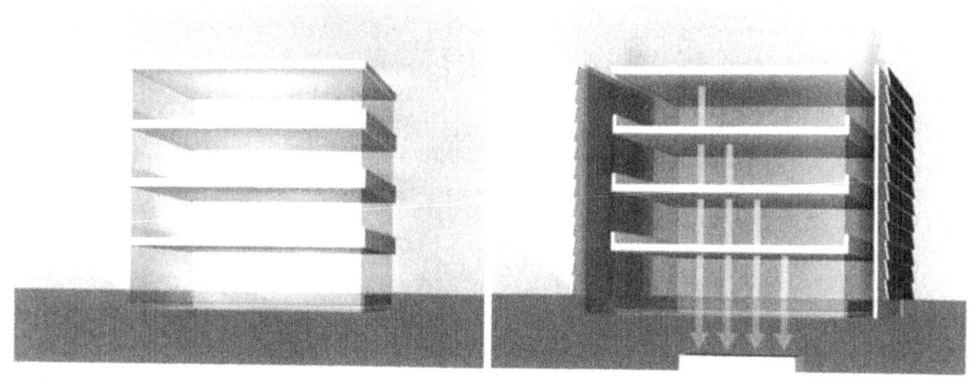


*Figure 11.11* Coupling a temperature-controlled mass to interior air temperature.

takes all the heat energy produced during a day by the users of the building into the systems circuit, and only a small temperature difference has to be added by such technical means as conventional warm water heating systems or solar collectors.

In one of our recent projects we employed rainwater in a reservoir (capacity = 250,000 l or 250 $m^3$) as a temperature (energy) buffer, situated under the fourth level of a deep garage. The building volume of this project is about 50,000 cubic meters, or 200 times larger than the rainwater reservoir, a very small fraction of the construction volume for such an important role.

## Controlled ventilation

Controlling the internal atmosphere of buildings has been an architectural theme for millennia. From the ancient buildings of Yemen to the renowned Unité d'habitation by Le Corbusier in Marseilles, France, "controlled ventilation" always was one of the design's leading aims. Modern air-conditioning systems are simply technically "upgraded" ventilation systems. But, as we know, cooling technology consumes much more energy than heating technology. When air-conditioning technology was not common a century ago, hospitals used a simple duct underneath the ground to cool air down in summer and to warm air up in winter. When air speed in this duct is limited to about 3 m/s, a duct length of 50 m delivers a cooling rate from about 3–5°C. Along with this cooling function, the dehumidification of the air transported through the duct occurs automatically. The surface temperature of the duct walls influences the cooling/warming capability.

Our "controlled ventilation" design (Figure 11.12) aims to improve the quality of the work space by dampening temperature variation and by adding higher-quality natural air to the interior. The design works without mechanical forcing of air movement. The only mechanics needed are pumps to circulate water in the ceiling. Air is drawn in by temperature differences (created by solar gain) through a ground duct and is pulled into the space between the two exterior façades. This air buffers the temperature of this intervening space and can be drawn into the building simply by opening a window. This design boosts the autonomy of the individual users, who can decide on their own how to regulate air intake. The system is robust enough to operate efficiently even when nudged

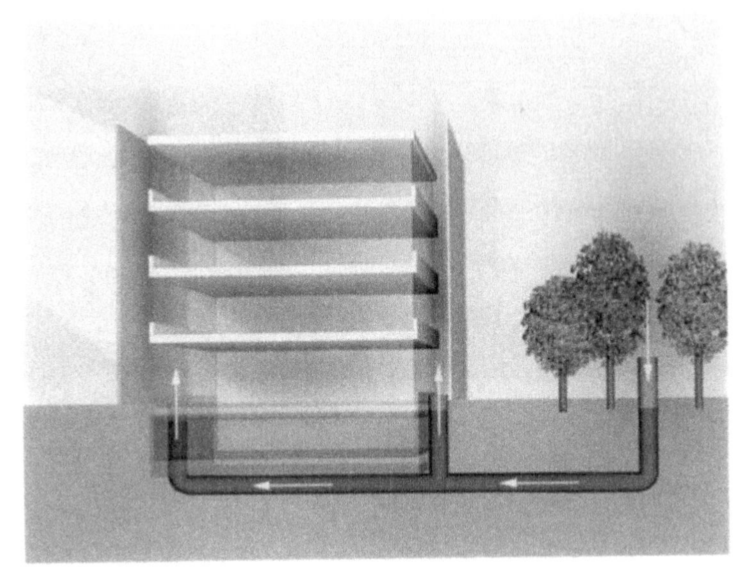

*Figure 11.12* Integration of "earth duct" and "ceiling activation" systems.

by so many individual decisions. This provides an atmosphere more agreeable than the callous and inflexible strictness of many "modern" buildings whose engineers and architects have removed personal choice by permanently closing windows.

Our experience has been that the whole architectural design process has to follow the site's climatic qualities to create an energy-efficient building. Of course, these buildings look different from conventional structures. It is not possible to simply adapt controlled ventilation, ceiling activation, and other techniques to a conventional design. However, these ideas will eventually be embraced by building design, given the enticing reductions in operational energy consumption and building materials. But, to be successful and really efficient, one cannot simply take these examples "off the shelf" and apply them in whatever ecosystem one lives. One has to engage in the whole design process, which has to follow the principles of design based on the "intelligence" of natural metabolism.

## Energy input reduction

Proper design principles achieve efficient interior climate control by using mass for support and energy storage, such as through our "ceiling activation" design described above. These principles reveal how interior temperature control is lost if heavy mass is placed on the building's exterior, where temperatures are dominated by climate. An exterior mass will store the temperature extremes that one tries to avoid. This may work for adobe structures in arid regions, where the sun's energy can be stored to damp the day's extreme heat and then released to warm the cold nights. But in moist, temperate zones it adds a burden that usually requires expensive countermeasures. These are unnecessary now that modern materials and façade design can reduce energy loss to the outside practically to the point of balance. Even building designs with very large glass facades can balance energy input/output when coupled with effective screening and shading technologies that can damp inputs. In moderate central European climates, we employ a "twin face" façade system (Figure 11.13) with double heat insulation glass, dynamic shading

*Figure 11.13* Twin-face façade with active shading.

adjustment and controlled ventilation. Such design requires additional heating inputs only in extreme winter conditions, but active (energy-expensive) cooling is not necessary.

## Selecting materials

We aim to minimize energy and material waste by replacing quantity with quality and durability. This is done in design by incorporating higher-quality materials and trimming material needs. A metabolically "intelligent" design process can have social impacts if its specifications of high-quality, "intelligent" materials mean more work for highly trained craftspeople. People with years of problem-solving experience form the pillars of "intelligent" industries. Investing in metabolic intelligence may open the opportunities for increasing the quality of our materials, working spaces, jobs, and business missions. Sustainable development should apply across all facets of society, and one's choice of bricks may symbolize the practical as well as philosophical axis of society.

History shows us how purely material choices that are not informed by sound principles can lead to society's decline. In ancient Nubia, the Sudan of antiquity, Naga persisted for centuries as a large city full of thousands of people. The paintings and graffiti on the surviving walls testify to a lush biosphere rich with trees, birds, water lilies, and lions. However, the demands of the transition to an iron culture and the need for wood and charcoal for smelting consumed the forests and caused the landscape to degrade to desert. The last pockets of that rich world can be found only in paintings on walls entombed in sand.

Neither our present glut of energy nor the passing of many centuries should be allowed to ease the bite of these historical lessons. First, we should notice the coincidence between catastrophic confrontations with the limits of Nature and the naive adoption of new paths that lack well-grounded principles like metabolic intelligence. New paths, ideas, and technologies will always open up. The question is: How and with what principles do we test them? Second, we should see how the quality of our science, engineering, and architecture relates to its aims and philosophical foundations, and how these interact with and affect all of society. Most people feel left out of decision-making processes in our society, humbly aware of their lack of political or technical connections. But it is still a social–philosophical process that all somehow participate in, and suffer from when wrong paths are chosen.

Our experiments with different materials in the pursuit of sustainable design have led to some general decision-making criteria:

1 Reduce material input – use a material only when absolutely necessary.
2 Reduce material diversity because it lowers the chances of reuse or recycling (for example, avoid composite construction).
3 Choose materials that are well known through empirical testing to be easy and simple to use. Avoid "high-tech" materials, so complicated in use that they are prone to waste, spoilage, or misuse.
4 Maximize your capacity to reconstruct with the materials you have in place.
5 Find designs and materials considered "traditional" for the region, learn why they work in the local context (climate, soils, hydrology), and use them to minimize transport distances and energy use.
6 Use the "ecological rucksack" as a measure of a material's impact and calculate that material's "standing" or turnover time.

A materials in the building structure last for decades, whereas B materials will turn over many times during the same period. Thus, one has to develop a factor with which to multiply the carpet's "ecological rucksack" in order to equate its impact with that of the building structure.

## Summary of approach

So far I have tried to show how applying basic principles of the "intelligence of natural metabolism" can create designs that lower our material and energy footprint. I would like to walk the reader through the entire design process to show how such efficiencies are achieved in practice.

### *Predesign phase*

Before any pencil is put to paper we try to gain a deeper understanding of the site by evaluating various quantities and qualities (Table 11.2) that our design must address. Different factors require different kinds of attention. Some factors can be exploited, for example constant ground temperatures can be used to reduce temperature variation. Other factors must be avoided, e.g. extremes of wind and temperature from certain directions. And some factors convey certain environmental or aesthetic qualities that should be preserved if not enhanced by design.

We discuss the planning team's structure with the client's staff while our field crew is making on-site measurements. A building climate simulation specialist is a crucial member of the team. In the predesign phase our office has the capacity to perform coarse simulations, but the need for predictions using sophisticated simulations increases as the project advances. We add engineering capacity to the team by searching for traditional competence coupled with an awareness of advances in related fields, particularly those pertaining to the "metabolic" efficiencies we are trying to achieve. For example, we seek

*Table 11.2* Data collected for site evaluation

| *No.* | *Data type* | *Description* |
|---|---|---|
| 1 | Temperature (hourly for several years) | Different heights above ground |
| 2 | | Underground structure and energy storage capability |
| 3 | | Ground water, level, temperature, speed |
| 4 | Wind | Direction and speed |
| 5 | Sun | Solar curve and constant |
| 6 | Ground water | Water table elevation |
| 7 | Shade | Buildings and vegetation |
| 8 | Local atmosphere | Air pollution |
| 9 | | Traffic noise |
| 10 | | Vegetation microclimate |
| 11 | Soil qualities | Physicochemical properties |
| 12 | | Pollution |

structural engineers who appreciate the fact that structural reduction is essential to reduce energy demands throughout a building's operation. We invite electrical engineers who understand the limitations of conventional suspended ceiling and façade designs and can work with us on new technologies for infrastructure, lighting, power sharing and transmission ("bus systems"). We also need mechanical engineers willing to work with alternative ideas and technologies for energy storage and transmission in the building foundation and structure.

The planning team also needs fresh perspectives from biological disciplines such as health and landscape architecture. We integrate into our designs the contributions of medical doctors who study the effects of work space climate and materials on users' health. We look for landscape architects who can design vegetation communities to regulate climate, particularly for "green roofs" in the extreme climates found on the top of buildings.

We orient the planning team with a start-up workshop to familiarize everyone with the overall design aims. The client and staff members are more than present: they are integrated into the discussion to maximize the transparency of the decisions such that they are confident that the design's functions achieve their standards. These proceedings help us to learn from one another, often re-examining and revising our standards.

Architects and engineers would immediately understand how conventional fee structures would not support a design process with such extended and intensive efforts both before and during the actual design. But the payback comes fast. The building and operational efficiencies of material and energy reductions give it a cost advantage that is complemented by the enhanced enjoyment of the building's users. Our clients' confidence increases as they appreciate the lower cost for a higher-quality work environment.

### *Design phase*

We commence the real design work once the site data are collected and analyzed and the planning team has established a "red line" or central "idea" that forms the thematic axis of the project. Design simply means transferring this idea to reality. First, we coarsely integrate the dimensions of the "room program" into a simple cube, the "standard model."

This model, including the internal load of the building, is modified iteratively by simulation of energy dynamics. In this way we learn the energy consequences of different manipulations of standard "causes" involving the addition or removal of various materials. We use these simulations to improve our intuition about the various interactions between site factors and materials. When it "feels right" in the light of the "red line," i.e. the enduring theme, we finally begin to design.

The building program is discussed first to broadly set the office layout. At this level we address a number of functions and integrate where possible. For example, we may de-emphasize individual territories in a shared office space that cuts building volume and costs and thereby tries to improve social quality, communication, and functional flexibility. Even at this early stage we begin integrating energy, communication, and water systems into the building's technical infrastructure, employing alternative approaches such as solar collectors and solar-driven absorption cooling systems where possible. Fundamental questions, such as " Why do we flush a toilet with drinking water?" are often encountered here.

Some of the elementary parts of the design emerge in a basic diagram in which we split the building's room and functional program into small patterns that describe all the subfunctions of any necessary part of the building. Even at the stage of setting the template on which the design will evolve, we often find opportunities to streamline and reduce the building's structure. This diagram also incorporates ideas about the quality, durability, and recyclability of different materials. The diagram is an arena to exercise ideas about space, functionality, and materials in light of the building's "red line." The patterns we experiment with in this diagram can be sketches, small technical drawings, or text phrases that describe the emotional quality of a detail or material or the light ambience, the mood of a room. To deepen our appreciation of the building's proportions we supplement the diagram with three-dimensional models in paper, wood, or acrylic glass. The diagram helps to secure our overview, the evaluation of the whole, within the design at the very beginning of its elaboration. This overview allows us to review how well the building's design is aimed at the individual human's sensibility. To do this it must be founded on symmetry so fundamental that it pervades our sense of space like a law. It must appeal to our intuitive sense of harmony like music. The "golden mean," Fibonacci numbers and the "Modulor" of the French architect le Corbusier come to mind. These are principles that are found throughout Nature – in plants and crystals.

The next design stage relies on a computer simulation of all energy-related building functions, including temperature, air stream, lighting, cooling, and heating. As described above, we employ software specialists to create the simulations. We further probe the building's potential qualities by integrating the findings of the computer simulations with the drawings through construction of a scale model that we test in a wind tunnel.

Design is not a steady elaboration of structure but a cyclical pulsing of creating followed by streamlining. We follow the simulation-inspired development of the design with a "make it simpler phase," a filter process for the whole concept. All details, all materials, all plans are brought to a "whole office" discussion that inevitably pares down the structure through reduction, simplification, or removal of various elements. Our resolve to simplify is reinforced by simultaneous discussions with the "practitioners": the craftsmen, construction engineers, and technicians of the product industry. We also maintain informal discussions with urban planners and building authorities to give them every opportunity to satisfy their doubts and become comfortable with our unorthodox designs. This effort pays off, for when they understand our aims they often assist in very positive and productive ways.

Before the entire progression of creation/reduction cycles brings the design phase to an end, we hold a final workshop as a "design quality check." This meeting engages all the stakeholders, both the creators and the users, in a closing review of the design. The aim is to sustain our integrated cooperation with the customer as we jointly examine how the building's fundamental "idea" has been faithfully extended from its forging in the predesign phase all the way to the design's full maturation.

## Lessons learned

A number of lessons have been learned along the way. Energy input to the building must be carefully controlled. Two successful approaches complement each other: shading, especially "active shading" and a façade that does not store energy but sheds it. Trees are excellent shading elements that add cooling by evapotranspiration and improve air quality. Additionally, their loss of foliage in winter can add solar gain and needed heat to a building in cold weather. Where trees are insufficient to do all the shading, then we design a façade with an active shading component. We attack energy input as the extreme outside layer because internal shading allows too much energy to accumulate inside. We avoid colored or reflecting glass because it lowers solar gain in winter and projects a hostile, secretive, and undemocratic image. With our buildings we use the openness and transparency of glass. We do not use colored glass because clear glass can be recycled and can advertise the living activity going on within. The building's vitality is further projected by the movement of the lamellae or blinds as the active shading structure adapts to hourly changes in solar input. The people and the building are dynamic and follow the shifting energy arc of the day. Active shading does require sensors to feed a computer-controlled feedback system that controls the lamellae electrically. Such control can be done subtly, such that extreme midday sun is bounced off the ceiling as indirect light, thereby lowering illumination energy needs.

## Limits and changes – an outlook

It is clear that we like to design buildings that use less energy than conventional buildings: less energy during construction, less embodied energy in the building materials, and less operational energy. We have discovered that materials with low embodied energy are almost always traditional building materials that are easy to recycle and whose biological influence on the building's users is well known. In the moderate climates of central Europe we have found that the building's interior climate can be created almost totally by intelligent design without technical (mechanical/electrical) back-up systems. The total energy consumption of our buildings is half that of conventional designs. We have also found that what we call "comfort" dramatically influences a building's capital and operational costs. When a client agrees that the maximum "in-room" temperature during the summer can be limited to 26°C, there is no need to add air-conditioning systems to the design. When we limit minimum interior daytime temperatures during the winter to 20°C and not to the conventionally accepted 22°C, we are also able to reduce building energy costs significantly. Around 40% of an office building's electrical power consumption is related to the use of artificial light. Intelligent building design and advanced illumination technologies are able to cut this energy consumption in half.

German building codes force us to use concrete for ceiling and staircase construction in large office and housing complexes. These heavyweight structures can be used easily and in a intelligent, multifunctional way to increase a building's sustainability qualities. In designing large industrial plants, we are able to design and build with steel. Both the structural and energy design of these buildings need a different approach.

Our basic intention is to develop designs that enable the building to operate largely with soft (renewable energy) technologies and that pay attention to second-law entropy considerations as well as first-law energy conservation possibilities. Only a client's extreme criteria or building code requirements for designing in extreme climate zones force us to install complex technical back-up systems. However, if the basic design is intelligent enough, these back-up systems can be reduced both in size and running time. Intelligent technical systems, that is alternative technologies (e.g. absorption cooling systems driven by solar processes; fuel cells), are not exotic systems and they are advanced enough to replace the conventional systems we have used for decades. But why do we still heat water using fossil fuels? We are unable to answer adequately even a simple question such as this.

Our limits, as I understand them, are created by ourselves. We should be careful in using the words "we used to ..." We should replace these words with "we used to think ..." The limits to design are caused by our understanding of comfort and by our building codes. Right now is not the time to create sustainable building codes because we are really just starting the learning process. What we need are responsible experiments, and each of our buildings is an opportunity to experiment. We should discuss our different approaches and we should all try to define construction metabolism as a "red line" or dominant theme for our thinking, as a philosophical approach to understanding. At present, we set our own limits by our thinking. Our opportunity is to recognize that these experiments and this new era of sustainability are in fact a great opportunity for change.

## Conclusions

I find it ironic that, while my approach to design may appear novel or modern, it has a very traditional basis. I simply reach back much further into tradition than do most architects. I seek traditions back at the foundations of human culture, when discussion about the aims of society encompassed all our philosophy of Nature and science. This classical kind of generalist and interdisciplinary approach broadens an architect's awareness such that design, raw materials, and products can be skillfully integrated in ways that benefit society and the environment. Architects who complain that their individual creativity is stifled by government-imposed environmental regulations should recognize that these breakthroughs came from my personal and individual reflection on the history of our cultures. The beauty and ingenuity of this new architecture may reflect a rare appreciation of the links to society and Nature, but it grows from a personal responsibility to learn from history.

My search for new approaches did not bring me to new, synthetic materials. My designs use materials that have been in use for decades or more, and I found that we rediscover lost traditions when we learn the reasons why these materials have worked successfully for so long in local climates. But we are just now building healthy new traditions as we develop construction ecology by applying our understanding of ecology and natural metabolism. Our new appreciation of materials cycling in Nature leads directly to sharpening our resolve to simplify design and to reduce material use. Understanding how the "intelligence" of natural metabolism has grown out of corrective feedbacks and interactions within and between Nature and society bring us to the next step. We look to apply that "intelligence" in sustainable buildings whose durability and flexibility of use and function make the design timeless – the pattern of interactions within the building can evolve even as the building can change form.

Our new tradition of successful sustainable construction shows us that the problem with "modern" architecture is that it is not modern enough; it does not address the real needs of our culture or society to adapt to the changing conditions of the planet. Change persists, but not in one direction; change itself changes, and so we have to continue to learn and learn again. However, flexibility and adaptability have little to work with unless one has a broad perspective to chose from. This disciplined, iterative changing of our outlook demands that we benefit from all branches of learning – the sciences (technical and natural), sociology, philosophy, politics, and the fine arts, the most sensitive seismograph of our culture. Responsibility means responding to these changes in both thought and action by learning, creating, and relearning from the way that people and the world respond to one's creations.

Peterson (Chapter 5) describes how Nature and society operate at many different scales of time and space. Our cycles of learning can be quick and local or slow and vast, and they are nested inside of one another. The local lessons cycle quickly inside the great, millennial awakenings. With each ecological building we gain immediate insights into materials and design. But I believe that construction ecology is part of a great process of closing one of the great loops of learning that is reconnecting Nature and science. With the Enlightenment, society took enormous leaps based on a perspective that split science from metaphysics. The wreckage of the past centuries has shown us that the power gained from separation from Nature has proven a destructive dead-end, a folly whose expense we only now begin to appreciate as the age of cheap energy wanes. Perhaps we are ready to learn from Giordano Bruno, the Italian monk and philosopher. He was burned 400 years ago during the Inquisition for his thesis that we must use both our intellect and intuition to discover that Genesis and Nature are identical. As an architect I resurrect Genesis with each creation, but my design integrates and reconnects with ecological processes and cycles. We are going beyond the calls for interdisciplinary thinking to the actions that integrate Genesis and Nature. And the signs are appearing at many levels, from the corporate logos on ecological structures to the satisfaction of the people who use them to healthier businesses, that society is weaving these actions into the heart of our culture. The architecture of our firm appears to be different and, in fact, it is different. Our primary goal is to minimize the life cycle use of energy and materials. The result is what the Japanese would call Wabi Shabi, i.e. materials and design in their purest form. Simply using sustainable building materials or new energy approaches and systems does not in and of itself create a new, sustainable architecture. The basic goal should always be to minimize. Minimization, I believe, is something our culture has to learn and integrate into all our designs, to include that of the built environment.

## Notes

1. Pythagoras, Tetraktys, *Golden Verses* V.50–54.
2. Platon, *Timaios, the Arts*, about 400 BC.
3. Vitruvii, *De architectura libri decem*, 22 BC, first printed in Italy in 1487.
4. René Descartes, *Discourse de la Méthode*, France 1637.
5. Carl von Linde, 1876, inventor of the compression cooling machine.

6 Henry Ford, Highland Fabrication Plant, Detroit, 1913.
7 C.P. Wolf, Box 587, Canal Street Station, New York 10013, USA.
8 Christopher Freeman, speech at the Conservatoire des Arts et Métiers, Paris, 1984.
9 The "ecological rucksack" describes the total materials that must be processed to produce a single unit of a given product. For example, the production of 1 g of gold requires the processing of 300 metric tonnes of material.
10 "Ceiling activation" is a system developed by the author.
11 For an average commercial building with a room height of 3.15 m and containing seven floors.
12 An average of 2.5 personal computers per engineer is the average data systems density for this project.

## References

Ayres, R.U. 1984. *The Next Industrial Revolution: Reviving Industry Through Innovation.* Cambridge, MA: Ballinger Publishing Co.
Ayres, R.U. and Simonis, U.E. 1994. *Industrial Metabolism. Restructuring for Sustainable Development.* Tokyo: United Nations Press.
Burghardt, L. 1985. Die Müllthеorie der Kultur (The garbage theory of culture). In *The Children Do Eat Their Revolution*. Cologne: Du Mont.
Ford, H. 1922. *My Life and Work*. Garden City, NY: Doublday, Page & Co.
Friedman, Y. 1975. *Utopies Réalisables*. Paris: Union Génerale d'Editions.
Gideon, S. 1948. *Mechanization Takes Command: a Contribution to Anonymous History*. New York: Oxford University Press.
Habermas, J. 1988. *Nachmetaphysisches Denken* (*Post Metaphysical Thinking*). Frankfurt.
Jungk, R. 1986. *Und Wasser bricht den Stein* (*And Water Breaks the Stone*). Freiburg: Herder Verlag.
Schmidt-Bleek, F. 1998. *Das MIPS-Konzept*. Munich: Droemer.
Steffen, D. 1996. *Welche Dinge Braucht der Mensch* (*What Things Do People Need?*). Germany: Anabas.
Toffler, A. 1980. *The Third Wave*. New York: Bantam Books.
Virilio, P. 1993. *L'art du Moteur*. Paris: Editions Galilée.

# 12 Green architecture

*Malcolm Wells*

Life on Earth is an endless experiment. Yet, to paraphrase Aldo Leopold, humans have become Earth's master tinkerers without really understanding all the parts and the ways that they interact. The Earth has been in existence for 5 billion years, but the human species has been in existence for only 3 million years. For the other 99.94% of the Earth's existence, humans were not involved in this grand experiment. Over the last 10,000 years (the agricultural revolution), the human population has increased from about 5 million to approximately 6 billion (Goudie 1994). It took 3 million years for the human population to increase from its beginnings to 5 million and it took only another 10,000 years for the human population to then increase 1,200 times over. If estimates of a population of 11 billion by the middle of the twenty-first century are realized, this will place a staggering burden on the Earth's capability to support the human species. Humans have the intelligence and creativity to develop tools and technology which can help alleviate our burden on the planet, but we still do not seem to have the wisdom to know how to use this creativity.

No matter where we go we find Nature's miracles on every hand. The Earth's current geologic age, the Cenzoic, is the latest in a series of major transitions in its evolution. The Earth has its own physical and biologic rules, and the human species continues to try to live apart from those laws. The consumption of resources and creation of wastes will eventually outstrip the capacity of the Earth to remain a living support system for the human species.

Humans produce hazardous and persistent substances, deplete soils, harvest and mine with little thought for the direct and indirect effects, overfish, and, relative to the construction industry, pave the Earth's surface with impervious buildings, with lifeless roads and parking lots, and with unnatural landscapes maintained by large quantities of potable water, pesticides, and fertilizers (Figures 12.1–12.3). In the USA alone, buildings are responsible for 35% of all annual $CO_2$ emissions.

Many have warned of the consequences of modern society's disregard for its actions. Rachel Carson's *Silent Spring* (1962), Paul Ehrlich's *The Population Bomb* (1968), and the Worldwatch Institute's yearly *State of the World* reports have unsentimentally reported on the issues and trends that most wish to ignore or pretend will somehow be miraculously solved by technological innovation. We might act with an environmental conscience in our personal and professional lives if we had a sense of the magnitude of the problems and a realization that these problems are not unrelated to personal and corporate decisions made every day. The life cycle components of the typical building – its materials, the land on which it sits, how it is designed to work with local temperature, rainfall, sun, wind, the humanity of the interior environment, the ability to adapt and change, and the

*Figure 12.1* Asphalt dominates the landscape at the Pentagon.

*Figure 12.2* Part of the world's largest building, the Boeing Assembly building at Everett, Washington.

*Figure 12.3* The flowering desert has been paved into oblivion at California's Disneyland.

preservation or reuse and recycling of the whole or the parts at the end of its functional life – are in fact controllable by the designers, builders, developers, and owners of buildings (Figure 12.4).

In spite of this ability, buildings continue to be built as though there are infinite resources to sustain the materials consumption and wastes and pollution that they "require." Suburban sprawl continues to remove productive forests and agricultural lands and create a landscape that numbs the soul. Only the most egregious examples are called to attention, but fortunately more and more attention is being paid to the effects that poor land planning is having on communities' quality of life.

The hard facts of "sustainable" development are that humans are totally dependent upon the green world – on living plants – for survival. Plants are the primary producers upon which all higher organisms are dependent. Soil and aquatic microorganisms and plants support biochemical and hydrologic cycles of the planet. Without this support there would be no food, few clothes, and, more to the point, no life. On the other hand, plants do not need higher organisms such as humans to survive. They can survive – thrive, actually – without us. One form of architecture that provides a possible means to provide shelter and a humane environment can be achieved by pulling sheltering blankets of earth over buildings (Figure 12.5).

Underground buildings on natural or man-made hillsides above the water tables and potential rising sea levels overlooking resurgent natural landscapes will answer the needs of both humans and the green plants in coexistence (Figure 12.6).

Humans are incredibly adaptive. In prehistoric times humans lived in caves that were organically produced by natural forces. Call it adaptive reuse, or living in synergy with

*Figure 12.4* The architect.

Nature. In modern society builders can learn to reuse and recycle almost anything. Computer models can calculate solar angles and heat flows and daylight levels and estimate the lifecycle monetary costs of capital and operating expenditures. Manufacturers of materials can calculate emissions and chemical off-gassing from synthetic compounds that are used to create interior environments. All of these efforts are but a pale shade of green in comparison with the dark green of the surface of the land. Continuing to pave over and remove natural environments can never be balanced by any number of "green" building strategies that do not address the fundamental abuse of productive ecosystems.

Our worldwide paving activities are not yet deadly enough to kill us. There is still enough green land left to keep some parts of the world enjoying a high quality of life. However, many parts of the world endure present deprivations and will experience increased deprivations long before others. These populations are typically those who can least afford any further erosion of access to resources. As urbanized and northern hemisphere societies abuse their own environments, they are going to depend increasingly on global exchanges to secure the natural resources of countries that have them in greater abundance. The picture is tilting ever more rapidly toward collapse, which will cross these divides of north and south hemispheres. When we add to that picture our ever-expanding population, our unashamed consumerism, our ruthless deforestation, and all the rest – pesticides, air pollution, overfishing – BUPATO (Figure 12.7) has got to be among the top ten threats to the world.

Fortunately, however, BUPATO is something we can begin to change right away, and since the change will be so highly visible a heartening message will start to appear on the American scene.

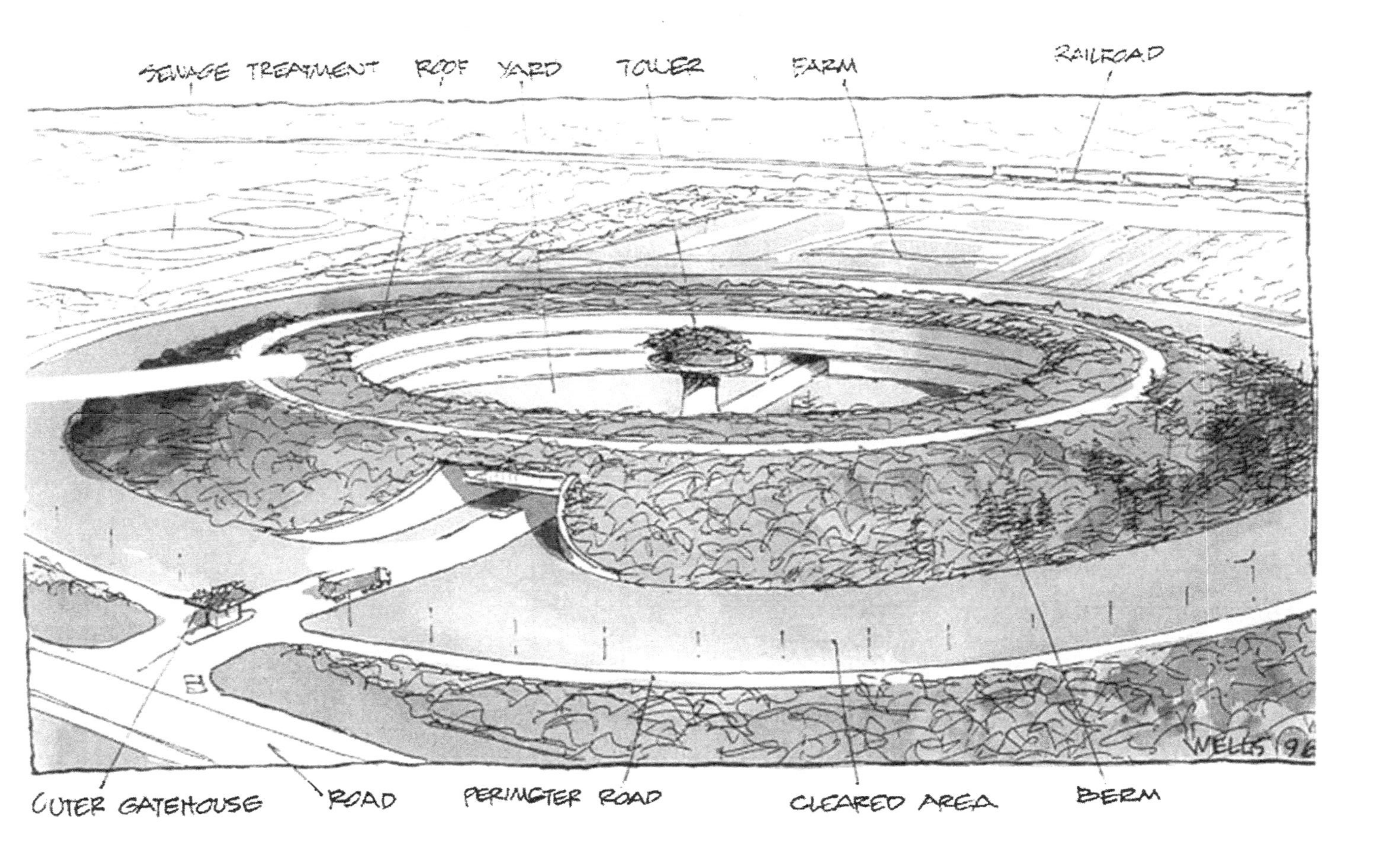


*Figure 12.5* Earth covered minimum-security prison design for southern Ontario.

*Figure 12.6* The author's office on Cape Cod.

*Figure 12.7* Buildings + Paving + Toxic Green Lawns (BUPATO).

Think of the message architecture sends to our kids right now: It is OK to build dead boxes on the surface of formerly living land. From generation to generation the silent idea we broadcast is that things as necessary as houses or shopping centers are more important than farms and forests and grasslands. How can our kids even imagine a different – a healthier – way if we continue to build so brutally?

Underground construction has long since been perfected (Figure 12.8). I have lived and worked in underground buildings for twenty-five years. There has never been a leak, never any dampness. Sunlight pours into the rooms (as it is doing as I write). The outdoor views are green and friendly, not gray and dead. Heating and cooling are a piece of cake. Exterior maintenance is minimal, nothing more than an occasional window washing. And the silence – the silence is wonderful – and, as is the case for many structures like mine, there is no need for fire insurance.

Watching first-time visitors shed their underground prejudices makes me smile every time. Visitors' expectations vary, of course, depending on the kinds of underground experiences they have had. Wet basements, caves, dark tunnels, subways: the visitors bring all such memories to my front door. They come expecting to find mold on the walls, expecting to need flashlights and rubber boots, expecting cobwebs and spiders.

The damp, dark fears they have carried for so long burst like bubbles when they see the sunny-day brightness that results when you are at home in the land (Figure 12.9). Some cannot be convinced that they are actually underground: others say, "You know, it feels *right* in here somehow." Chalk up a few more converts.

No special experience is needed to make successful use of earth sheltering. Structural techniques for resisting earth pressures are commonplace. Their designs are all in the

*Figure 12.8* Cityscape becomes landscape as a gentler architect moves in.

*Figure 12.9* Building partially recessed underground.

*Figure 12.10* The author's first underground building, his office at Cherry Hill, New Jersey, built in 1974.

engineering handbooks. The only secret ingredient needed is *great care*, and now it is no longer a secret. Just do it right. An earth-covered building will last for generations, even centuries, if the basics are done well. My underground buildings have been leak-free for the better part of thirty years, and the waterproofing, on examination, is just like new (Figure 12.10). But even if a leak should occur, finding and fixing it isn't that big a job.

Once the structure and the waterproofing have been done with great attention to detail all the rest is standard construction; little can damage the building. The principal

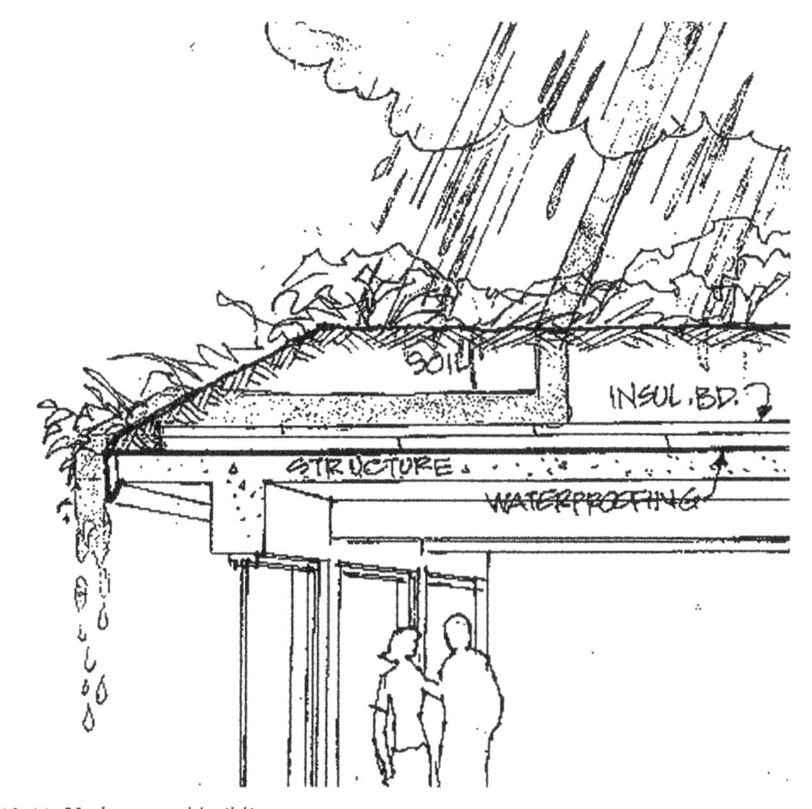


*Figure 12.11* Underground building.

*Figure 12.12* Plants, mulch, topsoil, subsoil, insulation board, waterproofing, and structure (1).

*Figure 12.13* Plants, mulch, topsoil, subsoil, insulation board, waterproofing, and structure (2).

enemies of building materials are sunlight, freezing rain, and baking heat, none of which can do much harm when the materials are protected by blankets of earth. Even acid rain loses much of its clout as it percolates slowly through the rooftop layers of plants, mulch, humus, roots, and subsoil before dripping slowly from the eaves into the surrounding gardens.

## How to build an underground building

The details are simple and straightforward (see Figures 12.11–12.14). These are typical. Notice that there is a commonsense sequence to the layers of materials. First comes the structure. Once it has been completed, thoroughly inspected, and its outer surfaces made fully smooth, the waterproofing sheets (or liquid membrane) are applied, making the building essentially weatherproof as far as construction activities are concerned.

(Rumors persist that concrete takes years to dry, that it will be damp for years. That is nonsense. During the rest of the construction period, before the building is enclosed, the concrete will become virtually dry.)

Now, moving outward from the waterproofed roof and walls, insulation board is applied. It not only slows the movement of heat in and out of the structure, but also acts as protection for the waterproofing. By insulating the outside of the concrete, the thermal mass of the building further adds to the temperature-leveling effect of the surrounding earth.

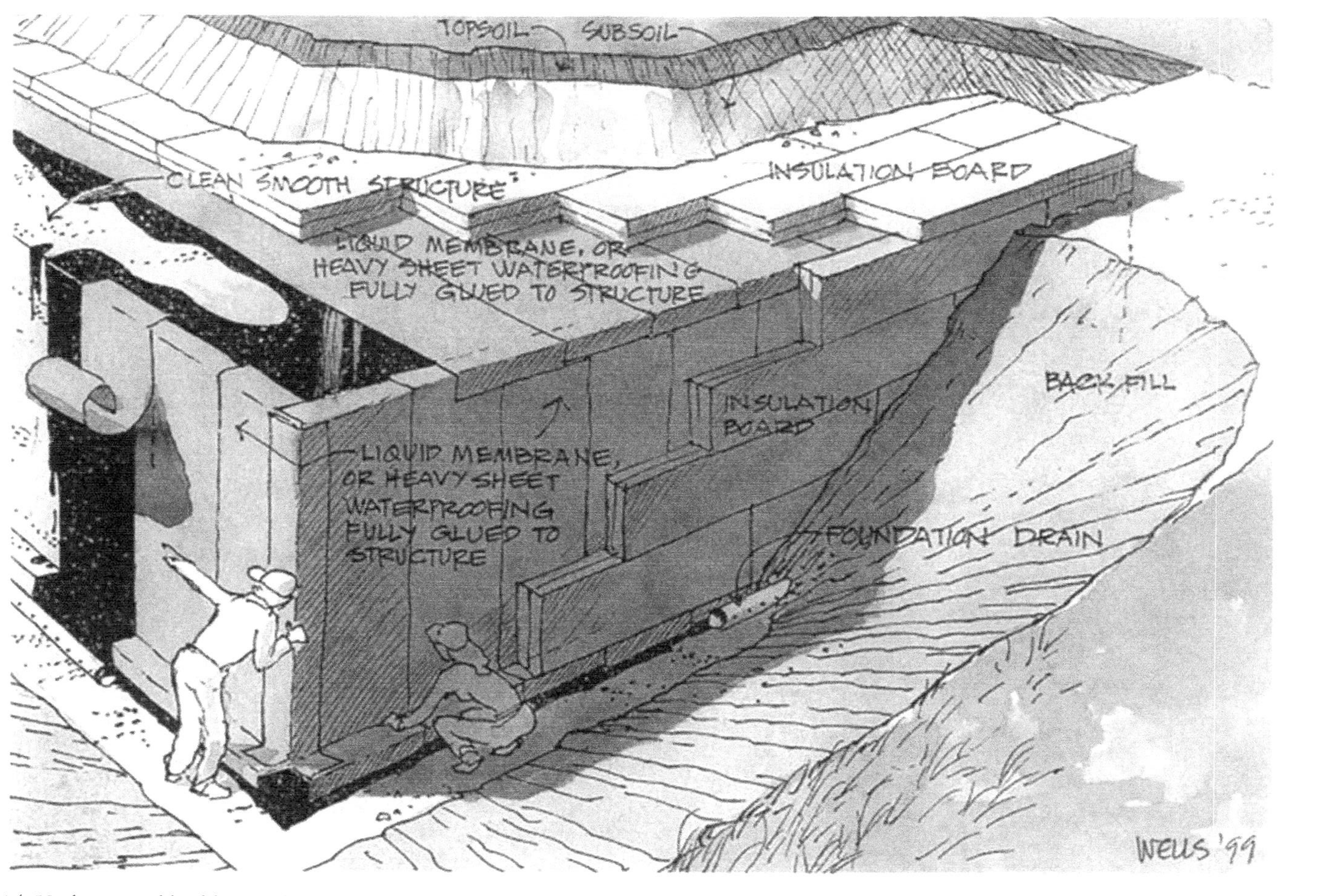


*Figure 12.14* Underground building (side view).

*Figure 12.15* Ground water is no problem when an underground building is built above it.

*Figure 12.16* View of underground building.

People ask if roots can damage these roofs. Roots are powerful creatures, and it would be foolish of me to say no, but in my experience they have done no harm. Roots follow water, most of which moves horizontally once it has percolated down to the insulation board. The roots fan out at that level, while a few thread-like roots follow the tiny amounts of water that flow through the joints in the insulation boards. Those roots form a lacy filigree of incredible thinness.

People also wonder if high groundwater levels make earth-covered buildings impractical in wet areas. Not if the buildings are fully above the high water level they don't. Simply build on a mound of the right height and then cover the building with earth, creating a little flowering hill with people living in it (Figure 12.15).

*Figure 12.17* View of underground building.

## Construction costs

People ask about construction costs. In dollar terms, you pay only a 10–15% premium for these long-lived, advantage-packed buildings. And if the designs are kept simple, there may be no premium at all.

In energy terms, fuel use is minimal because of the earth's natural temperatures and the soil's reluctance to change from warm to cool (or vice versa) very quickly. The energy used to extract, process, and assemble the materials used in earth-covered buildings becomes almost negligible because of the longevity of the structures. The use of concrete, which has a high energy "cost", is rightly condemned in less permanent buildings, but its negative impact almost disappears when the building lasts for centuries.

So why don't we see underground buildings all around us? I wish I could say that it is because they are hidden from us by their own landscapes, but the truth is, in most cases, a combination of two factors: the inertia of our tradition-bound construction industry and the negative reaction to the word "underground." Still, there are already 2,000–3,000 earth-covered buildings in the USA.

It may well be fifty years before earth buildings are commonplace. After all, much unthinking (and thinking!) resistance has to be overcome. And the virtual rebuilding of our civilization will involve an immense amount of work (Figures 12.16 and 12.17).

## The next step

Once we get used to the idea that we need not walk so heavy-footedly on the Earth we will be ready for the next step upward on the evolutionary spiral of architecture: huge, permanent earth ledges – giant shelves carrying deep layers of earth and native plants. Below each ledge, recyclable room components can then be plugged and unplugged at will, turning a hospital today into town houses tomorrow, or today's offices into tomorrow's factory, as the land itself lives on, undisturbed for centuries.

All of this could of course be little more than a pipe dream. A lot of very wise people pooh-pooh the whole idea. And we don't tend to feel much pity for plants and wildlife when things seem to be going well in our own lives.

Things aren't going well, though. The environment that sustains us is in real trouble in just about every category. We see more signs of it every day. As we waken more fully to this it will become a political movement, and once the political scales are tipped there'll be no stopping the movement towards the "greening" of the built environment.

## References

Carson, R. 1962. *Silent Spring*. Boston: Houghton Mifflin.
Ehrlich, P. 1968. *The Population Bomb*. New York: Ballantine Books.
Goudie, A. 1994. *The Human Impact on the Natural Environment*. Cambridge, MA: MIT Press.
Worldwatch Institute. Various years. *State of the World*. Washington, DC: Worldwatch Institute.

# Conclusions

*Charles J. Kibert*

The concept of construction ecology described in this volume is the result of a collaboration among ecologists, industrial ecologists, architects, and materials manufacturers. One of the products of this collaboration was a summary of general strategies for moving the green building movement from merely paying lip-service to ecological concepts to a much fuller integration of ecology into the creation, operation, and disposal of the built environment. In this final chapter I hope to provide a blueprint or roadmap for this shift in thinking. This roadmap is organized into three major sections: (1) "Recommendations and agreements"; (2) "Critical issues requiring further investigation"; and (3) "Additional observations." For the purpose of organizing this roadmap in a clear and systematic fashion, each of these sections is further categorized into up to six subsections: General, Materials, Design, Industrial ecology, Construction ecology, and Miscellaneous.

The first section that follows, "Recommendations and agreements", outlines the key points that the collaborators felt could be put forward to redirect industry in general, and green building in particular, onto a path much closer to the ideals of ecological integration. Additionally, the key points in this section are measures and conclusions where there is a high degree of agreement and confidence that they can be acted upon immediately. For example, the first point below under "Recommendations and agreements" was highlighted by the ecologists as the strategy of natural systems, the maximization of system effectiveness and *then* the optimization of efficiency. The ecologists noted that the efficiency of natural systems in converting solar energy to biomass was fairly low by human industrial standards, of the order of 5%. However, the effectiveness of natural systems in using solar energy exceeds 90%, far greater than that of industry, perhaps pointing the way to an appropriate shift in thinking for designers. At present, the consideration of effectiveness of energy or materials in the industrial system is virtually non-existent. Consequently, this point should be considered for implementation in the evolution of both industrial ecology, and one of its subsets, construction ecology.

The next section, "Critical issues requiring further investigation", highlights a range of issues about which our knowledge needs to be improved to be able to take effective action. The problem of defining precisely terms such as "synthetic" and "toxic" is one that needs to be resolved if all parties are to be able to communicate. Keeping materials in productive use implies that they have continued value, a concept that needs further development to make it useful in an industrial context.

The final section, "Additional observations", contains some of the major points of discussion among the collaborators that were not able to be categorized in the first two sections, and are intended to highlight some of the issues that are in play in the contemporary effort to create a stock of high-performance buildings. For example, the role of government is not recognizable to ecological systems other than the human one and, consequently, the collaborators felt compelled to comment on this institution and how it affects a shift toward a more ecologically sound economy.

## Recommendations and agreements

### *1 General*

1.1 Maximize second-law efficiency (effectiveness) and optimize first-law efficiency for energy and materials.

1.2 As is the case with natural systems, industry must obey the maximum power principle.

1.3 Be aware that the ability to predict the effects of human activities on natural systems is limited.

1.4 Integrate industrial and construction activities with ecosystems functions so as to sustain or increase the resilience of society and Nature.

1.5 Interface buildings with Nature.

1.6 Match the intensity of design and materials with the rhythms of Nature. In the built environment, move from the "weeds" stage to the "tree" stage for sites that are not frequently disturbed. "Weedy" structure (minimal built structure that is easily and cheaply replaced) may be much more adaptive for sites frequently disturbed by floods, storms, or fires.

1.7 Consider the life cycle impacts of materials and buildings on natural systems. Industry must take responsibility for the life cycle effects of its products to include take-back responsibility.

1.8 Address the consumption end of the built environment by integrating it with production functions.

1.9 Increase the diversity and adaptability of user functions in buildings through experiment and education.

1.10 Explore educational processes beyond academia that instruct through "learning-by-doing" by involving *all* stakeholders in processes that test different means by which the built environment is produced, sited, deconstructed, and resurrected.

1.11 Reduce information demands on producers and consumers by testing and improving the means by which materials, designs, and processes are certified as "green." This presupposes the development of a construction ecology based on Nature and its laws.

1.12 Systems analyses must look at system function, processes, and structure from different types of perspectives and at different scales of analysis.

1.13 Ecological thinking must be integrated into all decision making.

1.14 The precautionary principle should constrain and govern decision making.

### *2 Materials*

2.1 Keep materials in productive use, which also implies keeping buildings in productive use.

2.2 Use only renewable, biodegradable materials or their equivalent, such as recyclable industrial materials.

2.3 Materials created by the industrial system must be released only within the assimilative capacity of the natural environment.
2.4 Eliminate materials that are toxic in use or release toxic components in their extraction, manufacturing, or disposal. Focus first on materials not well addressed by economics, the intermediate consumables (paints, lubricants, detergents, bleaches, acids, solvents) used to create wealth (buildings).
2.5 Eliminate materials that create "information" pollution, e.g. estrogen mimickers.
2.6 Minimize the use and complexity of composites and the numbers of different materials in a building.
2.7 Not all synthetic materials are harmful and not all natural materials are harmless. Nature has many pollutants that are harmful. Natural fibers such as cotton are not necessarily superior to synthetic materials such as nylon.
2.8 The impacts of natural materials extraction can be high, for example as is the case with agricultural products or in forestry, in which pesticide use, transportation distances, processing energy, and chemical use are significant factors.
2.9 Plastics and other synthetic materials must be standardized based on recycling infrastructure and the potential for recycling and reuse.
2.10 Rather than being used for power generation, fossil fuels should be used for producing synthetic materials and renewable energy resources should be used as the primary power source.
2.11 It is not possible to rate or compare materials adequately based on a single parameter.

### *3 Design*

3.1 Model buildings based on Nature.
3.2 Make structures part of the geological landscape.
3.3 Design buildings to be deconstructable with components that are reusable and ultimately recyclable.
3.4 Design buildings and select materials based on intended use and then measure the outcomes of the design.
3.5 Incorporate adaptability into buildings by making them flexible for multiple uses.
3.6 Real savings can be realized by integrating the production, reuse, and disposal functions.
3.7 Focus on excellence of design and operation with "greenness" as a critical component. Exclusive focus on "greenness" can trivialize it as a marginal movement.
3.8 Investment in design that improves building function while minimizing energy use and the number of materials used will reduce the time and effort required to find and optimize new "green" materials.
3.9 At present it is critical that designs be revised to take into account major global environmental effects such as global warming and ozone depletion.
3.10 Green building design should allow for experimentation that produces structures that, like Nature, obey the maximum power principle.
3.11 Architects need to have a strong, fundamental education in ecology.
3.12 Performance-based design contracts would develop greener buildings and better architects.

### *4 Industrial ecology*

4.1 Changes needed to create an environmentally responsible industrial ecosystem must be intelligible to the population of the particular industry.
4.2 Owing to limits on time, knowledge, and resources, changes to the industrial system should focus on the clients and key stakeholders of the system. Major stakeholders include the educational system and insurance industry.
4.3 A new paradigm for industry is the collaboration of actors versus the possession of technical expertise.
4.4 Reducing consumption is more important than increasing production efficiency as the change agent for industrial ecology.
4.5 Ecological engineering must be incorporated into industrial ecology.

### *5 Construction ecology*

5.1 Construction ecology should balance and synchronize spatial and temporal scales to natural fluxes.
5.2 The corporations leading the way in the production of new, "green" building materials are a frontier species that may be creating a new form of competition that they are using to their advantage.
5.3 Green building can probably only be implemented as incremental change because of resistance and potential disruptions from the existing production and regulatory systems.

### *6 Miscellaneous*

6.1 Government officials and code-writing bodies need more education.
6.2 Performance standards for buildings and construction are needed to replace existing prescriptive standards. The performance standards need to include provisions for using green building materials.
6.3 The insurance industry is a major stakeholder in the built environment, and the threat of severe consequences from global warming should drive it toward promoting green building.
6.4 Certification has value as a starting point but should not be completely relied on for information on products.

## Critical issues requiring further investigation

### *1 General*

1.1 Application of systems theory to industrial ecology.
1.2 Precise definitions of key terminology – hazardous; harmful; toxic; biodegradable; green; and synthetic – to facilitate processes of certification, selection, production, deconstruction, reuse, and recycling.

*2 Materials*

2.1 Determining how to perform a quality analysis of materials.
2.2 The selection of products and production processes according to sustainable principles.
2.3 The process for introducing new chemicals by industry.
2.4 The definition of "synthetic materials" and how they are distinguished in terms of benefits and costs.
2.5 The scales (using different viewpoints to consider it, e.g. rucksack, footprint, hierarchy theory) of the materials and processes that control how, when, and where buildings are built, and how one integrates a view of these different scales.
2.6 The application of dematerialization to the built environment in light of requirements such as durability, strength, and thermal mass.
2.7 The impacts of materials at various scales.
2.8 The barriers to use of waste as materials.

*3 Design*

3.1 Insuring actors do not write their own standards.
3.2 Using performance standards rather than prescriptive standards.
3.3 Integrating risk assessments of long-term patterns (climate change, fossil fuel declines, water shortages) with processes that determine where, when, and how buildings are built, used, and recycled.
3.4 Improvement of design standards to eliminate unnecessary safety factors.
3.5 The scales for restoration because it is not possible to restore at all scales.

## Additional observations

*1 General*

1.1 There is still considerable misinformation about climate change, suggesting that burning more fossil fuels will create more carbon dioxide, which translates into more plants that can sequester carbon dioxide.
1.2 Government has a major role to play in developing and constraining markets, but it can also be an impediment if it does not encourage quality.
1.3 A structural change is different from a change in structure. These happen over different time and spatial scales.
1.4 Additive change is a technological route (increasing supply), whereas social change is a need-based approach (reducing demand).
1.5 There is no equivalent of photosynthesis in the industrial system.
1.6 Corporations are driven by competition, regulation, and customers.
1.7 Humans manage by default, and there is therefore a constant danger of a non-holistic perspective that will miss important conditions.
1.8 Long-term memory is weak in human systems.

*2 Materials*

2.1 Consumable goods such as hydrocarbons are burned and not a part of the final product. Disposal of the by-products degrades the environment, and there are tangible costs associated with lost assimilative capacity. This is depreciation of the environment that needs to be slowed down.

2.2 Synthetics pose a health risk when they are returned to the biosphere.

2.3 Hazardous materials can be created as long as they are retained in the industrial ecosystem.

### *3 Construction ecology*

3.1 Tools for change in the built environment are codes, standards, and "green" rating and labeling systems. Although systems such as the US Green Building Council (USGBC) Leadership in Energy and Environmental Design (LEED) standard and ISO 14000 are imperfect, they are good starting points to change design.

3.2 Contracts and bidding procedures are means to implement construction ecology and can reverse the efforts of designers and "green" building materials producers if not revised.

3.3 There are scale inefficiencies for life cycle assessment (LCA). *LCA may be appropriate for decision-making relative to commercial buildings but not for housing, for which the cumulative impacts of many structures are more important than the individual house.*

3.4 It is important that green building experiments are successful owing to the conservative nature of construction industry. A failed experiment occurring at a time when the whole industry is being asked to change could be catastrophic.

3.5 There is biodiversity of materials in green building materials.

3.6 Green building has no particular style and it addresses both human health and natural resources. It poses minimal burdens on the environment during construction, operation, and deconstruction.

## Closure

This initial effort to start a dialog among architects, ecologists, and industrial ecologists struggled at the onset to bridge the divides among these disciplines. It was clear from the start that attitudes, outlooks, and vocabularies differed significantly. Because ecology is a science, its terminology is precise, and the scientific method is the main tool for conducting research. Architecture, in contrast, is rooted as much in art as in science, and the tendency is to think of and articulate concepts in a totally different fashion. Industrial ecology faces an almost identical split between scientific and non-scientific considerations as it attempts to address the full range of sustainability issues, particularly its social aspects. In this respect, besides providing a bridge to ecology for built environment professionals, industrial ecology can also serve as an introduction to sustainable development for experts in natural systems.

Despite these cultural differences, there was much common ground and an often instant recognition of elements of problems that are common to the disparate range of disciplines represented by this group of authors. The problems of waste in construction were immediately recognizable to the industrial ecologists, who have developed both techniques and policy for closing materials loops. Similarly, the ecologists were able to address the practicalities of rematerialization and help identify approaches that would have a higher probability of success. Particularly worthwhile was the affirmation of many of the intuitive approaches that have been employed in greening the built environment. Future collaborations are, of course in order, and now is the time to begin experiments, in the form of buildings that test the common ground established by the participants in this collaboration. It is to be hoped that built environment professionals will recognize the need to deepen their knowledge of current ecological theory. If there is no other outcome from this collaboration, the realization by these professionals that their knowledge of ecology is critical to design would be reason enough to declare this initial foray into ecology and industrial ecology a tremendous success.

# Glossary

**Actor** A structure that is associated with the principal process in a system. By naming actors, one puts a scale on the formal model of the system.

**Adaptability** The capacity of a system (organisms, communities, landscapes) to respond to change in ways that preserve its ability to persist.

**Adaptive cycle** A model of the dynamics of systems (both human and natural) that maps changes through four phases (exploitation, conservation, release, and reorganization) that are linked in a loop.

**Biodegradable** Able to be decomposed by living organisms.

**Biodiversity** The complexity that contributes to the resilience of biological life as measured in the diversity of biological forms at many scales, from molecules to organisms to habitats to ecosystems.

**Biomass** The weight of organic matter, living or dead, in a given area.

**Biome** A natural system conventionally seen as covering a large area that is defined by its dominant vegetation (e.g. trees, grasses), a critical climate, and often an assemblage of animals that plays a central role in providing its particular structure.

**Biosphere** Zones of the Earth where living things exist, ranging from kilometers deep in the crust to the surface to kilometers high in the atmosphere.

**By-product** Material produced along with the primary product of the process.

**Carrying capacity** Amount of plant life, animal life, human life, and/or industry that can be supported on available resources in a given geographic area for significant time spans without any prolonged loss of system resilience.

**Change agents** Actors or substances that induce change in a system.

**Climax** A mature stage of succession in which the variability in diversity, biomass, and productivity is minimized by a set of dominant species in a presumed steady state that usually is the prelude to a catastrophic reorganization of the system.

**Community** An interacting group of organisms that is characteristic of a specific habitat.

**Complexity** A property of elaborate systems in which the elaboration is of organization. Complex systems are deeply hierarchical and exhibit simple behavior. It is not to be confused with elaborate structure, which is manifested by complicated systems.

**Composite materials** Industrial materials comprising multiple materials either in layers or intermixed in such a fashion that the separation at the end of their useful life is technically or energetically difficult.

**Conservation** Actions to sustain the character of systems as defined by their functions, cycles, flows, and actors (organisms).

**Construction ecology** A built environment (1) whose materials systems function in a closed loop integrated with ecoindustrial and natural systems; (2) that depends solely on renewable energy resources; and (3) that fosters the preservation of natural system functions.

**Construction metabolism** Resource utilization by the built environment that mimics natural systems by recycling materials and employing renewable energy resources.

**Consumers** Organisms or people that use the outputs of production.

**Culture** Shared information and values of a society.

**Cycle** A series of actions that repeatedly return to the initial state.

**Decomposer** Organism that consumes dead matter and returns nutrients to the system.

**Decomposition** The process of breaking down dead organic matter to simpler nutrients.

**Deconstruction** The disassembly of the built environment for the purpose of recovering and reusing materials and products.

**Demolishing** The tearing down and removal of the built environment, generally to landfills, with little or no regard for materials recovery.

**Depreciation** Loss of structure with the dispersal of material.

**Ecoefficiency** The delivery of competitively priced goods and services that satisfy human needs and bring quality of life while progressively reducing ecological impacts and resource intensity throughout the life cycle to a level at least in line with the Earth's estimated carrying capacity (World Business Council on Sustainable Development).

**Ecological design** Any form of design that minimizes environmentally destructive impacts by integrating itself with living processes.

**Ecological engineering** Environmental management, with part of the design resulting from human actions and part from attributes of Nature.

**Ecological footprint** An estimate of the resource consumption and waste assimilation requirements of a specific human population or economy in terms of a productive land area.

**Ecological rucksack** The quantity of materials that must be moved to produce a given product. For example, the ecological rucksack of a ring containing 1 g of gold is 300 metric tonnes of earth, the quantity of material that must be processed to extract the gold.

**Ecology** The study of the Earth's life support systems, of the interdependence of all beings on Earth.

**Economic web** The relationship of producers and consumers in an economic system which involves the exchange of money.

**Ecological system** A community of organisms in interaction with the environment.

**Ecosystem** In the vernacular, a generalized ecological system. In formal terms it is distinct from communities by the inclusion of the physical system inside the ecosystem boundary. It is a process-oriented view of ecological systems in which the parts are cyclical pathways of energy and matter fluxes.

**Effectiveness** Efficiency is about how well the quantity of flow is used. Effectiveness is how well the quality of the flow is used. When the flow is energy, quality is measured by exergy density. Effectiveness measures how much exergy is used in the system relative to a theoretical best-case scenario. (The theoretical best case, according to the second law of thermodynamics, is when all processes are performed reversibly and all the available work (exergy) is extracted from the energy; see Chapter 3.)

**Efficiency** The ratio of input to output energy or materials in a system.

**Emergence** Changes in system structure or organization that are too novel to have been predicted from understanding of its parts. It is distinct from evolution, in which change is merely behavior that is continuous and involves no new structures.

**Emergy** The energy that was required and used to make a product or service; its embodied energy.
**Emergy analysis** Calculation and comparison of emergy inputs and outputs of a system.
**Emjoules** Unit of emergy.
**End-of-the-pipe-technology** Technology that reduces emissions after they have formed.
**Energy** The ability to produce heat.
**Energy quality** The ability of a particular type of energy to be used in work, measured by its transformity.
**Entropy** The tendency of all systems to lose organization as they approach a steady state of thermal equilibrium with their environment.
**Environment** The context around a system that defines its opportunities to exchange energy, material, or information.
**Equilibrium** A stable state without available energy for change; dynamic equilibrium, a steady state.
**Evolution** A system of inspection of units so that their selection changes the state of the aggregate of the units, the population.
**Exergy** Represents the useful part of energy for a system in its environment, i.e. the maximum quantity of work that the system can execute in its environment. That part of energy that is convertible into all other forms of energy. Complex structures, rich in exergy and capable of reproduction, are formed by photosynthesis in the biosphere.
**Externalities** Impacts of an activity or product that are distributed to well beyond the area of its activity or production and are not paid for by the producer of the impact. Air pollution, for example, from an industrial complex affects a far larger area than the complex itself, and causes negative impacts that are not borne by the producers.
**First energy law (of thermodynamics)** Energy can change form but cannot be created or destroyed.
**Full-cost accounting** Including environmental costs with conventional economic costs in the assessment of the feasibility or outcomes of an activity or product.
**Gradient** Variation in concentration of energy or matter. Gradients are not in equilibrium. Energy or matter is dissipated down the gradient by movement.
**Green building** The creation and operation of a healthy built environment through resource efficiency and ecological principles.
**Greenhouse effect** Build-up of atmospheric heat caused by atmospheric carbon dioxide, other gases and water vapor, which absorb outgoing heat radiation.
**Hierarchy** Organization of objects or elements in a graduated series.
**Hierarchy theory** The theory of complexity wherein systems are expressed as a series of levels.
**Industrial ecology** The application of ecological theory to industrial systems.
**Industrial metabolism** The flow of materials and energy through the industrial system and the interaction of these flows with global biogeochemical cycles.
**Industrial symbiosis** The exchange of materials and energy among a cluster of companies that would otherwise be considered waste products.
**Information** Knowledge, facts, data (special use), parts and connections of a system
**Internalities** Environmental or ecological system impacts that have been assigned a monetary value.

**Level (two types)** Classes with asymmetry between (1) levels of organization, the relationship between which is a matter of definition; and (2) levels of observation which are ordered by spatial or temporal scale.

**Life cycle assessment (LCA)** Determination of the full energy and materials consumption and waste emissions (air, water, land) for a product.

**Materials intensity per service (MIPS)** The total quantity of materials, including waste produced during extraction, for example mine tailings, needed to create a given unit of a product. The Wuppertal Institute originated this approach and it includes the movements of soil, air, water, and biotic and abiotic material needed to produce a kilogram of material or a specific product.

**Maximum power principle** Principle that explains that the system designs that prevail are those that organize to use more energy.

**Metabolism** (1) The sum of the processes sustaining the organism: production of new cellular materials (anabolism) and degradation of other materials to produce energy (catabolism). (2) Processes of living organisms that utilize energy to maintain their structures and activities.

**Mode of analysis** There are two modes of analysis: structural and dynamical. The structural mode refers to an analysis of design. The dynamical mode refers to thermodynamic emergence, through positive feedback as a result of flow down a gradient.

**Net production** Gross production minus consumption over a certain time period.

**Organization** A set of limitations imposed by some design principle (evolution, God, a watchmaker).

**Positive feedback** A system whereby a perturbation in any one component causes a signal to go around the system components, returning to the original component and causing a further change in the same direction as the original perturbation. Positive feedbacks are unstable, and keep exhibiting amplified change until a controlling negative feedback is encountered.

**Primary consumer** Herbivore, the first consumer in a food chain, usually consuming an organism that captures energy from abiotic sources, such as the sun.

**Producer** Organism (green plant) that produces its own food from raw ingredients; a unit that combines input to form a new product.

**Production** The combination of two or more inputs to generate something new, as when plants produce food from inorganic materials or when industry generates a product from raw materials.

**Productivity** Rate of production, as of plant growth.

**Quality** There are two types: structural and dynamical. Structural quality is reliable and provides beneficial services. Dynamical quality is a force for improved change in structure or context.

**Recyclable material** A material that can be wholly or partially put back into productive use by a system by decomposing or disassembling the material to a more primitive form to allow its reassembly into new products.

**Recycle** Feedback of materials in a closed circle, as nutrients flow from decomposers back to plants.

**Renewable** A resource that can be replenished or renewed at a rate greater than or equal to its extraction rate.

**Resilience** There are several potentially conflicting definitions: (1) The speed with which a system returns to normal functioning. (2) The capacity of a system to return despite great displacement in state space.

**Resistance** The capacity of an ecosystem to oppose forces that would change its state.

**Reusable material** A material that is kept in productive use after one use or application by retaining its original form and without need for decomposition or disassembly to a more primitive state.

**Reuse** Utilizing products or components for anothr life cycle.

**Scale** A frame of reference for an observation. The frame is defined by the dimensions of space and/or time that are meaningful to the questions or ideas driving the observation. The frame has two components. The extent is the window in space and time that is large enough to capture the object of interest, for example the lifespan of an organism. The grain is the resolution (analogous to the pixel size on a computer screen) that is sufficiently fine enough that one can recognize the object for what it is.

**Second energy law (of thermodynamics)** The availability of energy to do work is used up in processes and in spontaneous dispersal of energy concentrations (depreciation).

**Secondary consumer** Animal that eats a primary consumer.

**Self-organization** (1) Set of limitations that result from historical happenings in the process of emergence of the self-organized entity. (2) Process by which various parts of a system connect themselves to work together. (3) Pattern formation processes in the physical and biological world including such diverse phenomena as grains assembling into rippled dunes to cells combining to create highly structured tissues to individual insects working to create sophisticated societies. What these diverse systems hold in common is the proximate means by which they acquire order and structure. In self-organizing systems, pattern at the global level emerges solely from interactions among lower-level components. Remarkably, even very complex structures result from the iteration of surprisingly simple behaviors performed by individuals relying on only local information (see Camazine, S., Franks, N. and Deneubourg, J. 2001. *Self-Organization in Biological Systems.* Princeton, NJ: Princeton University Press).

**Solar transformity** The quantity of solar emergy required to make 1 joule of another type of energy.

**Stability** (1) A property imposed by an observer indicating some degree of persistence of a system. (2) The resistance of a system. (3) The resilience of a system. (4) The capacity of a system description to survive changes in parameters (structural stability in mathematics).

**Stable state** A condition of a system at which the rate of change of its key parameters is zero or near zero.

**Steady state** System in which inflows equal outflows; dynamic equilibrium.

**Storage** A stock of matter, energy, or money.

**Succession** The process of self-organization of a system over time, resulting in a sequence of stages.

**Sustainability** All-or-nothing capacity to remain in the domain of a particular attractor for a specified amount of time.

**Sustainable** Capable of maintaining at a steady state or any state inside a given attractor.

**Sustainable construction** Creating and maintaining a healthy built environment based on ecologically sound principles.

**Synthetic material** A material created by industry that does not exist in Nature.

**System** A collection of parts related to each other directly or indirectly by interaction terms.

**System flip** A sudden change in a system that occurs when it exits one state and enters another with no prior warning.

**Thermodynamics** Science of heat and energy.

**Transformity** The ratio of energy of one type required to produce a unit of energy of another type.

**Turnover time** Time for things (water, organisms, systems) to be replaced.

**Urban ecosystem** A city and its environment.

**Wealth** Abundance of useful products.

# Index

CPSIA information can be obtained at www.ICGtesting.com
Printed in the USA
LVOW03s0332190814

399748LV00005B/96/P

9 780415 260923